An Introduction to
Mineralogy
for
Geologists

An Introduction to

Mineralogy
for
Geologists

W. J. & N. Phillips

Department of Geology
University College of Wales
Aberystwyth

JOHN WILEY & SONS

Chichester · New York · Brisbane · Toronto

Copyright © 1980 by John Wiley & Sons Ltd.

British Library Cataloguing in Publication Data:

Phillips, W. J.
  An introduction to mineralogy for geologists.
  1. Mineralogy
  I. Title   II. Phillips, N.
  549'.02'4553   QE363.2   79–42898

  ISBN 0 471 27642 1 (cloth)
  ISBN 0 471 27795 9 (paper)

Text set in 10/12 pt VIP Times printed and bound
in Great Britain at The Pitman Press, Bath

Dedicated to
Pam, Anne, Helen
and Shireen

Acknowledgements

The publishers and authors wish to thank the following for permission to publish new illustrations based on original work referred to in the text.

The Royal Society
Akademishe Verlagsgesellschaft, Wiesbaden
JCPDS. International Centre for Diffraction data.
Longman Group Ltd.

The authors wish to express their gratitude to Mr Howard Williams for the preparation of the photographic illustrations, and to Mr Julian Bird for constructing the model of the sphalerite structure. Thanks are also due to Mr Gordon Rattray for assistance with the mineral collection.

Contents

Symbols used in the text

Å	Angstrom units (10^{-8} cm).
a	Unit cell dimension in the **x** direction.
b	Unit cell dimension in the **y** direction.
c	Unit cell dimension in the **z** direction.
S.G.	specific gravity.
D	density in grams/cm^3
d	interplanar spacing.
H	hardness (Moh's scale).
m	mirror plane.
T	temperature.
x, y, u, z	the crystallographic axes.
Z	number of formula units per unit cell.
Z	atomic number.
K, L, M, N	quantum shells 1, 2, 3, 4 etc.
s, p, d, f	atomic suborbits.
α, β, γ	angles between the positive ends of the crystallographic axes $\widehat{y\,z} = α$, $\widehat{z\,x} = β$ $\widehat{x\,y} = γ$.
λ	wavelength.
v	volts.
e	charge on an electron.
i	angle of incidence.
r	angle of refraction.

Preface

Minerals are of great importance in developed societies because many of the industries that provide wealth, depend to a large extent on raw materials which are derived from minerals. The two important characteristics of a mineral are its chemical composition, and the 3-dimensional structural arrangement of the atoms which is the reason for the development of the attractive crystal faces and forms. These two characteristics of minerals together give rise to the potentially useful properties such as hardness, softness, refraction of light, pyroelectric, or piezo-electric effects.

Apart from the practical uses, natural minerals are worthy of serious study because they are visually beautiful objects which are the products of the operation of the laws of chemical bonding combined with the geometrical contraints of packing atoms together in 3-dimensional structures. Visual beauty may be defined as a combination of qualities such as colour, shape, proportion or form, which is pleasing to the eye, and this inevitably introduces the concept of symmetry. Consequently mineralogy involves the study of the symmetry of 3-dimensional repeated patterns, and the representation of 3-dimensional solids on 2-dimensional drawings, plans and stereographic projections. The applications of these skills are not only rewarding in themselves, but are also of great value for the understanding of a wide range of activities in industrial societies particularly in the manufacturing and constructional industries.

This introductory textbook has been developed from a course for first year university students of geology who have little previous experience of mineralogy. Consequently it includes much material that might be used in some schools, but it also introduces subjects which might not be studied until the second year of some university courses.

A special feature of the text is that it is designed as a series of studies of common naturally occurring minerals in order to introduce skills, concepts and principles of increasing complexity. The student may then more easily understand and use some of the many texts which deal systematically with crystal symmetry classes, mineral properties or crystal structures.

One fundamental aspect of geology is the study of the rock forming minerals. Since the most common rock forming minerals are silicates composed of combinations of various elements with silicon and oxygen, it is essential that the geologist appreciates the main chemical and crystallographic characteristics of the often complex silicate minerals. This is the principal objective for the student of geology, but in order to achieve this objective it is necessary to appreciate the principles of crystallography—the science of crystals.

In chapter 1 the characteristics of minerals are illustrated and a number of problems and objectives for study are defined. The principles of crystallography are introduced in chapters 2 and 3 with particular emphasis on some simple cubic minerals. Before extending the geometrical crystallographic studies to crystal forms of lower symmetry, the concepts of crystal chemistry and the methods used for the determination of crystal structures are explained in chapters 4, 5, and 6. Crystal structure analysis is introduced in chapters 7, 8, and 9 by considering the methods and reasoning used by W. H. Bragg and his son W. L. Bragg in their determinations of the structures of simple cubic minerals which were accomplished within a few years of Laue's discovery of the diffraction of X-rays by crystal structures.

In the present situation of overwhelming rates of data accumulation there is a danger that general interest is stimulated only by spectacular advances such as the solution of the structure of the DNA molecule. However, the most advanced studies depend on the reasoning involved in the original crystal structure determinations, and consequently it is very rewarding to study the original work and to recall the great public interest that was stimulated in the early years of the development of the new subject. Two books are of particular value because of their historical interest and the clarity of the explanations. An important first textbook by Sir W. H. Bragg explained the main principles of the new subject under the title, *An Introduction to Crystal Analysis* and it was based on a series of public lectures given at Aberystwyth in 1926. In the last book written by Sir Lawrence Bragg before his death in 1971, titled *The Development of X-ray Analysis,* he recalled the excitement and enthusiasm as each new insight into the understanding of the structure of matter was achieved, and shows how the methods were extended to allow the solution of the structures of large protein molecules.

The use of stereographic projections to illustrate the 3-dimensional angular symmetry of crystals is introduced in chapter 10 with reference to some cubic crystals. Then in chapters 11 to 15 the characteristics of the other crystal systems are developed from studies of selected geologically important minerals. The principles of crystal growth and twinning are considered in chapter 16. The principal objective for the geologist— the study of the structures of the important rock forming silicate minerals is the subject of chapter 17.

This textbook may be read through in a conventional manner but constructive independent work is of far greater value than passive reading. Consequently, questions and tasks are set at particular points in the text in order to test recall, or to lead the reader to solve problems independently. To assist the reader who may not have appropriate specimens, some crystals are represented by a series of plans. Estimates of interfacial angles may be obtained from these plans and the reader is invited to undertake certain crystallographic tasks independently, before reading on to check the work done. In most cases the solutions follow in the text so that the serious reader will have to apply a certain discipline in order to pause, and gain the benefit and satisfaction from obtaining the solution independently. It would be cheating to avoid serious consideration of the questions, and to avoid undertaking the simple mathematical and constructional tasks presented in the text, but only the reader would be cheated.

Aberystwyth   *John and Nahid Phillips*
April 1979

1 Introduction: The Nature of Minerals

The two characteristics of minerals which make them particularly attractive are the colours or transparency they exhibit and the plane, usually reflective faces which occur on the many different forms of crystals.

The purpose of this chapter is to illustrate the characteristics of selected minerals and to set a number of problems and objectives for study in subsequent chapters.

Contents of the sections

1.1 The colour of minerals

The colour of stones stimulated man's desire to possess pigments and these were used to produce cave paintings which formed part of the magic rites developed to ensure success in hunting and the regular supply of food and clothing. One of the commonest red pigments was obtained from the mineral **haematite** which is shown in plate 1. The name of the mineral is derived from the Greek word for blood. Haematite sometimes occurs in masses of black metallic radiating fibres which are terminated in rounded surfaces resembling the surface of a kidney as seen in the specimen in plate 1, so that this kind of occurrence of haematite is described as a **reniform mass** or as **kidney ore**. When there are sufficient quantities of a mineral for it to be mined the material is described as an **ore**. Haematite consists largely of iron combined with oxygen and the powdery form is bright red in colour. **Limonite** is an iron oxide which contains water and it is characterized by a brown or bright yellow colour also seen in plate 1.

Palaeolithic man learned how to make blades for tools and weapons by striking flakes of flint, and also used several different kinds of stones for ornamental purposes. The deep blue stone shown in plate 2 is called **lapis lazuli**, a name derived from the Persian word **lazhward**, meaning blue, and it often contains some bright brassy masses of the mineral **pyrite** which is known as 'fool's gold'. Lapis lazuli was greatly valued for stone ornaments and for its powder which produced the ultramarine pigment. The value set on precious stones in the early civilizations of the Middle East

1

resulted in exploration for mineral occurrences and in extensive trade. The small polished specimens of **turquoise** shown in plate 2 owe their beauty to the distinctive blue colour. Turquoise probably was the first precious stone to be mined extensively by the Egyptians. The finest turquoise comes from Persia where it is found between layers of limonite. In Persian the mineral is known as **firuse** and it is thought that the name turquoise means 'Turkish stone' referring to the trade route to Europe.

Brilliant green **malachite** is also shown in plate 2 and it often occurs as masses formed of distinct bands of slightly different colour. The banding is clearly seen in the polished stone. The naturally occurring group of spheroidal masses of malachite shown in plate 2 is described as a **botryoidal group** from the Greek for a bunch of grapes. Malachite is a hydrated copper carbonate which forms as a result of alteration of primary copper bearing minerals by weathering processes near the surface of the Earth. A less stable copper carbonate is known as **azurite** because of its azure blue colour, and in plate 3 one specimen consists of crystals of deep blue azurite on green malachite and a second specimen shows the more common bright blue colour of finely crystalline azurite also associated with some green malachite. The brightly coloured copper carbonate minerals are good indicators of the presence of copper in rocks below the surface. The third specimen in plate 3 is natural **metallic copper**.

Colour is one of the most attractive characteristics of minerals and it is usually the first property of a mineral that is observed. Minerals which have characteristic colours related to their composition are said to be **idiochromatic**, and the colour is useful as a means of identification. Lapis lazuli, turquoise, malachite, azurite, and copper shown in the first three plates are idiochromatic minerals with distinctive colours, but it will be seen that usually colour is not a reliable property for the identification of minerals.

The word **crystal** is derived from **krustallos**, the name the ancient Greeks gave to beautiful six sided quartz crystals like those shown in fig. 1.1. Beauty is defined as a combination of qualities such as colour, shape, proportion or form that delights the sight, and this introduces the concept of **symmetry of form** which is so important in the study of crystals. Ice crystals exhibit hexagonal patterns and so it was thought that the six-sided crystals of quartz were composed of water which had been frozen by extreme cold. This idea persisted until scientific crystallographic studies developed in the Middle Ages. Quartz has a simple chemical composition because it is composed of silicon (**Si**) and oxygen (**O**) in proportion of two atoms of oxygen for each atom of silicon so that its composition may be represented by the formula SiO_2. The forms of the quartz crystal are determined by the complex internal arrangement of the atoms and this will be considered in the final chapter.

Pure quartz is colourless but the presence of many inclusions of liquid produces a milky white colour. When quartz contains impurities it sometimes has the characteristic purple colour of **amethyst** shown in plate 4. The outer, later formed part of the quartz crystal contains the impurities which gives rise to the amethyst colour. **Amethyst** was one of the first precious stones used by man, and specimens of polished amethyst and a more modern cut gemstone are shown in plate 4. It was thought that gemstones could benefit the owner or could protect against evil so in addition to being status symbols, gemstones were worn as talismans and amulets. Amethyst was thought to give protection against drunkenness.

The lower specimen in plate 4 is part of a **vein** formed by crystallization in a fracture in rock. The outer part of the vein consists

Fig. 1.1. Six-sided columnar crystals of **quartz**. (×1)

of layers of the extremely fine grained variety of silica (SiO_2) called **chalcedony** which is known as **agate** when finely banded. The central part of the vein is filled with many quartz crystals which grew inwards from the margins until they coalesced in the centre. During the growth of the quartz crystals there was a period during which impurities were present in the solution from which the quartz crystallized, and incorporation of these impurities gave rise to the zones of amethyst colour.

Minerals which show a range of colours that are dependent on the presence of impurities or inclusions are said to be **allochromatic**.

The bright yellow metal **gold**, occurs primarily in veins of quartz as shown in plate 5 and because it is malleable and resistant to tarnish it became the ornament par excellence. Gold occurs usually as irregular shaped masses, but it is found in regular geometric forms such as the small cubes resting on the milky quartz in plate 5.

Pyrite, one of the minerals know as 'fool's gold', commonly occurs as cube forms as in plate 6. The name is derived from the Greek *pyrites lithos*, referring to the characteristic that pyrite is a hard brittle stone which produces sparks when struck with iron. Although superficially similar in colour and crystal form, gold and pyrite have very different physical properties indicated most obviously by the difference in hardness, but also by the presence of striations on the faces of the pyrite cubes shown in plate 6 and fig. 1.2.

It was customary to classify all materials into three groups, animal, vegetable or mineral so that the mineral group included rocks and soils. To the geologist a mineral is an inorganic substance which has two fundamental characteristics.

1. **A mineral possesses a characteristic chemical composition** which may range within certain limits.
2. **A mineral possesses an ordered arrangement of the atoms** of which it is composed and this results in the development of plane surfaces known as **faces**. If the mineral has been able to grow without interference, the faces may intersect to produce distinctive geometric forms known as **crystals.**

Gold (Au) and other elements such as **copper (Cu)**, **silver (Ag)** and **sulphur (S)** which occur naturally, are placed in the classification of minerals in the group known as **the native elements**. The form of the crystals of gold in plate 5 is determined by the manner in which the atoms of the metal are packed together and this subject will be considered in chapter 4.

Pyrite is composed of atoms of **iron (Fe)** and **sulphur (S)** in the proportion of two atoms of sulphur for each iron atom. Consequently the chemical composition of pyrite

Fig. 1.3. The pyritohedron crystal form of **pyrite**. (×2)

is represented by the formula FeS_2 and the mineral is classified as a **sulphide**. The atoms of iron and sulphur are arranged so that cube crystals often develop, but note that on each face of the pyrite cubes in plate 6 and fig. 1.2 there are distinctive striations parallel to one pair of edges of the cube face. These striations arise because of the manner in which the pairs of sulphur atoms are oriented in the pyrite crystal structure and this will be considered in chapter 8.

Pyrite also occurs as a crystal form called a **pyritohedron** illustrated by the specimen on the right side of plate 6 and in fig. 1.3. The pyritohedron consists of twelve pentagonal-shaped faces. The small crystal at the centre of plate 6 is composed of a **combination** of faces of the **cube** and the **pyritohedron forms**.

The **gemstones** mounted on the gold ring in plate 7 are a **diamond** flanked by two blue **sapphires**. **Diamond** is the hardest natural substance and the name is derived from the Greek word **adamas** meaning unconquerable. The gemstone was thought to endow the owner with purity and joy. Diamond usually occurs with octahedral forms consisting of eight triangular faces as shown by the impure diamond crystals on the right side of plate 7. Fig. 1.4 shows an **octahedron crystal form** of diamond with very pronounced growth layers on the faces.

Fig. 1.2. Cube crystal of **pyrite** showing conspicuous striations on the faces. (×2)

Until the 15th century only the natural surfaces of minerals could be polished. In plate 7 the **diamond** in the ring has been cut and polished to make the **gemstone**. Fig. 1.5 shows the pattern of polished surfaces on the top of the gemstone. This pattern is known as the '**brilliant cut**' because the polished surfaces are arranged to give the maximum internal reflection of light. The natural forms of crystals rarely show the exact geometric symmetry characteristic of the cut gemstones because the natural faces often develop to different sizes. However, it will be discovered in chapter 2 that the angular orientation of similar faces is constant and that **natural crystals possess characteristic degrees of symmetry**, which are direct consequences of the internal ordered arrangement of the constituent atoms. The crystals of a particular mineral may exhibit faces of different sizes but they always have the same degree of crystallographic symmetry. Consequently crystallography, the study of the forms and symmetry of crystals is an essential part of mineralogy.

Diamond is formed at high temperatures and pressures at depths possibly exceeding 70 km, and they are carried to the surface in volcanic pipes imbedded in a rock called **kimberlite**. Diamonds were first found in **alluvial deposits** consisting of sands and gravels in river beds where they had been concentrated due to the wearing down and washing away of the softer minerals. By tracing the diamond bearing gravels upstream the weathered rock on top of volcanic pipes was discovered, and **open pit diamond mines** were developed in South Africa after 1871. Expensive large scale mining is profitable even though the average yield is only 1 carat equivalent to 0.5 g of diamonds, from a ton of rock. This high capital investment is possible because 80% of the world production and trade is controlled by the Diamond Producers Association and the Central Selling Organization. About 20% of the diamonds are suitable for

Fig. 1.4. **Diamond** crystal showing prominent growth steps on the faces of the octahedron form. (×4)

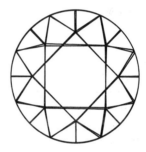

Fig. 1.5. The geometrically symmetrical pattern of polished surfaces known as the '**brilliant cut**', because it produces a large amount of reflection from the gemstone

gemstones and the remainder are used for cutting tools and polishing. Systematic prospecting for gemstones is limited to the search for diamond deposits because the impure diamonds have many industrial uses. Most other gemstones are discovered by accident and are commonly mined from alluvial deposits on a relatively small scale, by primitive methods.

The **sapphire gemstones** shown in plate 7 are usually cut from gem quality, water worn grains similar to the two rounded crystals obtained from an alluvial deposit. **Sapphire** is a transparent variety of the mineral **corundum** which is composed of

aluminium (**Al**) and oxygen (**O**) in the proportion **Al₂O₃**. Two very pale bluish crystals
of corundum are shown in plate 7. Plate 8
shows similar six-sided columnar crystals of
corundum but these are red in colour due to
the presence of chromium as an impurity.
The gemstone known as **ruby**, is cut from
the deep red, transparent variety of corundum.

The name **ruby** is derived from the Latin
word, *rubeus*, and refers to its deep red
colour. The name **sapphire** is derived from
the Greek word for blue and when used
alone it refers to the blue gemstone. Gem
varieties of corundum of different colours
are often given qualified names such as
white, yellow, or green sapphire. The group
of partly worn crystals of red corundum
shown with **ruby** gemstones in plate 9 still
show signs of the six-sided crystals forms.
The purple coloured ruby gemstone is
known as **oriental amethyst**. This range in
the colours of the crystals and gemstones of
corundum is due to the occurrence of different impurities, and it again emphasizes the
danger of reliance on colour for identification purposes, and the necessity for
the study of reliable characteristics such as
crystal form.

The red coloured gemstones in plate 10
are known as '**cape rubies**' but they are
transparent varieties of the mineral **garnet**.
The form of the almost black crystals of
garnet shown in plate 10 partly imbedded in
the rock within which it grew, is called a
rhombdodecahedron. The rhombdodecahedron is composed of twelve rhomb
shaped faces and it is represented in fig. 1.6.
Different views of garnet crystals with
rhombdodecahedron forms are shown by
the crystals at the top of plate 10. The three
very dark crystals of garnet in the lower part
of plate 10 consist of twenty four similar
faces and these make a form which is called
the **icositetrahedron** represented by fig. 1.7.
The crystallographic symmetry of the crystal
forms introduced in this section will be

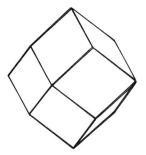

Fig. 1.6. The **rhombdo-
decahedron** crystal form

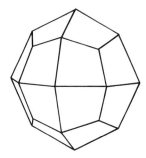

Fig. 1.7. The   **icositetrahedron** crystal form

studied in chapter 2, and the cubic forms
will be drawn in chapter 3 in order to demon-
strate some of the principles of crystallography.

The chemical composition of **garnet** is
complicated but the main constituent atoms
are silicon (**Si**) and oxygen (**O**) arranged as
groups represented by (**SiO₄**) in which each
silicon atom is surrounded by four oxygen
atoms. Garnet is classified as a **silicate**
mineral. Just as in pyrite each pair of sulphur atoms is balanced by an iron atom, in
the dark garnet every three (**SiO₄**)₃ groups is
associated with three atoms of iron (**Fe₃**)
and two atoms of aluminium (**Al₂**) so that
the average chemical composition can be
represented by the formula **Fe₃Al₂(SiO₄)₃**.
This is the composition of the common dark
red garnet which is called **almandine**. In
other garnets the places of the iron atoms
may be taken by atoms of calcium (**Ca**),
magnesium (**Mg**), or manganese (**Mn**),
while the places of aluminium atoms may be

taken by titanium (**Ti**), chromium (**Cr**) or trivalent iron. A pale green variety of garnet has a composition **Ca$_3$Al$_2$(SiO$_4$)$_3$** and it is known as **grossular**, a name derived from the botanical name for gooseberry.

The **garnet group** of minerals is an example of an **isomorphous series** referring to the similarity of crystal form of compounds with analogous chemical compositions. **Isomorphism** is common in minerals and is one of the bases of their classification.

Sphalerite (ZnS), the most important ore mineral of **zinc**, is almost colourless when pure. However, sphalerite usually contains iron and the colour ranges through browns to almost black for the most iron rich varieties as shown in plate 11. The specimens of sphalerite with lower amounts of iron have the appearance of resin. The general appearance of the surface of a mineral in reflected light is called its **lustre** and sphalerite is said to have a **resinous lustre**.

The colour of the fine powder of a mineral is known as its **streak** because the powder is conveniently produced by marking a piece of unglazed porcelain with the mineral. **Sphalerite** ranges in colour from light brown to black as shown in plate 11, but the streak of both specimens on the porcelain plate can be seen to be a light yellowish brown. The streak of minerals is usually constant although the colour of different specimens may be different, and therefore the streak is a more reliable property for mineral identification.

Chalcopyrite has a characteristic brass yellow colour due to reflection from the crystal faces, but it produces a black streak on the porcelain plate as shown in plate 12. Chalcopyrite is composed of copper, iron and sulphur with a composition approximating to **CuFeS$_2$**. The second specimen in plate 12 is **bornite**, a copper rich sulphide with a composition of **Cu$_5$FeS$_4$** which also produces a black streak. Bornite has a bronze colour but it tarnishes to purple colours and it is commonly known as **peacock ore**. Bornite and chalcopyrite are both **primary copper minerals** and they are usually altered to **secondary minerals** such as the carbonates, malachite and azurite in the weathered zone.

Olivine is an important silicate mineral with a composition represented by **(Mg,Fe)$_2$SiO$_4$**, and it is usually recognized by its yellowish green colour. The individual crystals of olivine in plate 13 show the range of colour, and the green gemstone variety of olivine is known as **peridot**. Distinctly yellowish green crystals of olivine can be seen in the large specimen of **lava rock** in plate 13. The smaller specimen in plate 13 consists of a mass of tiny green olivine crystals enclosed in a skin of lava rock, and it was formed when the mass of crystals in a blob of liquid lava was thrown out of a volcanic vent as a bomb so that the liquid cooled very quickly without crystallizing.

Beryl is a beryllium, aluminium silicate and it is mined for the beryllium which is used increasingly in industry. Green coloured, six-sided columnar crystals of beryl are shown in plate 14, and it can be seen that the large crystal is terminated by a face which is perpendicular to the side faces. The crystal forms of beryl will be studied in sections 15.1 and 15.2, and the nature of the crystal structure is considered in section 17.6. In plate 14 some pale green beryl crystals are seen partly imbedded in a specimen of a vein of rock called **pegmatite** because it is composed of relatively large crystals of feldspar. Beryl crystals are also found in dark rocks called **mica schists** which are composed of flake-like crystals of mica with some parallel alignment described as **schistosity**. The dark specimen in plate 14 is a mica schist which contains a green beryl crystal. The dark green transparent gemstone of beryl is known as **emerald**, and this colour is produced when some chromium and iron atoms occur in the mineral structure.

The physical properties of a mineral depend on its chemical composition and its crystal structure, and these determine the use to which the mineral may be put. Gemstones are rare minerals possessing certain special properties which are not found in other minerals, and which make them suitable for cutting and polishing. In addition to their transparency, brilliance, and colour, the most important physical property of the gemstones is their exceptional hardness and it is appropriate therefore to consider this property next.

1.2 The hardness of minerals

The resistance that a mineral offers to abrasion or scratching is called its **hardness**, and this can be determined in relative terms by comparing it with a sequence of ten minerals arranged by Mohs in increasing order of hardness. Care should be taken not to scratch good crystal faces. For the preliminary identification of common minerals a simplified relative hardness scale is adequate, and it consists of the finger nail (hardness 2) a knife blade (5) and a steel file

Fig. 1.8. **Talc** which is sometimes called soapstone, is easily carved and scatched with a finger nail. (×1)

(6). The minerals used in Mohs scale are described below.

1. **Talc** has a relative hardness of **1** on Mohs scale Talc occurs as grey compact masses composed of minute flake-like crystals, is greasy to the touch and is

Fig. 1.9. 'Satin spar' composed of very fine parallel fibres of gypsum. (×1)

easily carved as shown in fig. 1.8. Talc is a hydrous magnesium silicate, with a composition $Mg_3(Si_4O_{10})(OH)_2$ and it is used in such materials as paint and cosmetics. The lustre of talc is described as greasy and it is sometimes called **soapstone**.

2. **Gypsum** often occurs as well formed crystals illustrated in fig. 1.17. Veins of gypsum which are composed of very thin parallel fibres are known as **satin spar**, fig. 1.9. A massive variety composed of minute crystals of gypsum is called **alabaster** and is used for ornamental vases, etc. Gypsum is the commonest **sulphate** mineral, $CaSO_4$, $2H_2O$, and its main use is in the manufacture of plaster of Paris.

3. **Calcite** occurs as thick beds of limestone and also as well developed crystals of several different forms some of which are illustrated in figs. 1.22 and 1.24. Calcite is composed of calcium carbonate, $CaCO_3$, and fine grained varieties can be identified by the fact that it effervesces freely in cold hydrochloric acid. The most important use of calcite is in the manufacture of **cement**, but the crystalline varieties of limestone known as **marble** are used extensively as ornamental stones.

4. **Fluorite** occurs as crystals of cube form often showing interpenetration relationships. Its composition is calcium fluoride, CaF_2, and it is used mainly as a flux in steel manufacture. Fluorite crystals are illustrated in plate 15 and the structure will be considered in chapter 7.

5. **Apatite** is a calcium phosphate mineral, $Ca_5(F,Cl,OH)(PO_4)_3$, which occurs as six-sided columnar crystals shown in fig. 1.10. Apatite is not sufficiently hard to be used as a gemstone.

6. **Orthoclase** is a member of the most common group of rock forming minerals, the **feldspars**, a name derived from

Fig. 1.10. Crystals of **apatite** showing the characteristic six-sided columnar habit. (×1)

the German word for field. Orthoclase is a potassium aluminium silicate with a composition indicated by $K(AlSi_3O_8)$. Orthoclase crystals are illustrated in fig. 1.18 and are studied in section 13.1.

7. **Quartz** is composed of the oxide SiO_2 (silica) and it often occurs as six-sided columnar crystals terminated by pyramidal faces as shown in fig. 1.1.

8. **Topaz** is a colourless, yellow, or pale blue mineral often associated with tin ores. It is an aluminium silicate, $Al_2(SiO_4)(F,OH)_2$, and the transparent crystals are used as gemstones. A group of crystals is shown in fig. 1.11.

9. **Corundum** is composed of aluminium oxide, Al_2O_3, and its main use is as an abrasive. Some six-sided columnar crystals of corundum and the gemstone varieties of corundum are illustrated in plates 7 and 8.

10. **Diamond** consists of the native element carbon, C, and it occurs naturally as octahedral crystals, as shown in plate 7 and fig. 1.4. The brilliant transparent crystals are the most prized gemstones, but vast quantities of the grey crystals

Fig. 1.11. Crystals of **topaz**. (×2)

are used in cutting tools and as abrasives. The increase in the hardness of the selected minerals from 1 to 9 is approximately linear, but it is estimated that diamond has a hardness of about 42 compared with corundum. Diamond is more than four times harder than corundum.

The hardness of a mineral depends on its chemical composition and also on the structural arrangement of the atoms. The greater the bonding forces the greater the hardness of the mineral.

The mineral **graphite** also consists of carbon but in contrast to the exceptional hardness of the diamond crystals, graphite is a black mineral with a hardness of 1 and it easily marks paper. Graphite is shown in fig. 1.12 with a diamond crystal. Graphite has a metallic lustre but it is greasy to touch since it has a layered structure somewhat like that of talc. The great difference in the physical properties of diamond and graphite is due entirely to the manner in which the carbon atoms are arranged in the two minerals and this is considered in section 8.5. This is an example of **polymorphism**, the occurrence of two very different crystal structures, diamond and graphite, which have the same chemical composition, in this case pure carbon.

1.3 The fracture of minerals

In **glass** the atoms have a completely random orientation. When glass is broken the fractures are related to the point of application of the applied force and they tend to consist of curved surfaces which are described as **conchoidal** (shell-like) **fractures**, as shown by the specimen of natural glass called **pitchstone** in fig. 1.13.

The extremely fine grained massive variety of silica (SiO_2) is known as **chert** or **flint**. Early man discovered that chert can be struck to produce thin flakes bounded by conchoidal fractures which intersect to give very sharp cutting edges, as shown in fig. 1.14. **Quartz** is also composed of silica but the atoms are arranged with such regularity that the bonds between the atoms are very similar in magnitude in all directions. Con-

Fig. 1.12. A specimen of black **graphite** (hardness 1) with the octahedron crystal of **diamond** (hardness 10). Graphite and diamond are both composed of carbon atoms and this is an example of polymorphism—two very different crystal structures having the same chemical composition. (Graphite ×1; diamond ×4)

Fig. 1.13. A volcanic rock known as **pitchstone** is composed of glass, and is characterized by conchoidal (shell-like) fractures which arise because of the completely random distribution of the atoms of which it is composed. ($\times \frac{1}{2}$)

Fig. 1.14. A specimen of **chert** with a sharp edged flake. The small man-made blade is of Mesolithic age ca. 5000 BC. ($\times\frac{1}{2}$)

sequently there is no tendency for the mineral to break on any particular plane, and conchoidal fractures develop on quartz crystals as shown in fig. 1.15. The surfaces of fractures in transparent quartz crystals have a glassy appearance and this is described as a **vitreous lustre**.

Quartz is described as a tough **brittle** material. Soft brittle minerals are said to be **friable**. Soft minerals such as talc and graphite are described as **sectile** because when cut with a knife, curved shavings peel off.

1.4 The cleavage of minerals

When a mineral breaks along a definite plane surface it is said to possess a **cleavage**. A cleavage plane develops parallel to layers of atoms which are weakly bonded to the adjacent layers of atoms.

The pale coloured mica called **muscovite**

Fig. 1.15. A **quartz** crystal showing conchoidal fractures which arise because of the high degree of regularity of the structure. The small fragments have a vitreous (glassy) lustre. ($\times2$)

which sometimes occurs as flat six-sided crystals, has an exceptionally well developed plane of cleavage parallel to the flat side of the crystals illustrated in fig. 1.16. The thin transparent sheet of muscovite shown in fig. 1.16 contains lines of inclusions and is **flexible** and **elastic**. The name of the mineral is derived from Muscovy glass referring to its use in windows in Old Russia. **Muscovite** is a silicate mineral with a pronounced **sheet structure** which will be considered in section 17.10. The bonds between the atoms within the sheets are strong

Fig. 1.16. Hexagonal shaped crystals of **muscovite**. The crystal faces are almost perpendicular to a
well developed cleavage which produces thin, flexible transparent sheets. (×2)

while the bonds linking the adjacent sheets
are weak. The greater the contrast between
the strength of the bonds linking the atoms
in the directions parallel to the cleavage
plane, and the weakness of the bonds linking the atoms in the directions perpendicular to the cleavage planes, the greater the
tendency for the mineral to break along the
cleavage plane.

Since the growth of particular crystal
faces and the presence of planes of potential
cleavage are both determined by the character of the crystal structure it is important to
relate the orientation of the cleavage planes
to the crystal form.

Crystals of **gypsum** split along cleavage
planes which are parallel to the largest crystal faces as shown in fig. 1.17. The transparent sheets of gypsum are known as **selenite**.

Fig. 1.17. Crystal of **gypsum** illustrating the well
developed cleavage planes which are parallel to
the largest crystal faces. (×1)

Fig. 1.18. Crystals of **orthoclase** possess two prominent planes of cleavage which are perpendicular to each other and the name refers to this diagnostic feature of the mineral. (×1)

Fig. 1.19. **Halite** occurs as cube shaped crystals and has cleavage planes parallel to the three pairs of cube faces, (×1)

Many minerals possess more than one plane of potential cleavage. **Orthoclase** has two prominent planes of cleavage at right angles to each other and the name of the mineral refers to this diagnostic characteristic. The specimens in fig. 1.18 are arranged so that it can be seen that the two planes of cleavage are parallel to the two side faces of the crystals and to the faces which form the top and bottom of the crystal.

The mineral **halite** is known as **rock salt** because it is composed of sodium chloride, **NaCl**. Halite grows in cube-shaped crystals and cleaves easily parallel to all three pairs of cube faces, as shown in fig. 1.19. The question; *why do the sides fall off halite crystals?* will be considered in chapter 7.

Note that the cleavage planes can be observed usually without breaking the specimen, either by reflections of light from internal planes or by the intersection of cleavage planes with crystal faces.

Fluorite also grows as cube shaped crystals, but it possesses planes of cleavage which are oriented in such a way that the corners of the fluorite crystal tend to break away, as shown in plate 15. If this process is continued, the resulting fragment bounded by the four sets of cleavage planes has the form of an **octahedron**. The question: *why do the corners fall off fluorite crystals?* will be considered in chapter 8.

Barytes occurs as tabular shaped crystals which often have two faces perpendicular to

Fig. 1.20. Crystals of **barytes** showing three well developed planes of cleavage. One cleavage is parallel to the largest crystal faces and the other two are perpendicular to this face and parallel to the faces which form the pointed ends of the crystals. (×1)

Fig. 1.22. Transparent six-sided columnar crystals of **calcite**. (×2)

Fig. 1.21. A crystal of **barytes** resting on the large face to show the orientation of the two sets of cleavage planes which are perpendicular to the large face. (×1)

Fig. 1.23. Crystal of **calcite** showing three well developed cleavage planes which give rise to the rhombohedron form below. (×1)

Fig. 1.24. The rhombohedron cleavage form in transparent **calcite**. Note the double images of the edges and cracks on the lower surface. This is due to the refraction of light along two paths in the crystal. (×1)

the largest face and meeting to produce pointed ends as shown in fig. 1.20. Barytes has three well developed planes of potential cleavage, two of which are parallel to the crystals faces forming the pointed ends of the crystal as shown in fig. 1.20 and the other parallel to the largest crystal face on which the specimen rests in fig. 1.21. The crystal forms of barytes are considered in chapter 12.

Calcite crystals can be found with very many different crystal forms which are studied in section 15.3. Many are six-sided columnar crystals terminated by sloping faces as illustrated in fig. 1.22. Usually in all the crystal forms of calcite, cleavage planes with three different orientations can be recognized. These three planes of cleavage intersect to form masses bounded by six rhomb shaped surfaces. This form is called a **rhombohedron** and it is shown in figs. 1.23 and 1.24. Calcite may be identified by the rhombohedral form of cleavage fragments which have surfaces intersecting at 75° or 105°.

1.5 The habit of crystal forms

If allowed to grow without interference, crystals adopt shapes which are related to their internal structure. The general shape of the crystals of a mineral is called its **habit** and it is sometimes useful for the identification of a mineral. For example the lead sulphide, **galena (PbS)** and the antimony sulphide, **stibnite (Sb_2S_3)**, are minerals with a lead grey to black colour and streak, metallic lustre and hardness of 2, but they have very different habits as shown in fig. 1.25. Galena is easily recognized by its crystals of cube form and cleavages parallel to the cube sides. In contrast the crystals of stibnite are elongated in one direction and tend to have sharp edges like blades. Some of the common terms which are used to describe the habit of single crystals are listed below.

Columnar —elongated in one direction and like columns. The crystals of corundum shown in plate 8 and quartz in fig. 1.1 usually have columnar habits.

Prismatic —elongated in one direction like the andalusite crystals in fig. 1.44.

Tabular —elongated in two directions like the crystals of barytes in fig. 1.20.

Bladed —elongated in one direction and with thin edges like stibnite in fig. 1.25. Hornblende crystals usually have a bladed habit as illustrated in figs. 13.10 and 13.11.

Foliated —sheet-like, and easily split into folia like the muscovite shown in fig. 1.16.

Botryoidal—groups of globular masses, plate 2.

Reniform —radiating fibres terminating in rounded surfaces, plate 1.

Granular —the mineral consists of an aggregate of grains.

Massive —compact and irregular without any distinctive habit.

1.6 The density of minerals

The **specific gravity** of a mineral is a number that expresses the ratio between its weight and the weight of an equal volume of water at 4 °C. The **density** of a mineral is its mass per unit volume and it is necessary to specify the units used, for example kilograms per cubic metre or pounds per cubic foot. In this book **density** will be referred to in terms of **grams per cubic centimetre**, and since the

Fig. 1.25. Crystals of **galena** showing cube crystal forms and also cleavages parallel to the cube faces are seen on the left. **Stibnite** is also a soft, black mineral with a metallic lustre but its crystals have a very different habit and are elongated with sharp edges like a blade. (×1)

density of pure water at 4 °C is unity, density will be equivalent to specific gravity.

The density of a mineral may be obtained by the direct weighing of the specimen first in air and then in water. If the specimen is sufficiently large it may be suspended by a nylon thread from the hook on a **balance** beam as shown in fig. 1.26. The specimen is weighed in air and then is immersed in a beaker of water which rests on a bridge placed over the pan of the balance. The density is obtained by dividing the weight in air by the loss of weight in water.

$$\text{Density} = \frac{W_{air}}{W_{air} - W_{water}}$$

The relative weights of a mineral in air and in liquid can be obtained without direct weighing. With a simple **beam balance** shown in fig. 1.27 the mineral is first suspended in air from the shorter arm and is balanced by a counterpoise weight on the longer arm at position A. Then the mineral specimen is suspended in water as illustrated in fig. 1.27 and balanced by the counter-

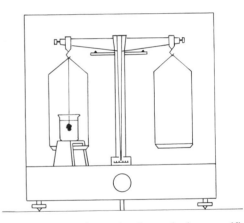

Fig. 1.26. The determination of the specific gravity of a mineral by weighing in air and water on a balance

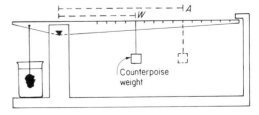

Fig. 1.27. A simple beam balance

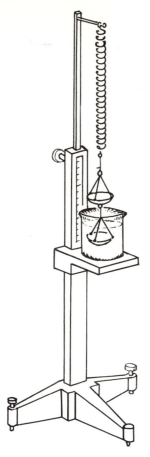

Fig. 1.28. The Jolly
spring balance

does not touch the sides or bottom. First record the extension of the spring when the mineral is on the upper pan and in air, *Ea*. Then record the extension of the spring when the mineral is on the lower pan and immersed in water, *Ew*. The specific gravity of the mineral is given by the ratio *Ea*/(*Ea* − *Ew*).

When determining the density of a mineral care must be taken to remove particles of any other mineral and any air bubbles that may remain on the surface of the specimen. Corrections may be made for variations in the temperature or the density of the liquid. Water is not the most suitable liquid for accurate density determinations because its high surface tension frequently results in air bubbles sticking to the mineral specimen when immersed. Pure organic liquids of known density such as carbon tetrachloride are more suitable.

The specific gravity of very small fragments of a mineral may be determined by the use of a **pycnometer**, a small bottle with a stopper within which there is a fine capillary tube. The bottle is weighed when dry and empty, then is weighed again with the mineral so that the weight of the mineral in air may be determined. The bottle containing the mineral is then filled with liquid and weighed. Finally the bottle is weighed with liquid only. in each case the liquid must fill the capillary tube and excess liquid must be removed from the outside of the bottle. The difference between the last two weighings is the weight of the mineral less the weight of an equal volume of water or other liquid.

The density is a fundamental property which is determined by the chemical composition and the structure of the mineral. The effect of the structural packing is well illustrated by the polymorphs of carbon, diamond and graphite. The closely packed, strong structure of diamond has a density of 3.5 grams/cm³, while the open structure of graphite has a density of 2.2 grams/cm³. An increase in density as a result of the increase in

poise weight in position *W*. The specific gravity is given by the ratio *A*/(*A* − *W*), and this number is equivalent to the density of the mineral given in grams per cubic centimetre.

One of the simple pieces of apparatus which has been devised for the determination of specific gravity of small specimens of a mineral is the **Jolly spring balance** shown in fig. 1.28, in which the relative weights of a mineral in air and in water are obtained by measuring the extension of a spring. Two small pans are suspended from the spring and a small platform on which stands a small beaker of water, is adjusted so that the lower pan is always immersed in water, but

the atomic weights of the constituent atoms is illustrated by the sulphate minerals listed below.

Element	Atomic weight	Mineral	Density gram/cm^3
Calcium	40.08	Anhydrite CaSO$_4$	2.9
Barium	137.36	Barytes BaSO$_4$	4.5
Lead	207.21	Anglesite PbSO$_4$	6.2

1.7 The magnetic and electrical properties of minerals

A few minerals are strongly attracted by a simple magnet and are said to be **magnetic**. The iron oxide, **magnetite Fe$_3$O$_4$** and the iron sulphide **pyrrhotite Fe$_{1-n}$S** are the only common magnetic minerals. A variety of magnetite called **lodestone** behaves as a magnet. When suspended, lodestone will orient itself in the Earth's magnetic field and was used in the earliest compasses.

All minerals are affected by a magnetic field. Minerals which are slightly attracted by a magnet are said to be **paramagnetic**, while the minerals which are slightly repelled are called **diamagnetic**. The different magnetic properties of minerals allows the separation of many minerals from mixtures when these are passed through special vibrating electromagnetic separators.

Variations in the Earth's magnetic field at the surface due to variations in the magnetic properties of minerals, can be detected by special magnetometers and this is an important geophysical method of mineral prospecting.

Minerals vary in their ability to conduct an electric current. The crystals of native metals and many of the sulphide minerals are good **conductors**, while minerals such as the micas are very good **insulators** because they will not conduct electricity. A change in temperature may induce an electric charge in some non-conducting minerals and this phenomenon is called **pyroelectricity**.

Piezoelectricity is the name given to the production of a charge on a non-conducting mineral due to the application of a directed pressure. Conversely when an electric field is applied to a piezoelectric crystal it is mechanically deformed. Quartz is a common mineral which exhibits piezoelectric properties, and plates cut parallel to particular planes in the quartz crystal structure are extensively used to provide oscillator plates for electronic equipment.

1.8 The nature of light

Sound waves and earthquake waves are transmitted through a medium and cannot travel through space. Light rays and X-rays form part of the **electromagnetic spectrum** and can travel through space.

Static electric charges give rise to electric fields of force in the surrounding space. In a similar way a permanent magnet gives rise to a magnetic field of force in the surrounding space and this can be demonstrated by placing iron filings on a piece of paper above a magnet. The iron filings become aligned parallel to the lines of force of the magnetic field. **Oersted** discovered that when an electric current flows along a helical wire which is known as a **solenoid**, a similar magnetic field is produced. Physicists then postulated that if electric currents produce magnetic fields, magnetic fields should produce electric currents. In 1831, **Faraday** demonstrated electromagnetic induction by recording electric currents in conductors when the orientation or intensity of a magnetic field surrounding the conductor was changed. **Maxwell** noted the symmetry of the two processes and considered whether they could combine to give a single electromagnetic process that would be self-sustaining.

A field of force depends on a source, and therefore changes in the source produces changes in the field. If a magnetic field was oscillating at a non-uniform rate, the in-

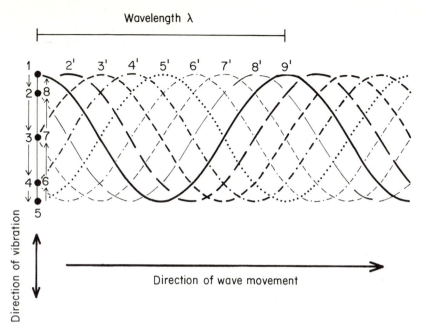

Fig. 1.29. A representation of the wave nature of light

duced electric field would change in a similar manner. This oscillating electric field would induce an oscillating magnetic field which in turn would induce an electric field and so on. Consequently under certain circumstances electric and magnetic fields may cooperate to form **electromagnetic waves** which once emitted from the source, are independent of the source.

The behaviour of electromagnetic radiation such as **light** can be understood in terms of **wave motion produced by the vibration of points in a direction perpendicular to the direction of transmission of the wave** as illustrated in fig. 1.29. When a point moves from position 1 to position 2 and then 3, the crest of the wave travels to position 2' and then 3' and so on. After one complete vibration through positions 1 to 5 and back to 1, the wave crest has moved to position 9'. The points are said to be in phase when they vibrate in unison, each maintaining the same relative position on the wave form. For example points 1 and 9' in fig. 1.29 are

in phase. The distance between any two points which are in phase is known as the **wavelength**. Wavelength is denoted by the Greek letter lambda, λ, and in this text it will be referred to terms of angstrom units, $1 \text{ Å} = 10^{-8}$ cm. Wave motion of this kind is known as **simple harmonic motion** and it will be considered further in section 6.12.

The time elapsing between the successive recurrences of some feature at a given point, for example the time interval between the recurrence of the vibration at its maximum displacement is called the **period T**. Vibrations or cycles per unit time is the **frequency** v, and $v = 1/T$.

The distance travelled by any particular feature such as the crest of the wave in unit time is the **wave velocity** v and $v = v\lambda = \lambda/T$ cm/ sec. The maximum displacement of the point of vibration is the **amplitude** of the vibration.

The portion of the spectrum of electromagnetic waves which contains the visible light waves is represented by fig. 1.30, and the ranges of the wavelengths of waves which are seen as individual colours are indicated. The most easily visible colour is yellow which is produced by waves with wavelengths ranging between 5700 Å and 5900 Å. White light consists of a mixture of all the colours produced by wavelengths extending from 3900 Å, the limit of visible violet light, to 7800 Å, the longest waves of red light.

Some minerals allow light to pass through them and are said to be **transparent** when objects can be seen through the minerals, or **translucent** when objects cannot be seen clearly. Colourless transparent minerals transmit all the wavelengths of white light. Coloured transparent minerals absorb certain wavelengths from white light and transmit the remainder which combine to give the observed colour. White non-transparent minerals reflect all the wavelengths of white light, while coloured non-transparent minerals reflect only the waves which give the observed colour. Black minerals absorb all wavelengths of visible light, and since they do not transmit any light they are said to be **opaque**.

1.9 The refraction of light

The velocity of light is greatest in vacuum and is approximately 299,739 km/sec. In all other substances the velocity of light is reduced. When a light ray passes obliquely from one medium, for example air, into another medium in which it travels with a lower velocity, for example water, its direction of transmission also changes, as indicated in fig. 1.31. The bending of a light ray at the boundary between two media is known as **refraction**. The ray of light in the first medium is known as the **incident ray**, and after it crosses the boundary into the second medium it is called the **refracted ray**. The incident and refracted rays always lie in a plane which is normal (meaning perpendicular) to the boundary surface. The angle between the incident ray and the normal to the surface is called the **angle of incidence** and is denoted by the letter i. The angle between the refracted ray and the normal to the surface is called the **angle of refraction** and it is denoted by the letter r as shown in fig. 1.31.

Consider a group of parallel light rays forming a beam of light of width represented by the line DE drawn perpendicular to the beam in fig. 1.32. DE may be regarded as part of a wave front which repre-

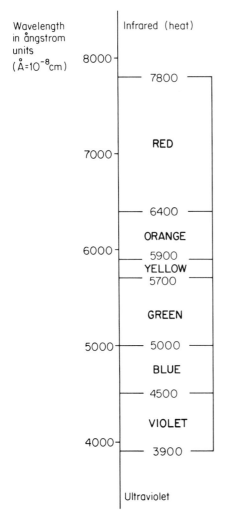

Fig. 1.30. The visible light portion of the electromagnetic spectrum

sents the position of a group of rays at a particular instant of time. The outside ray D will reach the boundary surface first and will be refracted to continue along the new path to G within the second medium. By the time that the outside ray E reaches the surface at H, the ray D will have travelled the distance DG at the lower velocity determined by the second medium. The line GH will then represent the wavefront of the beam in the second medium.

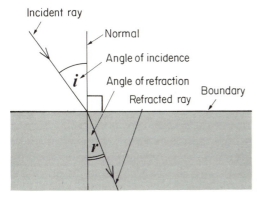

Fig. 1.31. The refraction of a ray of light at the boundary between two media

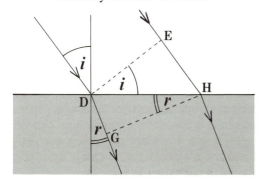

Fig. 1.32. The refraction of a beam of light

Since in the particular time interval, light has travelled the distance EH in the first medium and the distance DG in the second medium as represented in fig. 1.32, the distances EH and DG are proportional to the velocities of the light in the two media.

$$\frac{\text{Velocity in the 1st medium}}{\text{Velocity in the 2nd medium}} = \frac{\text{EH}}{\text{DG}}$$

but since $\sin i = \dfrac{\text{EH}}{\text{DH}}$, and $\sin r = \dfrac{\text{DG}}{\text{DH}}$

$$\frac{\text{Velocity in the 1st medium}}{\text{Velocity in the 2nd medium}} = \frac{\sin i}{\sin r}$$

The ratio of the velocity of light in the two media is a constant which is known as the **refractive index** and its value is obtained from the ratio **sin i/sin r**.

Since the velocity of light is greatest in vacuum this is the standard for comparison. The velocity of light is retarded in air but the refractive index of air is only 1.00029, so for convenience, air is commonly used as the standard for comparison. A light ray passing from vacuum into another substance is always refracted towards the normal and since $i > r$ the refractive index is always greater than 1. The greater the retardation of the velocity of light in a substance, the greater the refraction, and consequently the greater the refractive index of the substance. The refractive indices of some common substances are given in the list below.

Refractive index

Water	1.336
Fluorspar	1.43
Crown glass	1.53
Glass (imitation gems)	1.67
Almandine garnet	1.77
Diamond	2.42

1.10 The refractometer

The measurement of the refractive index is important for the identification of minerals, and it is useful for the detection of imitation gems since the refractive indices of glasses are less than the refractive indices of most gemstones. Consider again the rays of light passing through a medium of lower refractive index into a medium of higher refractive index as in fig. 1.33. The ray which is normal

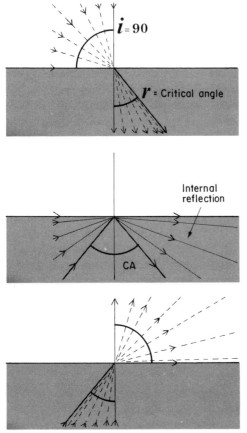

Figs. 1.33, 1.34, and 1.35. The refraction and internal reflection of light at the boundary between two media

to the surface goes through unrefracted, but all other rays are refracted towards the normal. For the ray nearly parallel to the boundary, i is nearly 90°. In this situation r is at its maximum value and this is called **the critical angle** CA.

A light ray which is just oblique to the boundary but which approaches the boundary through the lower medium, is internally reflected as shown in fig. 1.34. The position of the reflected ray is determined by the laws of **reflection**, one of which states that the angle of reflection is equal to the angle of incidence. As the angle of incidence decreases, so does the angle of internal reflection until the critical angle is reached because this is the minimum angle of internal reflection.

When the angle of incidence equals the critical angle (i = CA), or is less than the critical angle as shown in fig. 1.35, most of the light passes through the boundary and is refracted in the medium of lower refractive index in a direction away from the normal to the boundary.

The refractive index of some gemstones which have polished surfaces may be measured by a simple instrument known as a **refractometer**. The principle of the refractometer is illustrated in fig. 1.36. The gemstone is mounted on a glass hemisphere with a liquid of higher refractive index than the gemstone to exclude air from the contact surface. The hemisphere of special glass may have a refractive index as high as 1.88. Light internally reflected within the glass hemisphere illuminates part of a calibrated scale. The approximate refractive index of the gemstone is indicated by the position of the critical angle. Part of the scale is not illuminated because light is refracted through the gemstone. Clearly the use of the refractometer is limited to gemstones which have refractive indices less than that of the glass hemisphere.

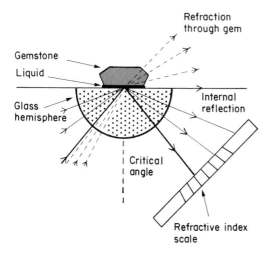

Fig. 1.36. The principle of the refractometer

1.11 Dispersion and absorption

Newton discovered that when white light is transmitted through a prism it is separated into a complex spectrum of colours. This phenomenon is known as **dispersion** and it arises because the amount of refraction of light is inversely proportional to the wavelength of light. The short wavelengths of violet light are refracted more than the long wavelengths of red light as illustrated in fig. 1.37. In diamond the refractive index is 2.407 for red light (6870 Å) and 2.465 for violet light (3970 Å). Consequently accurate measurements of the refractive index of a medium must be done with light which has a small range of wavelengths. This is called **monochromatic light** and the yellow light emitted by sodium is the most commonly used monochromatic light for refractive index determinations.

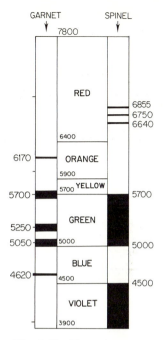

Fig. 1.39. The absorption spectra of red almandine garnet and red spinel

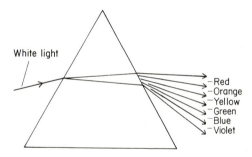

Fig. 1.37. The dispersion of light in a prism

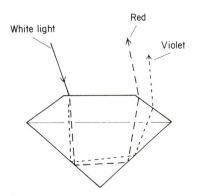

Fig. 1.38. The dispersion of light in a gemstone

Gemstones such as diamond, when rotated show flashes of spectral colours and are said to show a '**play of colours**' or to have '**fire**'. This is due to the high amount of dispersion of white light in the mineral, and the gemstones are cut in a way which accentuates the dispersion of the spectral colours. A side view of the gemstones shape which is known as the '**brilliant cut**' is shown in fig. 1.38, to illustrate the manner in which the 'play of colours' is produced.

When white light is passed through a colourless stone all the wavelengths are transmitted but with reduced intensity. When white light is passed through a coloured stone and then dispersed by a prism, the spectrum is seen to be incomplete. Certain colour ranges are absent and their positions are occupied instead by black bands. The black zones indicate that the rays have been absorbed by the mineral and consequently such an incomplete spectrum is

known as an **absorption spectrum**. The
wavelengths absorbed by the mineral are
characteristic of the mineral and may be
used to distinguished between certain gemstones by means of a simple instrument
known as a **spectroscope**.

For example the gemstones of red garnet,
almandine, and the red variety of the mineral **spinel**, have very similar crystallographic
and optical properties but they can be distinguished easily by their characteristic
absorption spectra which are represented in
fig. 1.39. Almandine garnet has narrow
absorption lines at 6170 Å and 4620 Å, and
wider more conspicuous absorption bands
centred on the wavelengths of 5700 Å,
5250 Å and 5050 Å between which green
light is transmitted. In contrast red spinel
absorbs all the green and violet wavelengths
and has narrow absorption lines at 6855 Å,
6750 Å and 6640 Å.

A mineral is said to be **irridescent** when it
shows a series of spectral colours usually
reflected from internal surfaces. A milky or
pearly reflection known as **opalescence** is
also due to internal reflections.

The general appearance of the surface of
a mineral in reflected light is called the
lustre. A mineral which has the brilliant
appearance of a metal such as the chalcopyrite in plate 12, is said to have a **metallic
lustre**. These minerals have very high refractive indices exceeding 3, but generally they
are opaque.

All minerals without a metallic appearance are said to have a **non-metallic lustre**.
Some of the terms used to describe the
lustre of non-metallic minerals are given
below.

vitreous	like broken glass, e.g. quartz in fig. 1.15.
resinous	like resin, e.g. sphalerite in plate 11.
greasy	greasy to the touch, e.g. talc (soapstone) fig. 1.8.
pearly	the lustre of pearls.
silky	fibrous structure of gypsum, satin spar fig. 1.9.
dull	like chalk.
adamantine	brilliant lustre due to a high index of refraction as in di-amond.

1.12 Fluorescent minerals

Fluorite is an example of a mineral which
shows a very wide range of colours without
any easily observable difference in composition or structure. The four specimens shown
at the top of plate 16 are yellow fluorite
occurring with black galena, and blue,
green, and purple fluorite crystals. Minerals
that become luminescent when exposed to
ultraviolet light radiation are said to be
fluorescent, a name derived from the mineral fluorite, many varieties of which
fluoresce. Three of the specimens of fluorite
of different colours shown in plate 16,
fluoresce giving the pale blue colour shown
in plate 17, but the yellow crystals do not
fluoresce. In some specimens the intensity
of the fluorescence varies in growth zones
which are parallel to the crystal faces. These
zones are related to periods of growth of the
fluorite crystals and suggest that slight differences in the composition of the fluorite
occurred during its crystallization.

The bright yellowish green mineral at the
centre of plate 16 is **autunite**, a hydrated
phosphate mineral containing uranium, and
it occurs as bunches of foliated crystals.
Autunite fluoresces with a brilliant green
colour as shown in plate 17.

Specimens from the zinc deposits of
Franklin, New Jersey are well known for
their fluorescent properties. The specimen
shown in the bottom right hand side of plate
16 is composed of the following minerals.

Franklinite	$ZnFe_2O_4$	black metallic crystals.
Zincite	ZnO	red coloured crystals.

Willemite $Zn_2(SiO_4)$ white crystals.

When excited by ultraviolet radiation the **willemite** is clearly seen in plate 17 by its green fluorescence.

Flourescence is a property that cannot be predicted because not all specimens of a particular mineral will fluoresce. However, in some mines certain minerals such as **autunite** can be located easily by their fluorescence. If the luminescence continues after the excitation radiation is removed, the phenomenon is known as **phosphorescence**.

1.13 The occurrence of crystals in rocks

The most conspicuous product of many volcanic eruptions is red hot lava which consolidates to form dark coloured rocks, within which crystals of certain minerals may be enclosed. A freshly broken surface of **lava rock** is shown in plate 13, and it can be seen that the yellowish green crystals of **olivine** occur with almost black crystals of the mineral **augite** embedded in a grey groundmass.

Lava is essentially molten rock which has travelled up to the Earth's surface, and a more general name for these liquids is **magma**. Rocks formed from high temperature magmas are classified as **igneous rocks**. A melt is a liquid within which there is an almost completely random arrangement of atoms in continuous motion. In contrast the geometric form and stability of a mineral arises because the atoms in the crystal structure are confined to definite positions about which they may vibrate but which they cannot easily leave.

When a melt cools it loses energy. As the translational energy of its atoms decreases it is less easy for individual atoms to move about and for the liquid to flow. The **viscosity** of a liquid is a measure of the resistence to flow, and clearly during the cooling of a lava the viscosity increases until the material becomes **rigid** like a rock. If the rate of cooling of a lava is relatively high, the viscosity may become so great that the liquid appears to be mechanically rigid at normal temperatures and over short periods of time. The melt is said to have consolidated as a **glass** and it can be regarded as a greatly cooled liquid with an extremely high viscosity. Rocks composed of glass such as the **pitchstone** illustrated in fig. 1.13, characteristically show **conchoidal fractures**. If a melt cools slowly, some clusters of atoms would begin to grow in size to form crystals of particular minerals. In some igneous rocks large individual crystals can be seen clearly in a fine grained ground mass as shown in plate 13, and by fig. 1.40. The manner in which the crystals occur in a rock is called the **texture**.

Lavas frequently contain cavities known as **vesicles** which were produced by the separation and expansion of small bubbles of gas consisting mainly of steam and the sulphur oxides. Subsequent crystallization on the margins of the vesicles and crystal growth inwards may produce globular masses known as **amygdales**, and the rock is then said to be **amygdaloidal**. Fig. 1.41 shows partly filled amygdales in lava rock called **basalt**, and well formed crystals of **chabazite** can be seen in the largest cavity. Crystals may form directly from the volcanic gases by a process known as **sublimation**, and the most common example of this is the growth of **sulphur** crystals around the volcanic vents from which volcanic gases escape.

Sedimentary rocks are produced from material deposited mechanically or chemically on the Earth's surface, and they are usually characterized by bedding structures. Due to the processes of erosion and transportation by flowing water, ice or wind, the crystals derived mechanically from other rocks have greatly modified shapes, like the water worn crystals of sapphire in plate 7. Certain minerals crystallize from aqueous solutions to fill the pore spaces of fragmen-

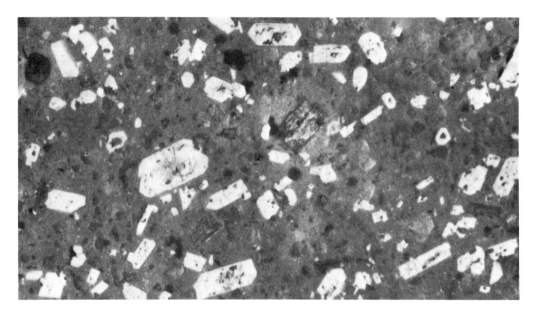

Fig. 1.40. An igneous rock composed of large crystals of white feldspar (plagioclase) and biotite enclosed in a fine grained ground mass. (×2)

Fig. 1.41. A lava rock called basalt containing gas cavities partly filled with crystals. The largest cavity contains crystals of **chabazite** on which the interfacial angles are close to 90°. The almond-like masses are called amygdales and the rock is called an **amydgaloidal basalt**. (× 1)

tary sedimentary rocks, or form continuous beds of chemically precipitated rock. The most common compound precipitated from aqueous solutions is **calcium carbonate ($CaCO_3$)** which most commonly occurs as the mineral calcite illustrated in figs. 1.22–1.24, but may also occur in a very different crystal form characteristic of the mineral **aragonite** shown in fig. 1.42. This is another example of the phenomenon of **polymorphism**. Calcite is the more stable mineral and it forms beds of limestone, and also the tapered cylindrical masses called **stalactites** suspended from the roof of caverns in limestone rocks and similar masses known as **stalagmites** which rise from the floors of caves.

Many marine organisms utilize calcium carbonate to build up hard protective shells,

Fig. 1.42. Crystals of **aragonite**, $CaCO_3$, showing the characteristic forms. The crystal on the right is viewed in the direction of the longest edges. ($\times$ 2)

and on the death of the animal the shell fragments may accumulate to produce limestone beds. Other animals construct their shells with aragonite crystals but later these change to calcite because this is the more stable crystal form of $CaCO_3$. Fig. 1.43 is a section of a large ammonite fossil in which the chambers have been completely or partly filled by the crystallization of calcite crystals that nucleated on the shell walls and grew into the cavities. Crystallization from solutions filling cavities or planar fracture openings in rocks frequently allows the growth of well-shaped crystal forms. When the cavity is not completely filled by crystallization and well developed crystals surround a space, the structure is known as a **vug** or **geode**.

Evaporation of large bodies of water in arid environments results in the saturation of the solution and then precipitation of certain minerals to form beds known as **evaporites**. The commonest minerals to form evaporites are **gypsum** (calcium sulphate) and **halite** (sodium chloride). When the evaporating body of water has been reduced to about 1.54% of its original volume, potassium chloride may precipitate as the mineral **sylvite**.

A variety of **apatite** known as **phosphorite** is an uncommon precipitate from sea water but it is found in beds of sedimentary rocks and provides an important source of phosphate fertilizer.

Crystals may grow within a rock by the **replacement** of other minerals. Plate 18 shows an ammonite shell which originally consisted of calcium carbonate but is now composed partly of the iron sulphide, **pyrite**. Through the medium of the aqueous solution occupying the pore spaces in the rock, the calcium carbonate has gone into solution and its place has been taken by the iron sulphide precipitated from solution. A completely replaced shell or crystal is known as a **pseudomorph** referring to the form of the original material.

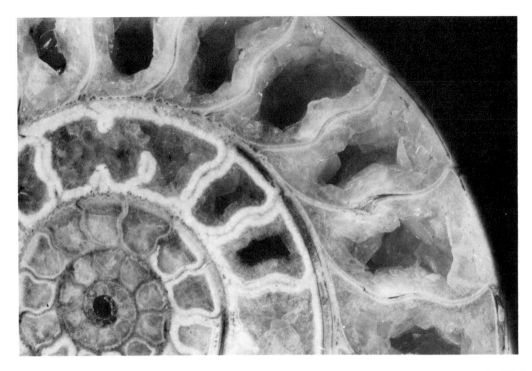

Fig. 1.43. Part of the interior of an **ammonite shell** showing the chambers completely or partly filled by the growth of calcite crystals from the walls. (×1)

The growth of crystals by the replacement of other minerals is particularly well seen in **metamorphic rocks,** which have been formed from other rocks as a result of an increase in temperature or by deformation. Fig. 1.44 is a photograph of a rock which was composed of compacted mud hence its name **mudstone,** but in this particular specimen prismatic crystals with diamond shaped end sections have grown by the replacement of the original material. The prismatic crystals within the mudstone are crystals of **andalusite**, an aluminium silicate which most frequently occurs in metamorphic rocks near large intrusions of igneous rock. The intrusion of a large mass of hot magma caused the original clay minerals of the mudstone to recrystallize largely in the solid state, to form the andalusite crystals. The larger crystal in fig. 1.44 shows the characteristic form of andalusite. If a mudstone suffers an increase in temperature accompanied by deformation it may recrystallize to form crystals of **biotite**, and the parallelism of adjacent biotite flakes produces the texture called **schistosity**. The almost perfect rhombdodecahedron crystal form of **garnet** in the rock shown in plate 10 and the **beryl** crystal in plate 14 grew by the replacement of the surrounding crystals of biotite and quartz in the biotite schists.

The minerals formed in igneous rocks and chemically precipitated sedimentary rocks, and the new minerals formed in metamorphic rocks, crystallized because they were the appropriate stable mineral phases for the particular conditions of temperature, pressure and chemical composition. Consequently careful studies of mineral assemblages will reveal information concerning the conditions of formation of the particular rock.

Fig. 1.44. Prismatic crystals of **andalusite** which have grown within a mudstone as the result of an increase in temperature. Note the characteristic cross section which is shown by the single large crystal on the left. During the crystal growth by replacement, material might remain on diagonal planes because the structure of the andalusite is less easily formed on the edges than on the faces. Andalusite containing diagonal planes of inclusions are known as **chiastolite** and a crystal is shown with two cross-sections. (×1)

When rocks are exposed to weathering processes near the earth's surface, the primary minerals may be altered to other minerals, known as **secondary minerals** which are stable under the new physical and chemical conditions. In temperate climates the most common products of weathering are the **clay minerals**. In tropical climates many elements are leached out of the weathered rock leaving a reddish brown residual deposit of iron and aluminium oxides which is called **laterite**. Under special circumstances the residual deposit may be rich in aluminium and it is then known as **bauxite**, the ore material of aluminium. The oxide minerals in many bauxites have crystallized in intricate, spherical concretionary structures known as **pisolites**, and the texture is illus-

trated in fig. 1.45. Some minerals such as the powdery limonite in plate 1, do not show regular crystal structures and they are said to be *amorphous*. When a mineral occurs as a mass of tiny grains which cannot be distinguished with a microscope but shows a good X-ray diffraction pattern it is said to be **cryptocrystalline**.

In this chapter many of the most attractive features of minerals have been illustrated and some of the methods of study have been introduced. In most cases the colour of a mineral although attractive, is not a reliable property for identification purposes, and the methods of study introduced here are of limited application. It is clearly apparent that the crystal forms and structures are the most fundamental characteristics of a mineral, and it is appropriate therefore to commence quantitative crystallographic studies so that many of the features introduced here can be understood and the problems that have been posed can be solved.

Fig. 1.45 **Bauxite**, the ore of aluminium, is often found composed of **pisolitic concretions** of aluminium and iron oxides and hydroxides

2 *The Symmetry of Crystals*

In the introductory chapter many of the most attractive characteristics of crystals were illustrated and some preliminary methods of study were described. However it is obvious that these methods are of limited value, and that progress in the study of minerals can be made only by appreciating the principles of crystallography—the science of crystals, which is concerned primarily with the description of the various crystal forms.

The objectives of this chapter are to gain an understanding of the aspects of crystallography which are listed below.

1. The angular relationships of crystal faces.
2. The elements of crystallographic symmetry.
3. The relationships between the elements of symmetry in geometrically symmetrical crystal forms.
4. The classification of crystals into symmetry classes.
5. The classification of symmetry classes into crystal systems.

Contents of the sections

2.1 Interfacial angle

A group of quartz crystals is shown in fig. 1.1 and it can be seen that they are six-sided columnar crystals with inclined pyramid like faces at the unattached ends. The characteristic cross sectional shapes of quartz crystals which are seen when the crystals are viewed along the longest dimension are illustrated in fig. 2.1.

The angle between two crystal faces measured in a plane which is perpendicular to both crystal faces, is called the **interfacial angle**. The interfacial angles on large crystals can be measured with a simple instrument known as a **contact goniometer** as

Fig. 2.1. Sections of quartz crystals perpendicular to the longest edges. (×2)

shown in fig. 2.2. This instrument can be made by attaching a straight arm so that it can rotate about the centre of the graduated semicircle of a protractor. The diameter of the contact goniometer and the arm are placed in contact with two faces so that the plane of the goniometer is perpendicular to the faces, and the interfacial angle can be read directly from the scale.

Either the external angle $D\widehat{B}C$ or the internal angle $A\widehat{B}C$ between two faces may be measured as shown in fig. 2.3. The external angle $D\widehat{B}C$ is equal to the angle $A\widehat{E}C$ between the normals to the two faces and this is the interfacial angle which is usually recorded. The interfacial angles between the side faces of quartz crystals are 60° irrespective of the size of the faces.

After studying sections of quartz crystals which consisted of faces of different sizes such as those illustrated in fig. 2.1 the Dane, **Nicolaus Steno**, established in 1669, the fundamental law of crystallography which states that **the angles between corresponding faces on the crystals of a particular mineral are constant regardless of the size or shape of the crystal**.

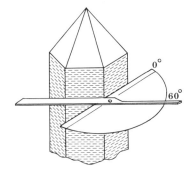

Fig. 2.2 A contact goniometer

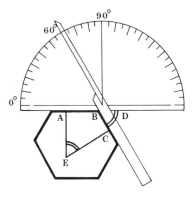

Fig. 2.3. The interfacial angle

For more accurate work particularly on small crystals, a **reflecting goniometer** is used in which a beam of light is reflected from faces of a crystal that has been mounted so that it can be rotated about an axis which is parallel to the line of intersection of the two crystal faces.

2.2 Zone and zone axis

The six faces which form the sides of the columnar quartz crystals intersect to form edges which are parallel as shown in fig. 2.4. A set of faces which intersect to produce parallel edges or which when produced would intersect along lines which are parallel, is said to constitute a **zone.** The direc-

tion of the lines of intersection of the faces of a zone is called the **zone axis**. The zone axis of the side faces of the quartz crystals is parallel to the longest dimension of the columnar crystals and the longest edges as shown in fig. 2.4.

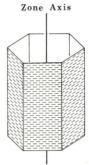

Fig. 2.4. A zone and zone axis

Fig. 2.5. Crystals of **gypsum.** (×2)

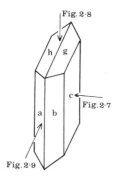

Fig. 2.6. An obli-
que view of a gyp-
sum crystal

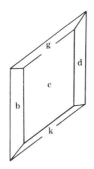

Figs. 2.7, 2.8, and 2.9. Views of the gypsum
crystal in the directions indicated in fig. 2.6

Study gypsum crystals similar to those
shown in fig. 2.5. A view of a gypsum crystal
perpendicular to the largest face is shown in
fig. 1.17.

Drawings of a gypsum crystal when seen
from four different viewing directions are
shown in figs. 2.6 to 2.9. On the drawings,
the faces are identified by letters. This is the
method commonly used when studying crystals and it is then necessary to answer the
following questions.

(1) *Does the gypsum crystal exhibit any
    zones?*
(2) *Which sequences of faces constitute each
    zone?*
(3) *What are the orientations of the zone
    axes?*
(4) *What are the approximate values of the
    interfacial angles in each zone?*

The sets of parallel edges indicate that the
gypsum crystal is composed of two zones.
One zone consists of the faces labelled **a, b,
c, d, e,** and **f,** and the intersections of these
faces produce the prominent vertical edges.
The second zone is made up of the faces
labelled **c, g, h, f, j,** and **k,** and the intersections of these faces produce the inclined set
of parallel edges shown in fig. 2.10. The
second zone axis is inclined at **52°** to the
vertical zone axis as shown in fig. 2.11.

Fig. 2.12 is plan of a gypsum crystal in a
plane perpendicular to the vertical zone axis
and the interfacial angles are indicated.
Note the symmetry of the interfacial angles
about a plane which is parallel to the vertical
zone axis. Fig. 2.13 is a plan of a gypsum
crystal in a plane perpendicular to the inclined zone axis to show the interfacial
angles of the inclined zone. Note the symmetry of the interfacial angles about a plane
which is parallel to the inclined zone axis.
When these two sets of observations of
interfacial angles are combined it can be
seen that the angular orientation of the faces
of the gypsum crystal are symmetrically

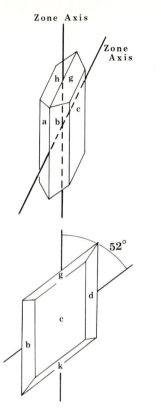

Fig. 2.10, and 2.11. The zone axes of the gypsum crystal

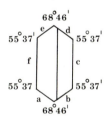

Fig. 2.12. The interfacial angles of the vertical zone

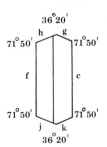

Fig. 2.13. The interfacial angles of the inclined zone

arranged with respect to a plane which is parallel to the two zone axes.

2.3 The symmetry of the gypsum crystal

The interfacial angles of the gypsum crystal shown in fig. 2.14 reveal that in both zones the crystal faces can be said to be symmetrically arranged in that they occur in pairs, one face on each side of the plane which contains the two zone axes. Each pair of faces has the same interfacial angle with respect to the pair of large faces and to an imaginary plane containing the two zone axes. This imaginary plane which is parallel to the two zone axes in the gypsum crystal is called a **plane of reflection symmetry**.

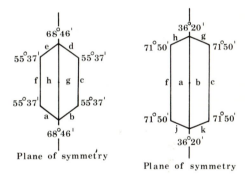

Fig. 2.14. The plane of symmetry of the gypsum crystal

A plane of reflection symmetry is defined as an imaginary plane that divides the crystal so that the angular orientation of a face on one side is the mirror image of a corresponding face on the other side of the reflection plane. One half of a gypsum crystal cut parallel to the pair of large faces has been placed on a mirror in fig. 2.15 to illustrate the concept of a plane of reflection symmetry. This crystallographic symmetry depends on the angular orientation of the similar crystal faces. The actual sizes of the crystal faces are not significant because the size depends on such factors as the position and conditions of crystal growth. Perfect

Fig. 2.15. A mirror reflection of one half of the gypsum crystal. (×2)

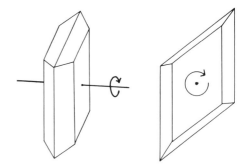

Fig. 2.16. The diad axis of symmetry

tions during a rotation through 360° about the axis. An axis of rotation which has this property is called an **axis of symmetry**, and since in this particular case the same orientation occurs twice during the rotation, it is said to be an axis of 2-fold symmetry or a **diad axis**. The axis of two-fold symmetry may be referred to by the symbol A_2 and in diagrams it is indicated by the symbol ◖. *Is the vertical zone axis in the gypsum crystal a symmetry axis?*

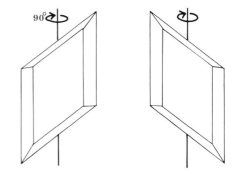

Fig. 2.17. Rotation about the vertical zone axis

geometrical symmetry is rare in crystals. Crystallographic symmetry is of fundamental importance because it is related to the internal structure of the crystal. The plane of reflection symmetry is sometimes called a **mirror plane** and it is represented by the letter M in the symbols of the symmetry of crystals.

If the gypsum crystal is rotated about an axis which is perpendicular to the two largest faces as shown in fig. 2.16, the same appearance is seen twice during a complete rotation through 360°. To be more precise it should be said that **exactly the same angular orientations of the faces occur in two posi-**

After rotation of 180° about the vertical zone axis, the gypsum crystal has quite a different orientation as show in fig. 2.17 and therefore this zone axis is not a symmetry axis.

A third element of symmetry is known as the **centre of symmetry** indicated by the symbol **C**. A symmetry centre is present if similar crystallographic features occur on opposite sides of a crystal. The gypsum crystal possesses a symmetry centre because each face is related to a similar parallel face on the opposite side of the crystal. **The operation by which the position of a face may be taken by the identical face which is on the opposite side of the crystal, is known as inversion through the centre of symmetry.**

In contrast, the tetrahedron form shown in fig. 2.18 consists of single faces and it has no symmetry centre. With the tetrahedron the inversion operation replaces a face by a point at which faces meet because there is no symmetry centre.

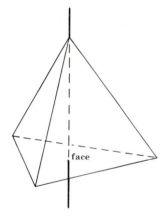

face

Fig. 2.18. The tetrahedron form illustrating the absence of a symmetry centre

The crystallographic symmetry of the gypsum crystal can be described by the symbols **M, A₂, C,** and it must be noted that the diad axis is perpendicular to the plane of reflection symmetry. In this way the symmetry of the angular orientations of the faces of a crystal are described in terms of the elements of symmetry.

In crystallography the term **form** has a special meaning. **A form consists of a group of faces all of which have the same angular relationship to the elements of symmetry.** For example the two large side faces of the gypsum crystal compose a form because they have the same orientation with respect to the plane of reflection symmetry which contains the two zone axes. *How many forms are present in the gypsum crystals illustrated above?*

It can be seen that the gypsum crystals are composed of three groups of faces which have identical angular orientations with respect to the plane of reflection symmetry. One form consists of the two faces labelled **c** and **f** which are parallel to the plane of reflection symmetry as shown in fig. 2.19. The second form consists of the four faces **a, b, d,** and **e** which are parallel to the vertical zone axis and have interfacial angles of

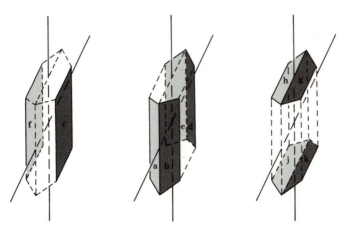

Fig. 2.19, 2.20, and 2.21. The forms present in the gypsum crystal

55°37' with the reflection plane, fig. 2.20. The third form is composed of the four faces, **g, h, j,** and **k** which are parallel to the inclined zone axis and have interfacial angles of **71°50'** with the reflection plane, fig. 2.21.

2.4 The symmetry of the cube form

Crystals of halite and fluorite with the cube form are shown in fig. 1.19, and plate 15, respectively. Although crystal symmetry refers to the angular arrangement of the faces without consideration of the actual size of the faces, it is convenient to study the different degrees of crystallographic symmetry by using models of crystals on which similar faces have equal sizes. The models possess geometrical symmetry as well as crystallographic symmetry.

Consider the cube and determine whether it is composed of zones and also how many forms are present.

The cube exhibits three sets of parallel edges and therefore it is clear that the cube is composed of three zones. The three zone axes are mutually perpendicular. All six faces of the cube are identical since they are each parallel to two zone axes and perpendicular to the third zone axis. Consequently the cube consists of one form of six faces which completely enclose space. The cube form is termed a **simple form.**

Determine the kinds of axes of symmetry possessed by the cube and also determine the numbers of each kind of symmetry axis.
Consider the rotation of a cube about an imaginary line which is parallel to the vertical edges as shown in fig. 2.22. *Can this axis be regarded as an axis of symmetry?*

Since a similar view of the cube is observed in four different positions during rotation, an axis which is parallel to the vertical edges is an axis of 4-fold symmetry and it is called a **tetrad axis**. A tetrad axis may be referred to by the symbol A_4 and in diagrams is indicated by a square ■. In the cube there are three tetrad axes ($3A_4$), each axis lying parallel to a zone axis as shown in fig. 2.23.

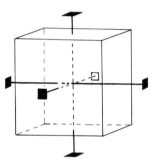

Fig. 2.23. The tetrad axes
of the cube

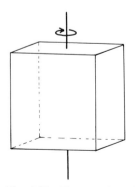

Fig. 2.22. Zone axis of
the cube

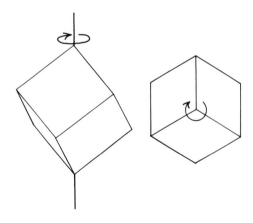

Fig. 2.24. A rotation axis in the cube

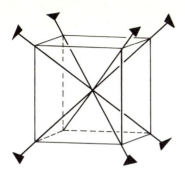

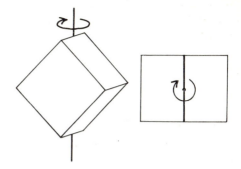

Fig. 2.25. The triad axes of the cube

Fig. 2.26. A rotation axis in the cube

Consider the rotation of cube about an axis joining opposite corners, as illustrated by the oblique view and the plan in fig. 2.24. *Is this an axis of symmetry? If so how many axes of this kind are present?*

Since the faces of the cube have the same orientation in three positions during a complete rotation, the axis joining the corners of a perfectly symmetrical cube can be described as an axis of 3-fold symmetry or a **triad axis**. Since the triad axes join pairs of opposite corners of the cube there must be four triad axes, (**4A₃**) indicated by triangles in fig. 2.25. The triad axis lies at equal angular distances from the three tetrad axes.

Are the axes which are each perpendicular to a pair of opposite edges of the cube as illustrated by the oblique view and the plan in fig. 2.26, axes of symmetry?

When rotated on an axis perpendicular to a pair of opposite edges the appearance of the cube is repeated twice so that the axis is one of 2-fold symmetry (**A₂**), and it is called a **diad axis**. Since there are six pairs of opposite edges on the cube there must be six diad axes as illustrated in fig. 2.27.

How many planes of reflection symmetry are present in the cube?

The plane which is perpendicular to a tetrad axis is a plane of reflection symmetry because it divides the crystal so that each half is a mirror image of the other, as shown in fig. 2.28. There are three planes of sym-

metry each perpendicular to one tetrad axis, and these are sometimes called the **axial planes of symmetry** because each is parallel to two of the three mutually perpendicular tetrad axes which later will be used as reference axes for describing the orientation of the faces.

The diagonal plane joining two opposite edges of the cube is a plane or reflection symmetry which is perpendicular to one of the diad axes and parallel to one diad and one tetrad axis as shown in fig. 2.29. Since there are six diad axes there are six **diagonal planes of symmetry** which bisect the angles between the crystallographic axes.

The cube has a centre of symmetry because the faces occur in pairs one on each side of the centre. The complete elements of symmetry of the cube may be described by the symbols listed below:

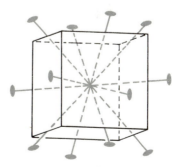

Fig. 2.27. The diad axes of the cube

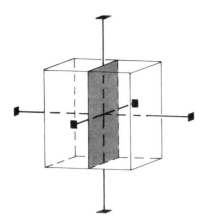

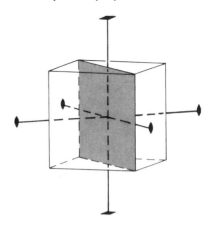

Fig. 2.28. An axial plane of symmetry in the cube

Fig. 2.29. A diagonal plane of symmetry in the cube

**$3A_4$, $4A_3$, $6A_2$,
9M** described as 3 axial planes each parallel to two tetrad axes and perpendicular to
the third tetrad axis, and 6 diagonal
planes each perpendicular to a diad axis.
C.

2.5 The symmetry of the icositetrahedron, octahedron, and rhombdodecahedron forms

Garnet crystals often occur as a form called
the icositetrahedron which consists of twen-

Fig. 2.30. Diagrams of the icositetrahedron form as seen from three particular directions which are
labelled b, c and d on the oblique view. Corresponding views of the garnet crystal are shown by the
photographs below. ($\times 1$)

ty four similar faces and two different views of the form are shown in plate 10. Fig. 2.30 shows the icositetrahedron form when seen from an oblique direction and along three particular directions which are labelled with the letters **b, c** and **d**. *Study the diagrams of the icositetrahedron to determine the complete elements of symmetry of this simple form and consider whether it shows the same degree of crystallographic symmetry as the cube form.*

It can be established from fig. 2.30 that the direction labelled **b** is a tetrad axis of symmetry and that two planes of reflection symmetry intersect along the tetrad axis. The operation of these elements of symmetry imply that there are three tetrad axes which are mutually perpendicular and these will be used as reference axes later. There are also three planes of reflection symmetry which are parallel to two tetrad axes and these are referred to as the axial planes of symmetry.

The direction **d** in fig. 2.30 is a diad axis and it can be seen that two planes of reflection symmetry intersect along the diad axis. The diad axis bisects the angle between two tetrad axis as shown in fig. 2.31. Therefore there will be six diad axes of symmetry in the icositetrahedron form. The reflection planes which are parallel to the diad axes can be described as diagonal symmetry planes, to indicate their orientation with respect to the reference axes.

The direction labelled **c** in fig. 2.30 of the icositetrahedon form is a triad axis. The triad axes are at equal angular distances from the tetrad axes and their orientation is best visualized by relating the icositetrahedon to the cube as in fig. 2.31.

The icositetrahedron possesses the elements of symmetry indicated by the symbols **3A₄, 4A₃, 6A₂, 3M** (axial), **6M** (diagonal), **C**. It can be seen that the icositetrahedron and the cube forms possess identical combinations of elements of symmetry. When studying models of these forms it is instruc-

tive to hold them with similar orientations so that the arrangement of the symmetry axes in the icositetrahedon can be related to the symmetry axes of the cube.

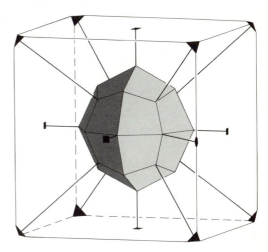

Fig. 2.31. The icositetrahedron form related to the cube

Fig. 2.32. **Magnetite** crystal showing growth steps on the octahedron faces. (×1)

The diamond crystals shown in plate 7 and in fig. 1.4 exhibit the octahedron form and an oblique view of a similar crystal of magnetite (Fe_3O_4) with prominent growth steps is shown in fig. 2.32. A diagram of an oblique view of a perfect octahedron and views of crystals of cuprite (Cu_2O), along three particular directions which are labelled **b, c,** and **d** are shown in fig. 2.33. *What*

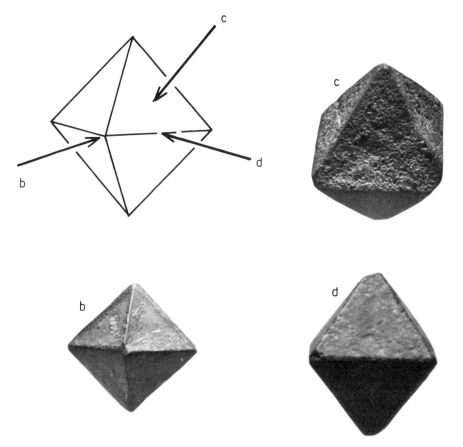

Fig. 2.33. The octahedron form of **cuprite** crystals seen from three particular
directions which are labelled b, c, and d. (×2)

*are the elements of symmetry exhibited by the
octahedron form?*

The garnet crystals shown in plate 10 also
exhibit the rhombdodecahedron form which
consists of twelve rhomb shaped faces.
*Which elements of symmetry can be identi-
fied from the different views of the rhombdodecahedron form shown in fig. 2.34?*

2.6 The symmetry of the pyritohedron form

Pyrite crystals exhibiting a form commonly
known as the pyritohedron are illustrated in
plate 6, and fig. 1.3. The **pyritohedron** form
shown by the model and crystal in fig. 2.35 is
also known as the **pentagonal dodecahedron**

and it is composed of twelve pentagonal
shaped faces. It can be seen that the pyritohedron form has three pairs of parallel
edges. *Study the different views of the pyritohedron form which are shown in fig. 2.36
and determine the complete elements of symmetry possessed by the form.*

The direction labelled **b** in fig. 2.36 is
perpendicular to a pair of parallel edges of
the pyritohedron and it is a diad axis of
symmetry. Two planes of reflection symmetry intersect along the diad axis. There
are three mutually perpendicular diad axes
($3A_2$) and three planes of reflection symmetry (**3M**) each perpendicular to a diad
axis.

Fig. 2.34. The rhombdodecahedron form of **garnet** crystals seen from three particular directions which are labelled b, c, and d. (×2)

Fig. 2.35. Crystal of **pyrite** and a model showing the pyritohedron form. (×2)

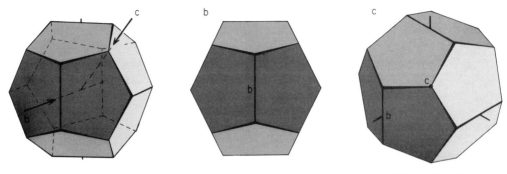

Fig. 2.36. The pyritohedron form as seen in two particular directions which are labelled
b and c

The direction **c** in fig. 2.36 is a triad axis and since this is at equal angular distances from the three diad axes there are four triad axes (**4A₃**) in the pyritohedron form.

It can be seen from the fig. 2.36 that the pyritohedron exhibits a centre of symmetry but the diagonal plane parallel to one diad axis and bisecting the angle between the other two diad axes is not a plane of reflection symmetry. The complete elements of symmetry of the pyritohedron form can be represented by the symbols **3A₂, 4A₃, 3M** (axial), **C**.

The pyritohedron can be related to the cube since both forms possess four triad axes as shown in fig. 2.37, but the pyritohedron has a lower degree of symmetry as shown below.

Cube   **3A₄, 4A₃, 6A₂, 3M(axial), 6M(diagonal), C**
Pyritohedron **3A₂, 4A₃,   3M(axial)   C**

Note that the three mutually perpendicular symmetry axes in the pyritohedron are diad axes. Although pyrite often crystallizes as cubes these usually show conspicuous striations as in fig. 2.38. The striations reveal that the axes perpendicular to the face of the pyrite cube are diad axes rather than tetrad axes since after a 90° rotation the orientation of the striations is different.

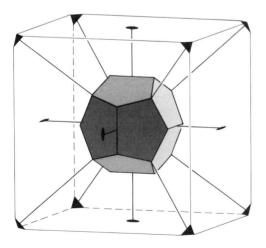

Fig. 2.37. The pyritohedron form related to the cube

Fig. 2.38. A **pyrite** crystal showing the cube form with striated faces. (×2)

Similarly the striations reveal that the pyrite cubes do not possess diagonal planes of symmetry.

2.7 The symmetry axis of rotary inversion of the tetrahedron form

The mineral sphalerite (**ZnS**) is illustrated in plate 11 and a group of crystals is shown in fig. 2.39. Sphalerite crystals often exhibit the **tetrahedron** form which is composed of four equilateral triangular faces. In fig. 2.40 the first diagram shows an oblique view of the tetrahedron form and the other two, show views along particular directions which are labelled **b** and **c**. *Use fig. 2.40 to determine the elements of symmetry of the tetrahedron form.*

The tetrahedron does not possess a symmetry centre because each face of the tetra-

Fig. 2.39. Crystals of **sphalerite** exhibiting the tetrahedron form. (×4)

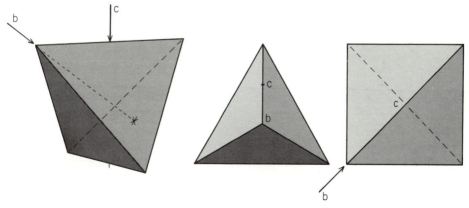

Fig. 2.40. The tetrahedron form seen from two particular directions which are labelled
b and c

hedron is opposite a point at which faces meet. The direction labelled **b** in fig. 2.40 is a triad axis perpendicular to a face, and since there are four faces, there are four triad axes in the tetrahedron form. The direction labelled **c** in fig. 2.40 is a diad axis and since there are three pairs of opposite edges, the tetrahedron has three diad axes which are perpendicular to each other.

It can be seen also from fig. 2.40 that the planes which are parallel to one diad axis and bisect the angle between the other two diad axes of the tetrahedron, are planes of reflection symmetry as shown in fig. 2.41. Consequently the complete elements of symmetry of the tetrahedron are $3A_2$, $4A_3$, and **6M** (diagonal).

Fig. 2.42 shows the tetrahedron related to the cube so that the four triad axes coincide. Note that the three diad axes of the tetrahedron coincide with the tetrad axes of the cube.

It is appropriate at this point to introduce a composite symmetry element known as the **symmetry axis of rotary inversion**. This composite symmetry axis combines two operations.

1. **rotation about an axis**, and
2. **inversion** before the new position is obtained.

Consider the diad axis shown in fig. 2.43 and note the position of the upper edge. If the crystal is rotated through 90° on the diad

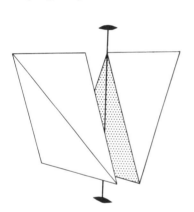

Fig. 2.41. A diagonal plane of symmetry in the tetrahedron

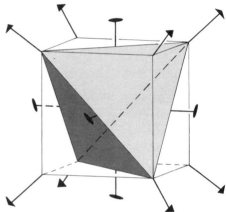

Fig. 2.42. The tetrahedron form related to the cube

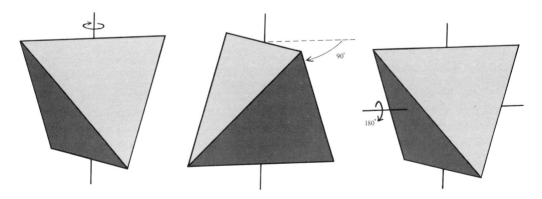

Fig. 2.43. The operations of a rotary inversion axis

axis and then inverted the new position is the same as the original one.

These two operations will produce the same crystal orientation four times during a rotation through 360° and therefore the axis may be described as a tetrad axis of rotary inversion or **tetrad inversion axis**. Note that crystals which exhibit a tetrad inversion axis do not possess a symmetry centre.

A centre of symmetry is no longer used because it can be represented by an inversion axis giving a rotation of 360° combined with inversion. A diad inversion axis is equivalent to a reflection plane of symmetry.

2.8 Symmetry class: The point group

It was demonstrated in sections 2.4 and 2.5 that the cube and icositetrahedron possess identical combinations of elements of symmetry. Careful study of figs. 2.33 and 2.34 should have revealed that the octahedron and the rhombdodecahedron also possess the same combination of elements of symmetry. The combinations of symmetry elements of the cube are tabulated below together with the symmetry elements of the pyritohedron, the tetrahedron and the gypsum crystal which was studied in section 2.3.

Cube	$3A_4$	$4A_36A_2$	3M(axial)	6M(diagonal)	
Pyritohedron	$3A_2$	$4A_3$ —	3M(axial)	—	C
Tetrahedron	$3A_2$	$4A_3$ —		6M(diagonal)	—
Gypsum	$1A_2$	— —	1M	—	C

The pyritohedron, the tetrahedron and the gypsum crystal all exhibit lower degrees of crystallographic symmetry than the cube. The elements have been arranged so that the three mutually perpendicular symmetry diad axes of the pyritohedron and tetrahedron, and the single diad axis of gypsum are in the same column as the tetrad axes of the cube.

Crystals may be classified according to the degree of symmetry which they exhibit. The cube, icositetrahedron, octahedron and

rhombdodecahedron forms may be grouped together in a particular **symmetry class** which is given the name hexaoctahedral class.

The pyritohedron, the tetrahedron and the gypsum crystal possess different combinations of symmetry elements and therefore they belong to different symmetry classes. The importance of the classification of crystals into symmetry classes, each class characterized by a particular combination of elements of symmetry was realized at the beginning of the nineteenth century.

The concepts of the elements of crystal symmetry have been introduced in the preceding sections by considering the external forms of some crystals which can be classified into four different symmetry classes. The reverse procedure in which the repetition of a particular crystallographic feature as the result of reflection, rotation, or rotation combined with inversion, is of greater value. For example, consider the orientation of one triad axis in the cube. The presence of three mutually perpendicular reflection planes determines that there must be four similar triad axes. If only one reflection plane and one symmetry axis is allowed then the axis which may be diad, triad, tetrad, or hexad must be perpendicular to the reflection plane. By reasoning in this way the German crystallographer, Hessel, demonstrated in 1830 that there could be **only 32 different symmetry classes**. At the time only a few of these were known to be represented by natural crystals but since, representatives of all but one of the symmetry classes have been discovered. Table 2.1 lists for later reference the elements of symmetry of the 32 symmetry classes. The symmetry class which is characterized by a centre of symmetry only, is said to possess a one-fold axis of rotary inversion. The symmetry class with no symmetry is said to have a one-fold axis, meaning that the same orientation is attained only by rotation through 360° to the original position.

The operations **reflection, rotation,** and

rotation combined with inversion, are called the **symmetry operations**. The group of symmetry elements which is possessed by a particular crystal is described as a **point group** because it describes symmetry about a single central point, which is indicated by the intersection of the zone axes. The point group is a fundamental property of the crystal which is independent of the external form and it expresses the relationship of all symmetrically disposed properties of the crystal to each other.

2.9 The Hermann–Mauguin symbols

Up to this point the elements of symmetry have been summarized by symbols composed of the capital letters **A** (axis), **M** (mirror plane) and **C** (centre) and the symmetry classes are described in this way in the reference table 2.1.

By international agreement the essential symmetry characteristics of the symmetry classes are summarized by the Hermann–Mauguin symbols and these are also listed in the table. In the Hermann–Mauguin notation the symmetry elements are represented by the symbols explained below.

2, 3, 4, and **6** represent respectively the diad, triad, tetrad, and hexad symmetry axes.

3̄, 4̄, and **6̄** (pronounced bar three, etc.) represent triad, tetrad and hexad inversion axes.

1 is a one-fold symmetry axis which represents asymmetry, meaning only one occurrence of a particular orientation during a complete rotation.

1̄ is a one-fold axis of rotary inversion which represents a symmetry centre, meaning inversion through a point.

m indicates a mirror plane which represents a plane of reflection symmetry, and it is equivalent to $\bar{2}$.

In the Hermann–Mauguin notation the symmetry of the principal axis of the crystal is given first and then its relationship to a plane of symmetry. For example a diad axis with a symmetry plane parallel to it is represented by the symbol **2m**, and a diad axis perpendicular to a symmetry plane is represented by the symbol **2/m**.

The symmetry of the other axes and their relation to a plane of symmetry are then given in turn. For example the cube has the symmetry elements $3A_4$, $4A_3$, $6A_2$, 9M, C. The Hermann–Mauguin symbol for the symmetry class to which the cube belongs is **4/m3̄ 2/m**.

4/m indicates that there is a tetrad axis with a perpendicular plane of symmetry. $\bar{3}$ is a triad inversion axis which has the same effect as a triad axis combined with a centre of symmetry. If necessary refer back to fig. 2.25 which illustrates the triad axis in the cube. **2/m** represents a diad axis with a perpendicular symmetry plane. The operation of the symmetry elements indicated by the Hermann–Mauguin symbol produces the complete combination of symmetry elements of the cube.

Recalling the meaning a triad inversion axis from section 2.7. *What is the Hermann–Mauguin symbol likely to be for the symmetry class represented by the pyritohedron (section 2.6) which is characterized by the symmetry elements $3A_2$, $4A_3$, 3M, C?*

The Hermann–Mauguin symbol which represents the symmetry of the pyritohedron is **2/m3̄**. This indicates that there is a diad axis perpendicular to a symmetry plane, and the triad inversion axis represents a triad axis associated with a symmetry centre.

What is the Hermann–Mauguin symbol of the symmetry class represented by the tetrahedron which has the symmetry elements $3A_2$, $4A_3$, 6M, in which a characteristic feature is that there are symmetry planes parallel to the triad axes?

The Hermann–Mauguin symbol which represents the symmetry of the tetrahedron is **4̄3m**. The tetrad inversion axis (4̄) represents the occurrence of a diad axis with no centre of symmetry as explained in section 2.7, **3m** refers to a triad axis with a parallel symmetry plane.

2.10 The crystal systems

Inspection of the elements of symmetry of the cube, pyritohedron and tetrahedron will indicate that these three symmetry classes contain one common element of symmetry. The common element of symmetry is the occurrence of **four triad axes** and because of this, the three symmetry classes are grouped together in a crystal system which is called the **cubic system**. It can be seen in table 2.1 that the 32 symmetry classes have been grouped into seven major divisions called the **crystal systems** which are **characterized by the occurrence of particular symmetry axes** indicated in the last column. The gypsum crystal belongs to the monoclinic system because it possesses a single diad axis. Another symmetry class in the monoclinic system is characterized by a single reflection plane which can be represented by a diad inversion axis ($\bar{2}$). The crystal class which exhibits the highest degree of symmetry within a crystal system is called the **holosymmetric class.** In the table the crystal systems have been arranged in the order in which representative crystals of each system will be studied in later chapters.

Before studying some crystals of the other systems, the use of axes for the description of crystals in the cubic system is explained in chapter 3. Then the nature of the internal structure of some cubic minerals and the methods involved in the determination of their structures will be developed in chapters 4 to 9.

Books for further reading.

A systematic treatment of all the symmetry classes is given in the following text books.

C. S. Hurlbut, **Dana's Manual of Mineralogy**. John Wiley and Sons. London, 1959.

F. C. Phillips, **An Introduction to Crystallography**, Longmans. London, 1957.

A. C. Bishop, **An Outline of Crystal Morphology.** Hutchinson, 1967.

L. G. Berry and Brian Mason. **Mineralogy**. W. H. Freeman and Co., 1959.

2.11 Questions for recall and self assessment

These questions are designed to draw attention to the main concepts and facts that should be fully understood and easily recalled. Because most of the questions require explanations of concepts, scoring will be difficult, but it is important to compare each attempt with the answers given in appendix A very carefully and critically. In the answers the key words or phrases are indicated by bold type.

1. What is the definition of interfacial angle?
2. What is the name of the instrument which is used to measure interfacial angles?
3. What is the most fundamental law of crystallography which was discovered by Nicolaus Steno in 1669?
4. What are the definitions of zone and zone axis?
5. What is the definition of a plane of reflection symmetry?
6. What is the characteristic of an axis of symmetry?
7. Which names are used to indicate the degree of symmetry shown by the symmetry axes?
8. What is the characteristic of a symmetry centre?
9. What is the name of the symmetry operation which relates the orientation of a crystal face to a parallel face on the opposite side of the crystal?
10. What does the word form mean in crystallography?
11. Which symmetry operations are associated with an axis of rotary inversion?

12. Which symmetry element is equivalent to a diad inversion axis?
13. Which symmetry axis represents a symmetry centre?
14. What is the characteristic of a symmetry class?
15. How many symmetry classes are possible?
16. Name the three symmetry operations.
17. What is the definition of a point group?
18. In which way are the symmetry classes classified into systems?
19. Which element of symmetry is characteristic of the cubic system?
20. What is the name given to the symmetry class which exhibits the highest degree of symmetry in each system?

Table 2.1. The thirty-two crystal classes

Elements of symmetry of the symmetry classes				Hermann–Mauguin Symbols		Crystal System	Characteristic symmetry element of the crystal systems
$3A_4$	$4A_3$	$6A_2$	9M	C	$4/m\bar{3}2/m$	Cubic	4 Triad axes
$3A_4$	$4A_3$	$6A_2$	.	.	432		
$3A_2$	$4A_3$	.	6M	.	$\bar{4}3m$		
$3A_2$	$4A_3$	.	3M	C	$2/m\bar{3}$		
$3A_2$	$4A_3$	.	.	.	23		
$1A_4$	$4A_2$		5M	C	$4/m2/m2/m$	Tetragonal	1 Tetrad axis or
$1A_4$	$4A_2$		.	.	422		1 tetrad inversion
$1A_4$	.		4M	.	$4mm$		axis $(\bar{4})$.
$1A_4$	$2A_2$		2M	.	$\bar{4}2m$		
$1A_4$			1M	C	$4/m$		
$1A_4$	.			.	4		
1 Tetrad inversion axis					$\bar{4}$		
$3A_2$			3M	C	$2/m2/m2/m$	Orthorhombic	3 diad axes or
$3A_2$			.	.	222		1 diad axis and
$1A_2$			2M	.	$mm2$		2 inversion diad $(\bar{2}) = m$
$1A_2$			1M	C	$2/m$	Monoclinic	1 diad axis or
$1A_2$			.	.	2		1 inversion diad or
.			1M	.	$m = (\bar{2})$		both
.				.	C $\bar{1}$	Triclinic	1-fold axis of rotary
No symmetry					1		inversion or 1-fold axis
$1A_6$	$6A_2$		7M	C	$6/m2/m2/m$	Hexagonal	1 Hexad axis or
$1A_6$	$6A_2$		.	.	622		1 hexad inversion axis
$1A_6$	.		6M	.	$6mm$		axis
$1A_3$	$3A_2$		4M	.	$\bar{6}m2$		
$1A_6$	.		1M	C	$6/m$		
$1A_6$	.		.	.	6		
$1A_3$	.		1M	.	$\bar{6}$		
$1A_3$	$3A_2$		3M	C	$\bar{3}2/m$	Trigonal	1 Triad axis
$1A_3$	$3A_2$		.	.	32		
$1A_3$	.		3M	.	$3m$		
$1A_3$	.		.	C	$\bar{3}$		
$1A_3$	.		.	.	3		

3 Drawing and Indexing Cubic Crystals

The principle objectives of this chapter are listed below.

1. To enable the student to understand the representation of the angular relationships of crystal faces in terms of their intercept ratios on three mutually perpendicular axes of reference.
2. To provide practice in the labelling of crystal faces by use of Miller indices which are the reciprocals of the intercept ratios.
3. To provide exercises in the representation of 3-dimensional geometrically symmetrical solids in 2-dimensional diagrams.
4. To provide exercises which require accuracy in draughtsmanship that is rewarded by the satisfaction of producing geometrically exact drawings of the crystal forms.

The objectives can be achieved by the construction of drawings of crystals which belong to three symmetry classes of the cubic system. The method of construction involves the drawing of an axial cross to represent the crystallographic axes on which the unit intercepts are marked, and then locating the following features.

(a) the points at which a crystal face intersects the axes; these points are determined by the intercept ratios.
(b) the lines of intersection of the extension of the crystal face with the axial planes of symmetry.
(c) the lines representing the intersection of adjacent crystal faces.

Contents of sections

3.1 Coordinate axes

The position of a number of points on a plane, such as this sheet of paper, may be recorded by stating their position with reference to two axes drawn at right angles as in fig. 3.1. The point is located by coordinates which are the perpendicular measurements to each axis, for example 2 cm from the y

axis gives the coordinate in the direction of the x axis, and 3 cm from the x axis gives the coordinate parallel to the y axis.

The position of a line on a plane may also be recorded by giving the points at which it intersects the two reference axes. In fig. 3.2 a line intersects the x axis at a distance of 2 cm and it intersects the y axis at a distance of 1.15 cm. These coordinates (2, 1.15) are known as the **intercepts** of the line with the x and y axes respectively. The ratios of the

intercept on the y axis to the intercept on the x axis for the two parallel lines shown in fig. 3.2 are 1.15/2 and 2.31/4 respectively, and in each case the intercept ratio = 0.5774. The ratio 0.5774 is the tangent of the angle of inclination of the line to the horizontal axis x, and it is known as the gradient or slope of the line. It can be seen from tables of natural tangents that 0.5774 is the tangent of an angle of 30°. The actual intercepts produced by the line which intersects the reference axes are not important because they depend on the units of measurement. The angular position of the line is important and it can be given either as the angle between the line and the x axis = 30°, or as the tangent 0.5774, which is the ratio of the intercept on the y axis to the intercept on the x axis.

In solid geometry the position of a plane in space can be given by the intercepts that the plane makes on three imaginary axes, x, y, and z which are usually arranged at right angles to each other. The size of the face is not important, only its angular relationship to the three axes, and this can be represented by the ratios of the intercepts on the three reference axes as in fig. 3.3. The tangent of the angle $\widehat{oyz} = oz/oy$ will provide the ratio of the intercept on the z axis relative to the intercept on the y axis. If $\widehat{oyz} = 45°$, tan $\widehat{oyz} = 1$ so that $oz = oy$. The tangent of the angle $\widehat{oyx}$ will provide the ratio of the intercept on the x axis relative to the intercept on the y axis. If $\widehat{oyx} = 45°$ then $ox = oy = oz$ and the intercept ratios on the reference axes x, y, and z are $1:1:1$.

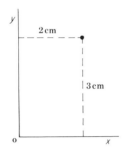

Fig. 3.1. Coordinates
of a point

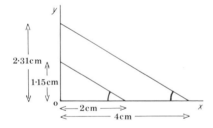

Fig. 3.2. The intercept ratios

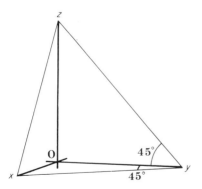

Fig. 3.3. The intercept ratios of a
plane

3.2 Crystallographic axes and intercept ratios

The French crystallographer, Haüy, introduced the concept of axes of reference in crystallography. The orientations of crystal faces are described in relation to axes which are arranged parallel to axes of crystallographic symmetry whenever possible.

The octahedron has three axes of tetrad symmetry at right angles to each other as in fig. 3.4 and these can be used as the crystallographic axes. The crystallographic axes illustrated in fig. 3.5 are always referred to in the following order.

x front (positive) to   −*x*   back (negative)
y right (positive) to   −*y*   left (negative)
z top (positive) to   −*z* bottom (negative)

Study the octahedron by looking along the *z* axis as in fig. 3.6. *Measure the angle between the x axis and the edge extending between the x and y axes and determine the intercept ratio y/x.* The angle between the *x* axis and the crystal edge is 45°. Therefore the intercept ratio is 1 and the intercepts on the *y* an *x* axes are equal. *Will the intercepts of any face of the octahedron be equal on all three crystallographic axes?* Since any of the tetrad axes of symmetry may be placed in the position designated the *x* axis, it follows that all the edges of the octahedron will intersect two crystallographic axes at 45°, and consequently each face of the octahedron will have equal intercepts on all three crystallographic axes. The intercepts are always given in the order *x,y,z* and are given as ratios with the shortest intercept taken as unity. The **intercept ratios** are also known as **parameters**. The position of each face of the octahedron can be described by stating the intercept ratios with the positive or negative signs of the axes.

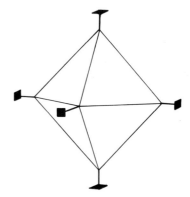

Fig. 3.4. The tetrad axes of the octahedron

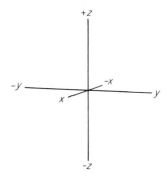

Fig. 3.5. The crystallographic axes

3.3 Crystal drawings

Normally when we attempt to illustrate an object which contains planes and straight edges, we draw it in perspective with all parallel lines converging to a point at infinity as in fig. 3.7. Since the parallelism of sets of edges on a crystal is one of the most important features, it is necessary that crystal drawings should show this parallelism as in fig. 3.8.

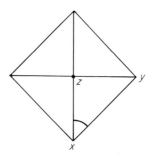

Fig. 3.6. The intercept ratios of the octahedron face

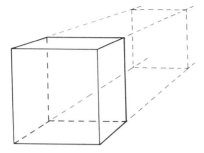

Fig. 3.7. A cube seen in perspective

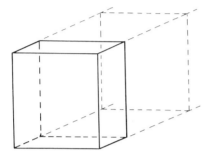

Fig. 3.8. Drawing indicating the parallelism of the cube edges

One method would be to represent two of
the crystallographic axes parallel to the
plane of the paper with the third axis perpendicular to the paper. A drawing can then
be made by parallel projection on to the
paper. A cube would be illustrated as a
square, by parallel projection along the x
axis as in fig. 3.9. However, three dimensional representation is usually much more
instructive. If the cube had been rotated
through an angle θ about the z axis, one side
of the cube would also be shown and the x

axis would appear as a greatly foreshortened
line, as in fig. 3.10.

A clinographic projection is developed by
raising the point of view through an angle ϕ
so that the top of the cube is also seen as in
fig. 3.11. The method of crystal drawing
adopted here involves drawing an axial cross
on which the unit intercepts are marked and
then superimposing the crystal faces until a
three dimensional drawing of the model is
achieved. The construction of drawings of
the reference axes is given in appendix B,
but a simplified construction of the axial
cross is illustrated in fig. 3.12 and is explained below.

1. Draw a vertical line and a line normal to
    it, OK.
2. Select the distance OK so that it is
    divisible by 3 repeatedly, e.g.
    OK = 8.1 cm.
3. Mark off OL so that OL = $\frac{1}{3}$ OK.
4. Draw perpendiculars at K and L.
5. Mark off LQ = $\frac{1}{3}$ OL (downwards).
6. Mark off KP = $\frac{1}{3}$ LQ (upwards).
7. Join OP = unit intercept on $-y$.
8. Join OQ = unit intercept on $+x$.
9. Draw a line from O through Q to R,
    OR = length of the unit intercept on
    the vertical axis and this distance can be
    marked off along the vertical line.

Remember that the unit intercepts are the
same on all three axes, but in the drawing
the unit distance on the x axis, OQ, appears
to be shorter because we are looking nearly
along the axis. The unit of measurement on

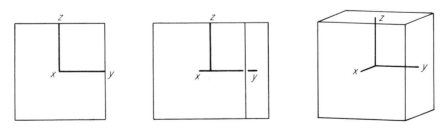

Fig. 3.9, 3.10, and 3.11. The reorientation of the cube for clinographic projection

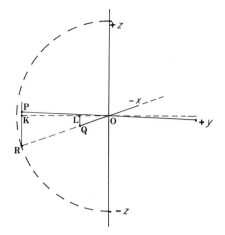

Fig. 3.12. A simplified construction of the crystallographic axes produced by clinographic projection

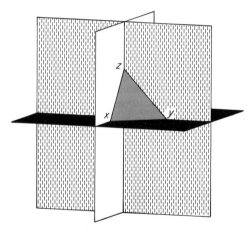

Fig. 3.13. The intersection of the octahedron face with the axial planes of symmetry

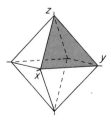

Fig. 3.14. The octahedron drawn on an axial cross

the *y* axis, OP, is also slightly shorter than OR on the *z* axis because the positive end of the *y* axis is turned slightly towards the observer.

10. Complete the axial cross by producing the axes and marking off the appropriate unit intercepts.
11. Erase the construction lines.

Consider the problem of drawing the octahedron by representing the positions of the inclined crystal faces in relation to a drawing of the three mutually perpendicular crystallographic axes on which the units of measurement have been marked. The lines joining the unit intercepts *x* to *y*, *x* to *z*, and *y* to *z* represent the intersection of a face with the three planes of symmetry which are each parallel to two of the crystallographic axes as illustrated in fig. 3.13. The operation of the symmetry elements result in the appearance of similar faces in all eight sections bounded by the planes containing the crystallographic axes. The drawing of the octahedron can be completed by joining all the points of unit intercept as in fig. 3.14 but remember that the lines represent edges which are formed by the intersection of two crystal faces.

3.4 Drawing the icositetrahedron

The icositetrahedron is a form composed of 24 trapezium shaped faces. Garnet crystals exhibiting the icositetrahedron form are illustrated in plate 10, and the elements of symmetry of the icositetrahedron form were considered in section 2.5.

The three mutually perpendicular axes of tetrad symmetry are selected as the crystallographic axes. Study the icositetrahedron by viewing it along one of the crystallographic axes which can be labelled *y* as in fig. 3.15. The angle that the front edge makes with the *x* axis is approximately 63°26′ as shown in fig. 3.16. *What is the ratio of the intercept*

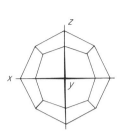

Fig. 3.15. The icosi-
tetrahedron viewed
along a tetrad axis

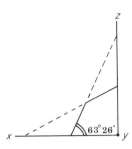

Fig. 3.16. The intercept
ratios of the edges of the
icositetrahedron face

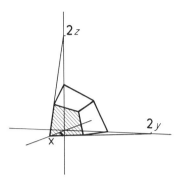

Fig. 3.17. The intercept ratios
of the icositetrahedron face

of this edge on the z axis compared with the intercept on the x axis?

The ratio $z/x = \tan 63°26' = 2$ and therefore if the intercept on the *x* axis is unity, the intercept ratio on the *z* axis is 2 as shown in fig. 3.17. The edge just considered is part of a face which also intersects the *y* axis. The edge joining the *x* and *y* axes also lies at an angle of 63°26' to the *x* axis, and therefore the intercept ratio on *y* is also 2, as shown in fig. 3.17. Consequently the intercepts of the front face are *x*, *2y*, *2z*, and the intercept ratios are 1,2,2.

The operation of the symmetry elements of the icositetrahedron result in the occurrence of 23 other faces which have intercept ratios similar to 1,2,2. Each face can be regarded as intersecting one axis at unit distance and the projection of the face intersects the other two axes at intercepts which are twice the unit distance from the origin of the axial cross. Note that if one

reads the intercepts on the three axes in an anticlockwise direction starting from the point of unit intercept, then the intercept ratios are always 122. Haüy recognized that in a particular crystal, the faces intersected the axes of reference at simple rational multiples of certain lengths as illustrated above. This is one aspect of the **law of simple rational intercept ratios** established by Haüy.

Consider the problem of drawing the icositetrahedron on an axial cross. *How can the face which has intercept ratios of x, 2y, 2z, shown in fig. 3.18 be represented on the drawing of the axial cross?*

The line joining *x* to *2y* represents the line of intersection of the crystal face with the plane of symmetry which is parallel to the *x* and *y* axes as shown in fig. 3.19. Similarly the lines joining *x* to *2z* and *2y* to *2z* represent the lines of intersection of the extension of the face with the two vertical axial planes of symmetry. The face with intercepts of *2x*, *2y*, *z* can be represented in a similar way as in fig. 3.20.

In the vertical plane of symmetry the two edges of the icositetrahedron extending from *x* to *2z* and from *z* to *2x* intersect at a point midway between the *x* and *z* crystallographic axes as indicated in fig. 3.21. This is the point of emergence of a diad axis of symmetry. *At which other point are the two faces with intercepts of x, 2y, 2z and 2x, 2y, z*

Fig. 3.18. The icosi-
tetrahedron face

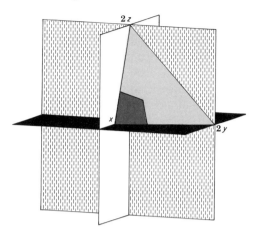

Fig. 3.19. The intersection of the icosi-tetrahedron face (*x, 2y, 2z*) with the axial planes of symmetry

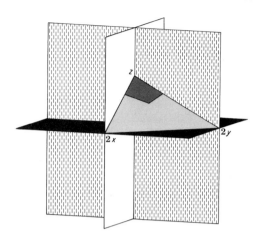

Fig. 3.20. The *2x, 2y, z* face of the ico-sitetrahedron

known to intersect?, and which line on the drawing will represent the line of intersection of the two faces?

Inspection of these intercept ratios shows that both faces intersect the *y* axis at the same intercept ratio which is *2y*, and in fig. 3.22 it can be seen that the line which joins the intercept *2y* to the point of intersection of the edges in the vertical plane of sym-metry (black arrow), represents the crystal edge between the faces having the intercept ratios *x, 2y, 2z* and *2x, 2y, z*.

The drawing of the icositetrahedron is easily completed by determining the inter-cepts of the faces, locating the lines of intersection of the faces with the axial planes of symmetry and then identifying the lines of intersection of pairs of faces. The symmetrical arrangement of the planes re-presenting the crystal faces is brought out by the completed diagram of the ico-sitetrahedron, fig. 3.23.

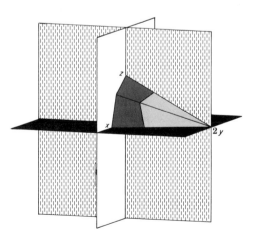

Fig. 3.22. The intersection of the (*x, 2y, 2z*) and (*2x, 2y, z*) faces

Fig. 3.21. The (*x, 2y, 2z*) and (*2x, 2y, z*) faces

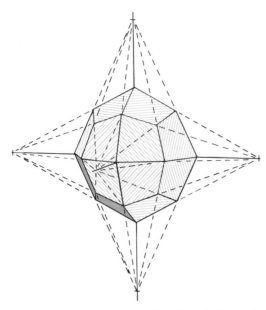

Fig. 3.23. The completed drawing of the icositetrahedron

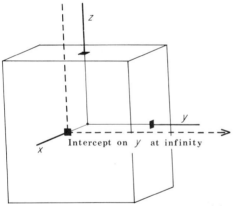

Fig. 3.24. The intercepts of the cube face

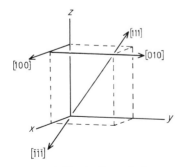

Fig. 3.25. The indices of directions in the cube

3.5 Miller indices

In the preceding sections the orientations of the crystal faces of the octahedron and the icositetrahedron have been identified by means of intercept ratios on reference axes. In the case of a crystal face which is parallel to a crystallographic axis, the intercept of the face on that axis is regarded as at infinity, and the intercept ratios will include the sign for infinity ∞. In the cube the tetrad axes of symmetry which are parallel to the edges are selected as the crystallographic axes as shown in fig. 3.24. The front face of the cube intersects the x axis but it is parallel to the y and z axes so that the intercept ratios of this face are x, ∞y, ∞z. The intercept ratios are known as the Weiss system of crystallographic notation.

However, the system of crystallographic notation in general use was popularized by W. H. Miller. In Miller's system there are three features to remember.

1. Since **the axes are always referred to in the same order**, x, y, and z, the labels of the axes are omitted.
2. The **reciprocals of the intercepts are used** so that 2 becomes $\frac{1}{2}$.
3. Any **fractions are cleared**.

For example consider one face of the icositetrahedron.

The intercept ratios are x, $2y$, $2z$.
The reciprocals are $\mathbf{1}$, $\frac{1}{2}$, $\frac{1}{2}$.

When the fractions are cleared the Miller indices of the face are **(211)**, pronounced two, one, one.

Note that in the Miller indices 1 denotes the largest intercept of a face, while 2, 3, etc. denote shorter intercepts which are the reciprocal fractions of the longest intercept. A face which is parallel to a crystallographic axis is said to have an intercept at ∞, and the reciprocal of infinity is 0 (nought). Since the front face of the cube is parallel to the y and *z* axes, the Miller indices of this face are (**100**) pronounced one, nought, nought.

The Miller indices are in general use because they avoid the mathematical difficulties which arise with parameters at infinity. Because of the regularity of atomic structures of the cubic minerals the indices of any face are either nought or are small rational numbers usually 1,2,3,4, or 6. This is the law of simple rational ratios of indices; an important concept which will be extended in later chapters.

Intersections at the negative ends of the axes are indicated by a bar over the index, for example $\bar{2}$, pronounced bar two. When indices are used to give the orientation of planes they are enclosed in simple brackets, for example (**100**) represents the front face of the cube. **Indices may also be used to indicate directions and these are enclosed in square brackets.** In fig. 3.25 [**100**] is the direction of the *x* axis and [**111**] and [**$\bar{1}\bar{1}\bar{1}$**] are the directions from the origin along one of the cube diagonals.

Return to fig. 3.23 and work out the Miller indices of the visible faces of the icositetrahedron and the indices of the directions of the triad axes. The Miller indices of the faces of the icositetrahedron are given in section 3.10, solution 1, and are indicated on fig. 3.38.

(**100**) are the Miller indices of the front face of the cube but since the cube is a simple form consisting of 6 identical faces, the simple form can be represented by the indices {**100**}. **Note that the indices of the form are enclosed in curved brackets which are known as braces.**

Directions related to symmetry are called directions of form, and a set of these may be represented by the indices of one of them enclosed in angular brackets⟨ ⟩. For example the triad axes of the cube may be represented by ⟨**111**⟩.

Remember that in the Miller indices **0** (nought) denotes that a face is parallel to the crystallographic axis, while **1** indicates

the largest intercept ratio of a face. The numbers **2,3,4,6,** are shorter intercept ratios which are the reciprocal fractions of the longest intercept ratio that might be observed on the particular axis. *List the indices which would describe the point group of the icositetrahedron form* {211}. The answers are given in section 3.10, solution 2.

3.6 Drawing the rhombdodecahedron

In order to become familiar with the use of Miller indices, construct a drawing of the rhombdodecahedron on an axial cross. Garnet crystals exhibiting the rhombdodecahedron form are seen in plate 10 and its elements of symmetry were considered in section 2.5. The rhombdodecahedron is a simple form consisting of 12 rhomb shaped faces and it is shown in fig. 3.26. The three mutually perpendicular tetrad axes of symmetry are the most appropriate crystallographic axes. The interfacial angle between the two faces which are parallel to the vertical axis and cut the *x* and *y* axes, is 90°. *What are the intercept ratios and Miller indices of the face which intersects the positive ends of the x and y axes?, and how can the orientation of this plane be represented in*

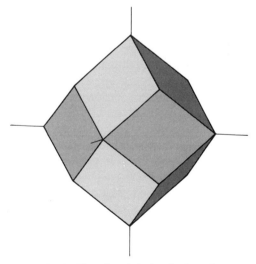

Fig. 3.26. The rhombdodecahedron form

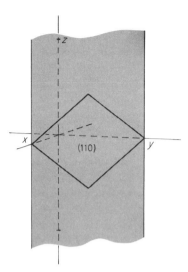

Fig. 3.27. The intersection of the
(**110**) face with the axial symmetry
planes

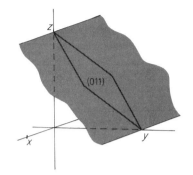

Fig. 3.28. The intersection of
the (**011**) face with the axial
symmetry planes

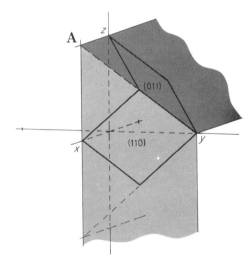

Fig. 3.29. The intersection of the (**110**)
and (**011**) faces

*relation to the axial cross and the axial planes
of symmetry?*

The intercept ratios of the face which
intersects the positive ends of the x and y
axes are x, y, ∞z and the Miller indices are
(**110**). The (**110**) plane is parallel to the z
axis, consequently lines drawn parallel to
the z axis through the unit intercepts x and y
represent the intersection of the (**110**) face
with the vertical axial planes of symmetry as
shown in fig. 3.27.

The face which is parallel to the x axis but
cuts the y and z axes has the indices (**011**), so
in a similar way lines drawn parallel to the x
axis through the unit intercepts y and z
represent the intersection of the (**011**) face
with the horizontal plane of symmetry and
the vertical plane of symmetry containing
the x and z axes as shown in fig. 3.28. *At
which points in the axial planes of symmetry
do the* (**110**) *and* (**011**) *planes intersect?*

The (**110**) and (**011**) faces intersect at the
unit intercept on the y axis as indicated by
the common index. The faces also intersect
in the vertical axial plane of symmetry which
contains the x and z axes. The point of
intersection **A** in fig. 3.29 is given by the line

through unit intercept x and parallel to the z
axis, and the line through unit intercept z
and parallel to the x axis. Join the two points
of intersection, y and **A**, and the position of
the edge between the (100) and (011) faces
is obtained.

Complete the drawing of the rhombdodecahedron by locating the lines of intersection of the faces with the axial planes of
symmetry and with each other. The constructions may be checked by studying solution 3, figs. 3.39 and 3.40 in section 3.10.

3.7 Drawing the pyritohedron

The pyritohedron form of pyrite is shown in plate 6 and the symmetry has been considered in section 2.6. The three diad axes of symmetry which are perpendicular to each other should be selected as the crystallographic axes. The front faces of the pyritohedron indicated in the fig. 3.30 intersect the x and y axes and are parallel to the z axis. A section of the crystal in the plane parallel to the x and y axes is shown in fig. 3.31. *Measure the interfacial angle between the two front vertical faces of the pyritohedron and determine the intercept ratios and Miller indices of the face which cuts the positive ends of the x and y axes.*

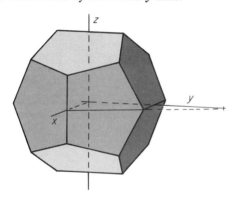

Fig. 3.30. The pyritohedron form

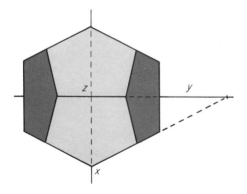

Fig. 3.31. The pyritohedron viewed along
the z axis

The interfacial angle between the two front faces of the pyritohedron is $53°08'$, and therefore the angle between the face indicated in fig. 3.31 and the x axis is $63°26'$. The ratio $y/x = \tan 63°26' = 2$. Therefore the intercept on the y axis is at 2, and the intercept ratios are $x, 2y, \infty z$. The reciprocals of the intercepts are 1, 1/2, 1/∞ and after clearing the fractions the Miller indices are **(210)**. The operation of the symmetry elements determined in section 2.6 require that there are twelve faces having similar indices in the pyritohedron form.

Determine the indices of the faces shown in fig. 3.30, and then draw the pyritohedron by locating the intersections of the faces with the axial planes of symmetry and with each other, in a similar manner to that followed in the drawing of the rhombdodecahedron. Alternatively the drawing may be completed by studying the symmetrical arrangement of the planes about the axes and planes of symmetry. The construction of an oblique drawing of the pyritohedron form is explained in section 3.10. solution 4, figs. 3.41, 3.42, and 3.43.

3.8 Drawing a combination of the cube and octahedron forms, and the tetrahedron

On the cube form {**100**} the axial planes of symmetry intersect the faces and indicate the position of the cube edges as shown in fig. 3.32. The octahedron form {**111**} can be drawn within the cube simply by joining the points of unit intercept.

The cube crystals of fluorite illustrated in plate 15 are recognized because they have a well developed cleavage parallel to the octahedron form so that the corners break away easily. When constructing oblique drawings of combinations of cubic forms it is necessary to choose different intercept lengths on the axial cross for each form.

Construct a drawing of the combination of the octahedron and cube forms on an axial cross on which the intercept units for the octahedron are 2.5 times the intercept units for the cube.

A drawing of the combination of the cube and octahedron is shown in fig. 3.33 and the construction lines for obtaining the lines of intersection of the forms are indicated on the (1$\bar{1}$1) face. This construction illustrates clearly the relationship between the cube

{**100**} and octahedron {**111**} forms which is important in the study of the atomic structures of the cubic minerals. The relationship between the tetrahedron and the cube is even more important.

An oblique drawing of the tetrahedron is most easily constructed by commencing with a cube as shown in figs. 3.34 and 3.35. In fig. 3.34 the tetrahedron is developed by the extension of the (**111**) face and this is the orientation which is usually shown. The

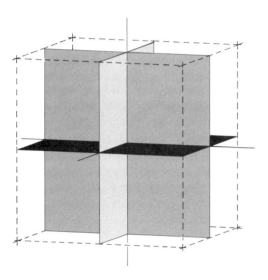

Fig. 3.32. The axial symmetry planes of the cube

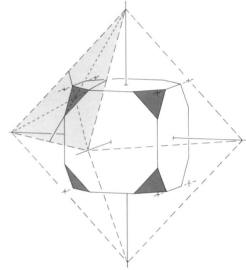

Fig. 3.33. The combination of the cube {100} and octahedron {111} forms

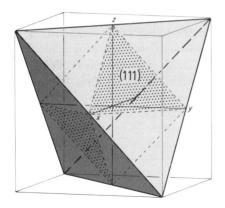

Fig. 3.34. The tetrahedron form {111}

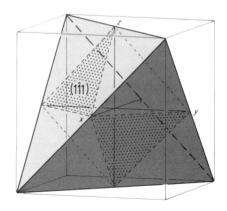

Fig. 3.35 The tetrahedron form {1$\bar{1}$1}

faces present are (**111**), (**1Ī̄Ī**), (**Ī̄Ī1**) and (**Ī̄11**). In fig. 3.35 a tetrahedron has been developed from the (**1Ī̄1**) face and in this alternative orientation the faces represented are (**1Ī̄1**), (**11Ī̄**), (**Ī̄11**) and (**Ī̄Ī̄Ī̄**).

3.9 Some problem exercises in drawing cubic crystals

Crystals sometimes exhibit faces which are characteristic of different simple forms. For example in fig. 3.36 the two crystals of garnet labelled c and d are composed of combinations the faces of the icositetrahedon {**211**} and rhombdodecahedron {**110**} forms. Construct a oblique drawing of the combination of the icositetrahedron and rhombdodecahedron forms on an axial cross on which the unit intercepts for the rhombdodecahdron are 1.1 times the unit intercepts for the icositetrahedron. A completed drawing of this

Fig. 3.36. **Garnet** crystals viewed along a diad axis with a tetrad axis vertical in the plane of the photographs (×2)

(a) icositetrahedron form {211},
(b) rhombdodecahedron form {110},
(c) and (d) combinations of the icositetrahedron and rhombdodecahedron forms.

combination is given in section 3.10, solution 5, fig. 3.44.

In fig. 3.37 the crystal of pyrite shows a combination of faces of the cube and pyritohedron {**210**} forms. Even if no striations occur on the pyrite cube faces the symmetry of the combined forms is that of the lowest member, in this case the pyritohedron. Construct an oblique drawing to show combination of the cube and pyritohedron forms using unit intercepts for the pyritohedron which are 1.1 times the unit intercepts for the cube. A completed drawing is shown in section 3.10, solution 6, fig. 3.45.

For comparison construct a drawing to show the combination of the cube and rhombdodecahedron using unit intercepts for the rhombdodecahedron which are 1.5 times the unit intercepts for the cube. The completed drawing is given in section 3.10, solution 7 and fig. 3.46.

As a final exercise construct an oblique drawing of the simple form {**321**} and determine its point group. A completed drawing of the {**321**} form and the symmetry is given in section 3.10, solution 8, fig. 3.47.

Fig. 3.37. **A pyrite** crystal showing a combination of faces of the cube {100} and pyritohedron {210}. (×4)

3.10 Solutions of the problem exercises

1. The **indices of the faces of the icositetrahedron** and the directions of the triad axes are shown in fig. 3.38.
2. The list of **indices which would describe the point group of the icositetrahedron** is given below.
   Tetrad axes ⟨100⟩
   Triad axes ⟨111⟩
   Diad axes ⟨110⟩
   Reflection planes {100}
   Reflection planes {110}
   Symmetry centre.

3. **Completing the drawing of the rhombdodecahedron.** In fig. 3.39 the orientation of the (**101**) face is represented by lines drawn parallel to the y axis through the unit intercepts x and z. These are the lines of intersection of the (**101**) face with the axial planes of symmetry. The (**101**) face intersects the (**011**) and (**0Ī1**) faces at unit intercept z (common index), and at the points **B** and **C** respectively within the horizontal plane of symmetry. The points **B** and **C** are given by the intersections of the lines parallel to the x axis passing through unit intercepts y and $-y$, and the line parallel to the y axis passing through unit intercept x. The lines through points **B** and **C** to unit intercept z give the positions of the edges between

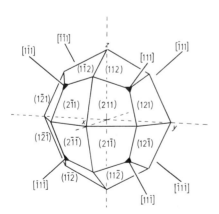

Fig. 3.38. The indices of the faces of the icositetrahedron form and the triad axes

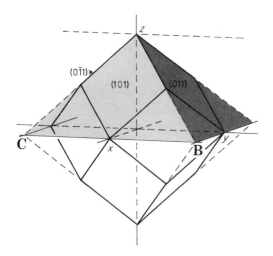

Fig. 3.39. The intersection of the (101) face with the (011) and (0$\bar{1}$1) faces

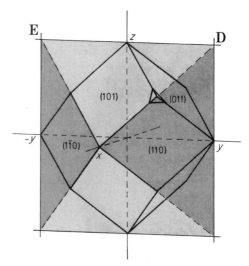

Fig. 3.40. The intersections of the (101) face with the (110) and (1$\bar{1}$0) faces

the **(101)** face and the **(011)** and **(0$\bar{1}$1)** faces respectively. *How can the lines of intersection of the (101) face and the (110) and (1$\bar{1}$0) faces be located?*

The **(101)** face intersects the **(110)** and **(1$\bar{1}$0)** faces at unit intercept x (common index), and at the points **D** and **E** in the vertical plane of symmetry containing the y and z axes, as shown in fig. 3.40. The points **D** and **E** are given by the intersections of the lines parallel to the z axis passing through unit intercepts y and $-y$, with the line parallel to the y axis passing through unit intercept z. The lines through the points **D** and **E** to unit intercept x give the edges between the **(101)** face and the **(011)** and **(0$\bar{1}$1)** faces respectively.

The three lines representing the edges between the **(110)**, **(011)** and **(101)** faces should intersect at a point as shown in fig. 3.40. This is the point of emergence of one of the four triad axes of symmetry. The drawing of the rhombdodecahedron may be completed by locating lines of intersection of the faces with the axial

planes of symmetry and with each other, or by studying the symmetry of the construction.

4. **Completing the drawing of the pyritohedron.** The method of construction is illustrated by considering the **(210)**, **(021)** and **(02$\bar{1}$)** faces shown in fig. 3.41. The **(021)** and **(02$\bar{1}$)** faces intersect to produce the edge which is parallel to the x axis and passing through unit intercept y. The **(210)** face intersects this edge at point **A** which is found by drawing the line from x to **2y**. The extensions of the **(021)** and **(02$\bar{1}$)** faces intersect the vertical plane of symmetry which is parallel to the x axis, along lines parallel to the x axis and passing through the intercepts **2z** and $-2z$. The **(210)** and **(2$\bar{1}$0)** faces intersect the same vertical plane of symmetry along the line through unit intercept x and parallel to the z axis. The intersection **B** in fig. 3.41 is the point at which the extensions of the **(210)**, **(2$\bar{1}$0)**, **(021)** and **(0$\bar{2}$1)** faces meet. The edge between the **(210)** and **(021)** faces is obtained by drawing the line **AB** as shown in fig. 3.41.

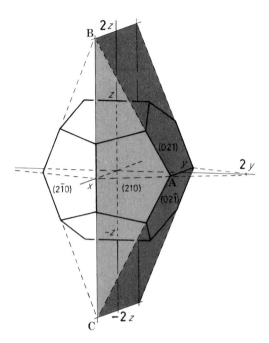

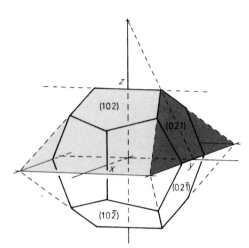

Fig. 3.42. The intersection of the (021) and (102) faces

Fig. 3.41. The intersections of the (210) face with the (021) and (02$\bar{1}$) faces

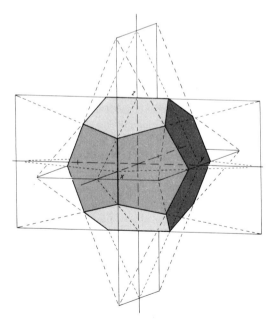

Fig. 3.43. The completed drawing of the pyrito-hedron

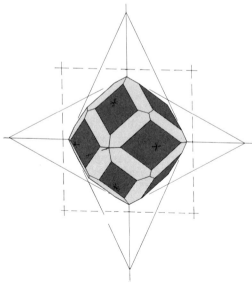

Fig. 3.44. The combination of the ico-sitetrahedron {211} and rhombdodecahedron {110} forms

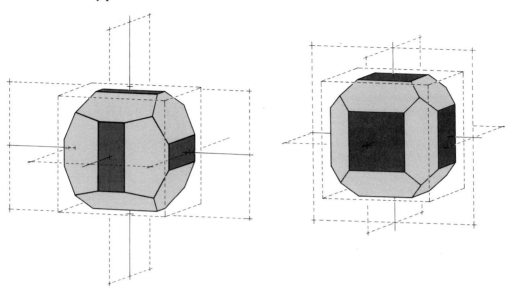

Fig. 3.45. The combination of the cube and pyritohedron {210} forms

Fig. 3.46. The combination of the cube and rhombdodecahedron {110} forms

The corresponding faces on the lower part of the form meet at the point **C** in fig. 3.41.

The line of intersection of the (**021**) face with the (**102**) face is found in a similar way which is illustrated in fig. 3.42. The lines of intersection of the (**210**) and (**102**) faces is shown in the completed drawing of the pyritohedron fig. 3.43.

5. **A drawing of the combination of the icositetrahedron and rhombdodecahedron forms** produced when the unit intercepts for the rhombdodecahedron are 1.1 times the unit intercepts for the icositetrahedron, is shown in fig. 3.44.

6. **A drawing of the combination of the cube and pyritohedron forms** which is produced when the unit intercepts for the pyritohedron are 1.1 times the unit intercepts for the cube is shown in fig. 3.45. It can be seen that the symmetry of this combination is that of the pyritohedron, the form with the lowest symmetry.

7. Fig. 3.46 is **a drawing of the combination of the cube and rhombohedron forms** which is produced when the unit intercepts for the rhombdodecahedron are 1.5 times the unit intercepts of the cube. The combination of the cube and rhombdodecahedron shown in fig. 3.46 has the symmetry of the cubic holosymmetric class to which both forms belong.

8. Fig. 3.47 is **an oblique drawing of the {321} form which is known as the diploid.** The indices (**321**) represent the intercepts 1, 1.5, 3 and the method of construction is very similar to that of the icositetrahedron considered in section 3.4, but the diploid has lower symmetry. It will have been discovered from the construction that there are axial planes of symmetry but no diagonal planes of symmetry. The crystallographic axes are diad axes of symmetry and there are four triad axes of symmetry which are characteristic of the cubic system. The point group of the diploid is summarized below.

Diad axes	⟨100⟩
Triad axes	⟨111⟩
Reflection planes	{100}
Symmetry centre.	

The diploid has the same point group as the pyritohedron which was considered in section 2.6. An identical shape but with the alternative orientation is produced by the form {**312**}.

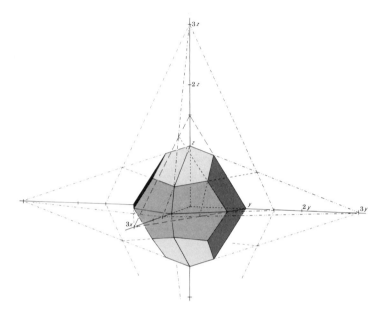

Fig. 3.47. The diploid form {321}

4 The Internal Structures of Metals

A mineral possesses an ordered arrangement of the atoms of which it is composed and this results in the development of the crystal face and cleavage forms which have been considered in the first three chapters. In this chapter the concepts involved in understanding the nature of crystal structures are introduced with reference to the structure of metals which consist of atoms all of the same kind. The principal objectives of this chapter are listed below.

1. To consider briefly some early explanations for the constancy of form of the crystals of a particular mineral.

2. To understand the classification of 2-dimensional repeated patterns into five kinds of 2-dimensional lattices composed of two sets of equally spaced parallel lines, and to note that patterns with greater symmetry are indicated by lattices with special characteristics.

3. To apply the concepts obtained by the study of 2-dimensional patterns to 3-dimensional patterns and to introduce the 14 kinds of space lattice identified by Bravais.

4. To understand the structure of metals in terms of the packing arrangements of spheres of equal size.

5. To determine the appropriate 3-dimensional pattern (space lattice) for the description of the structures of pure metals.

6. To index the principal planes of points in the cubic space lattices, and to determine the ratios of the spacings between the planes in preparation for the study of the structures of simple crystals by X-ray methods in chapters 6 and 7.

Contents of the sections

4.1 Early ideas concerning the constancy of crystal form

Why do the crystal faces and cleavage planes in crystals of a particular mineral always show the same angular relationships?

Speculations concerning the reasons for the external regularities of crystals were made by the earliest crystallographers. In 1611 the astronomer **Kepler**, suggested that the regularity of 'hexagonal snow' was probably due to a regular geometrical arrangement of minute building units.

The French crystallographer, **Haüy**, introduced the concept of crystallographic axes and established the law of simple rational intercept ratios. In 1781 Haüy published an essay on the theory of crystal structure. After studying the relationship between the cleavage fragments and crystals of calcite and galena, Haüy put forward the hypothesis that all crystals are made up of small units bounded by cleavage planes which were characteristic of the mineral. Haüy illustrated his hypothesis with calcite and some cubic minerals such as galena. Galena breaks into cube shaped fragments and its crystals are cubes or less commonly octahedra. Haüy's concept of the formation of cube and octahedron crystal forms from the cube shaped structural units is illustrated in fig. 4.1.

The mineral fluorite also occurs as cube shaped crystals, but the cleavage planes of fluorite are parallel to the {111} form. Con-

sequently the cleavage fragments of fluorite are octahedra as shown in plate 15. *Can Haüy's hypothesis be used to explain the structure of fluorite, in other words can octahedra be packed together to completely fill space?*

Octahedral shaped structural units cannot be packed together to completely fill space, and some tetrahedral shaped spaces remain as illustrated in fig. 4.2. This test of Haüy's hypothesis shows that it is not completely satisfactory. However, the general concept that the crystal faces and cleavage planes are related to a regular packing of some fundamental unit was correct. Sometime before Haüy published his hypothesis, **Huygens** in 1678 postulated that the crystal form, cleavages, and directional variation of hardness and optical characteristics of calcite, might be explained if the crystal was made up of minute ellipsoid shaped particles packed in a regular manner as illustrated in fig. 4.3. This hypothesis may be considered as forming the basis of modern ideas of crystal structures. Basic to the concepts of both Huygens and Haüy is the idea that a crystal is composed of a fundamental unit which is repeated to form a three-dimensional pattern.

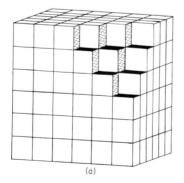
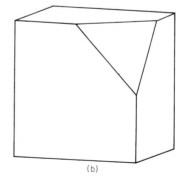

(a)   (b)

Fig. 4.1. (a) Haüy's concept of the packing of cube cleavage forms to produce cube and octahedron crystal faces which are shown in 4.1(b)

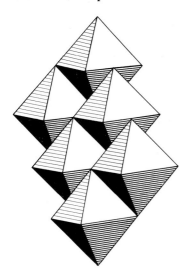

Fig. 4.2. Octahedron forms packed together but not completely filling space

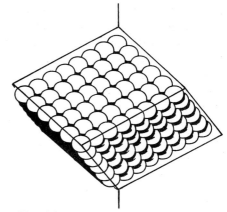

Fig. 4.3. Huygen's concept of the structure of calcite

Fig. 4.4. Linear patterns

As the ideas of modern physics and chemistry developed it became evident that the crystal units could not be solid particles but were groups of atoms which have a periodic arrangement in three dimensions.

4.2 Two-dimensional patterns

The simplest representation of a periodic arrangement is a straight row of equally spaced points or patterns. Two different linear patterns are shown in fig. 4.4 and it can be seen that each pattern is repeated indefinitely along each line.

Most wallpaper and fabric patterns are repetitions of a unit pattern in two dimensions. In fig. 4.5 a pattern of this kind is produced by repetitions of the letter P. The manner in which a complex pattern is repeated can be understood easily in terms of the **net** or **lattice** composed of two sets of equally spaced parallel lines which outline the selected unit pattern. The two sets of equally spaced lines intersect at identical points on the pattern and in fig. 4.5 they

intersect at the base of each letter P. Each line represents a row of identical points with equal spacing. The smallest area outlined by the lines can be regarded as a **unit cell** or **unit pattern** as shown in fig. 4.5. It can be seen that there are a number of ways of arranging the 2-dimensional lattice so that the lines intersect at identical points in the pattern, and three different arrangements are shown in fig. 4.5. In each case the unit cell is a parallelogram and the repetition of the unit pattern in two dimensions generates the complete pattern.

The way in which the unit pattern is repeated can be represented by the **2-dimensional lattice** with points at the intersections. Three 2-dimensional lattices which define the repetition of the pattern are shown at the bottom of fig. 4.5. The three 2-dimensional lattices are of the same kind because they show the same degree of symmetry which is limited to symmetry about a centre. Note that this symmetry refers only to the network of points. The pattern is represented by a repetition of the

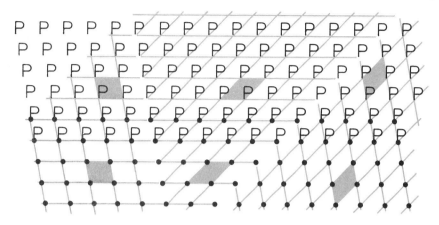

Fig. 4.5. A 2-dimensional pattern produced by repetitions of the letter P. The pattern can be defined by a 2-dimensional lattice composed of two sets of parallel lines which intersect at identical points. Three arrangements of a two-dimensional lattice are illustrated and in each the unit cell is a parallelogram

letter **P** which does not possess a symmetry centre. The orientation of the unit pattern will be considered in chapter 9.

In two-dimensional lattices of lines and points, the **parallelogram** can be regarded as the shape of the unit cell in the general case.

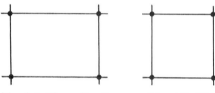

Fig. 4.6. The unit cell of a 2-dimensional lattice is a rectangle

Fig. 4.7. The unit cell is a square

Particular arrangements of the points give rise to networks with greater degrees of symmetry. If the two sets of parallel lines are perpendicular to each other, the unit cell has the shape of a **rectangle** (fig. 4.6), and the network is symmetrical about each row of points. If in addition the points have equal repeat distances on both sets of lines the unit cell is a **square** as shown in fig. 4.7 and the network is then also symmetrical about lines which are diagonal to the unit cell.

If the points have equal separations on each set of lines but the lines are not perpendicular to each other, the unit cell has the

Fig. 4.8. The unit cell is a rhombus

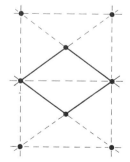

Fig. 4.9. The rhombus represented as a centred rectangle

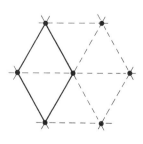

Fig. 4.10. The hexagon unit cell

shape of a **rhombus**, as in fig. 4.8. This unit cell could be described as a rectangle with a point at the centre or a **centred rectangle**, fig. 4.9.

In the special case where the points have equal separations on each set of lines and the sets of parallel lines intersect at 60°, the unit cell is a parallelogram. However the special characteristic of this 2-dimensional lattice may be described by a unit cell having the shape of a **hexagon** with a point at the centre as shown in fig. 4.10.

There are three important features to be noted.

1. All 2-dimensional patterns may be described in terms of one of only five types of unit cell which have been illustrated above.

2. The 2-dimensional lattices with the lowest symmetry are described in terms of a unit cell which is a parallelogram, but lattices with greater degrees of symmetry are described by one of four unit cells which are the rectangle, square, rhombus or centred rectangle and hexagon.

3. In the case of the centred rectangle and the hexagon additional points have been used in the unit cell in order to emphasize their higher degrees of symmetry.

4.3 The primitive space lattices

Imagine space to be divided by three sets of parallel equally spaced planes. In this way space is divided into cells, each identical in size, shape, and orientation as shown at the top of fig. 4.11. The intersection of two planes can be regarded as a row of equally spaced identical points which are located where the third set of planes intersect the line. In this way a three-dimensional array of points is produced as shown in the lower part of fig. 4.11.

A regular three-dimensional array of points in space is known as a **space lattice**. All the points in a space lattice have identi-

cal surroundings. The unit cell of the space lattice is the smallest portion of the space lattice from which the whole space lattice could be built up by translation movements of the unit-cell with no rotations. The space lattice shown in fig. 4.11 consists of three mutually perpendicular lines of equally spaced points and the unit cell of this space lattice is a simple cube with a point at each corner.

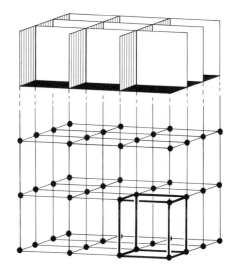

Fig. 4.11. The cubic space lattice P

In the cubic space lattice shown in fig. 4.11 each point belongs equally to eight adjacent unit cells and counts as 1/8th so that there is one point for each unit cell. A space lattice defined by a simple unit cell which has a point at each corner is called a **primitive space lattice** and it is denoted by the symbol **P**. The space lattice shown in fig. 4.11 is called the primitive cubic space lattice.

In section 4.2 it was demonstrated that all possible two-dimensional point arrays may be defined by one of five kinds of two-dimensional lattices. In a similar way, the French crystallographer **Bravais**, demonstrated in 1848 that only fourteen kinds of unit cell are required to describe all possible three dimensional space lattices.

In the reference table at the end of chapter 2 the thirty-two symmetry classes are grouped into seven crystal systems which were characterized by a certain element of symmetry. The fourteen kinds of space lattice can also be grouped into crystal systems with the crystallographic reference axes placed parallel to rows of identical points.

The **primitive cubic space lattice** and the crystallographic axes of the cubic system labelled *x*, *y* and *z* are illustrated in fig. 4.12. The repeat distance on all three axes is the same and is denoted by the letter *a*. The crystal forms which were drawn in chapter 3 all belong to the cubic system.

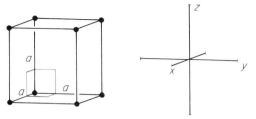

Fig. 4.12. The cubic unit cell P

The other main kinds of primitive space lattices may be regarded as modifications of the primitive cubic space lattice. In the **primitive tetragonal space** lattice shown in fig. 4.13 the repeat distance in the *x* and *y* directions is the same and is labelled *a*, but the repeat distance in the *z* direction is either greater or smaller than *a* and it is labelled *c*. Through each point the vertical axis *z* is a tetrad axis and the *x* and *y* axes are diad axes of symmetry.

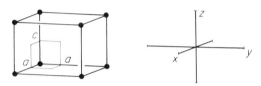

Fig. 4.13. The tetragonal unit cell P

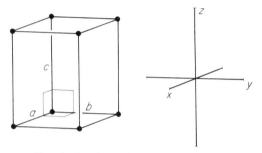

Fig. 4.14. The orthorhombic unit cell P

In the **primitive orthorhombic space lattice**, shown in fig. 4.14 the repeat distances of the points are different in the three mutually perpendicular directions *x*, *y*, and *z*, and the repeat distances are labelled *a*, *b*, and *c* respectively. Through each point in the orthorhombic space lattice diad axes of symmetry lie parallel to the rows of identical points.

In the **primitive monoclinic space lattice**, shown in fig. 4.15 the repeat distances *a*, *b*, and *c* in the directions *x*, *y*, and *z* are all different and in addition the *x* axis is not perpendicular to the *z* axis. The angle between the vertical axis *z* and the positive end of the *x* axis is denoted by the symbol β and the axes are arranged so that β is greater than 90°. Through each point, the *y* axis is a diad axis of symmetry.

In the **primitive triclinic space lattice** shown in fig. 4.16 the three rows of identical

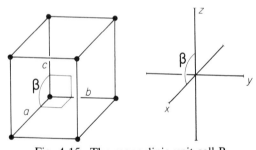

Fig. 4.15. The monoclinic unit cell P

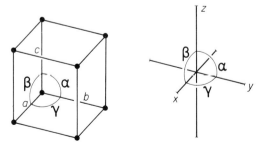

Fig. 4.16. The triclinic unit cell P

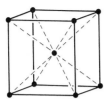

Fig. 4.17. The
body-centred
cubic unit
cell I

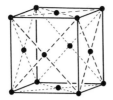

Fig. 4.18. The
face-centred
cubic unit
cell F

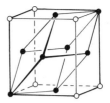

Fig. 4.19. A
primitive
unit cell in
a cubic
face-centred
lattice

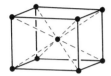

Fig. 4.20. The
body-centred
tetragonal
unit cell I

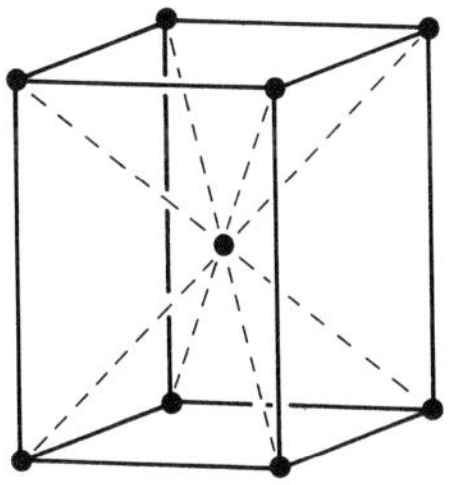

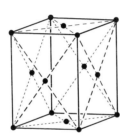

Fig. 4.21. Body-
centred orthorhombic
unit cell I

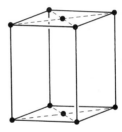

Fig. 4.22. Face-
centred orthorhombic
unit cell F

Fig. 4.23. Base-
centred orthorhombic
unit cell C

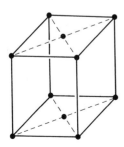

Fig. 4.24. Base-
centred monoclinic
unit cell C

points are not perpendicular to each other, and the repeat distance of the points is different in all three directions. The angles between the reference axes are labelled $\hat{y}\,z = \alpha$, $\hat{z}\,x = \beta$, $\hat{x}\,y = \gamma$ as shown in fig. 4.16. A primitive unit cell characterized by the presence of a single triad axis is described later with a hexagonal unit cell.

4.4 The multiple point space lattices

The cubic system is characterized by the presence of four triad axes of symmetry. In addition to the primitive cubic space lattice there are two other ways of arranging points with identical surroundings so that the whole array displays four triad axes of symmetry. In one arrangement a point occurs in the centre of each primitive cube cell denoted by the letter **I** (from the German Innenzentrierte), fig. 4.17. Since in the **body-centred cubic space lattice** there are two points for each unit cell, the space lattice is said to be doubly primitive.

In the other arrangement a point is placed at the centre of each face of the primitive unit cell and it is described as a **face-centred cubic cell** denoted by the letter **F**, as shown in fig. 4.18. In the face-centred cubic cell each of the six face-centred points is shared with another cell so that they each count as 1/2. Consequently in the face-centred space lattice there are four points for every unit

cell and the lattice is said to be quadruply primitive.

It is possible to select a primitive unit cell to define a face-centred cubic space lattice as shown in fig. 4.19. However this primitive unit cell is a rhombohedron which does not possess the symmetry characteristics of crystals of the cubic system. It is much more informative to describe the symmetry of a crystal structure by a multiple point lattice rather than use a primitive cell which has a very different symmetry.

A **body-centred tetragonal space lattice** is possible as shown in fig. 4.20 because all the points in this lattice have identical surroundings and it is different from the primitive space lattice. There is no face-centred tetragonal space lattice because this arrangement may be described in terms of a body-centred space lattice with a new orientation.

A **body-centred orthorhombic space lattice**, fig. 4.21 and a **face-centred orthorhombic space lattice** fig. 4.22 are both different from the primitive lattice and in each all the points have identical surroundings. Points at the centres of one pair of opposite faces of the primitive orthorhombic space lattice produce a different orthorhombic space lattice. Points at the centres of either of the other pairs of opposite faces produces a similar kind of space lattice but with a different orientation. Usually the points are placed in the centre of the top and bottom

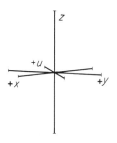

Fig. 4.25. The hexagonal crystallographic axes

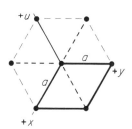

Fig. 4.26. Plan of the hexagonal unit cell

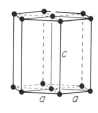

Fig. 4.27. Oblique view of the hexagonal unit cell

Fig. 4.28. The
trigonal
unit cell R

{001} faces, and the space lattice is then called a one-faced or **base-centred orthorhombic space lattice** and it is denoted by the letter C.

A **base-centred monoclinic space lattice** shown in fig. 4.24 is different from the primitive monoclinic space lattice. Crystals like those of beryl shown in plate 14 are characterized by the presence of a hexad symmetry axis. Crystals which exhibit a hexagonal symmetry axis are grouped in the **hexagonal system** and they are described by reference to four crystallographic axes. The vertical axis z is placed parallel to the hexad axis and three similar axes perpendicular to z which intersect each other at 120° are labelled x, y, and u as shown in fig. 4.25. An appropriate **hexagonal space lattice** is shown in figs. 4.26 and 4.27. The hexagonal unit cell contains a primitive unit cell which can be regarded as a special kind of primitive monoclinic cell in which the repeat distances in two directions intersecting at 60° are equal but different from the repeat distances in the perpendicular direction as shown in fig. 4.27. However the hexagonal space lattice is more appropriate for the description of hexagonal crystals because it possesses the same symmetry characteristics.

Crystals which belong to the **trigonal system** exhibit a unique triad axis of symmetry. The crystallographic axes used for the description of trigonal crystals are similar to those used for hexagonal crystals, and the z

axis is placed parallel to the triad axis. The unit cell of the **trigonal space lattice** shown in fig. 4.28 is a rhombohedron composed of rhomb shaped sides and it is denoted by the letter **R**.

All fourteen space lattices possess the maximum symmetry of the respective crystal systems, and Bravais realized that the occurrence of crystal forms with lower symmetry such as the pyritohedron and the tetrahedron in the cubic system, must be due to the arrangement of the groups of atoms which are represented by the points in the space lattice. The space lattice type determines the crystal system, but the particular symmetry class within that system is determined by the symmetry of the arrangement of the atoms forming the real crystal structures. This subject will be considered in chapters 7, 8, and 9.

The first 20 questions for recall in section 4.11. refer to the preceding sections.

4.5 Hexagonal close packing of spheres

Metals consist of atoms of the same kind and consequently the structures of metals are determined by the close packing of spheres of equal size. There are three packing arrangements.

Consider a layer of spheres represented by the heavy circles in fig. 4.29. It can be seen that it is not possible to pack more than six spheres around a single sphere when they are all of equal size. Therefore this is a layer of closest packed spheres, and the spheres cover over 90% of the area. Consider a second layer of closest packed spheres represented by the light shaded circles in fig. 4.29 in which each sphere rests in a hollow between spheres of the lower layer. It would be possible to arrange a third layer of spheres so that they rested in hollows and were also directly above the spheres of the first layer. The third layer of spheres is a repeat of the first layer of spheres and this is illustrated in fig. 4.30.

The structure is built up of the repetition of two layers and because it possesses an axis of six-fold symmetry the structure is called **hexagonal close packing**. The sym-

metry of the structure may be represented by the hexagonal space lattice as shown in fig. 4.30. Note that the spheres of the middle layer are not represented by points in the space lattice. Some metals of the platinum group are found in nature with hexagonal close packed structures.

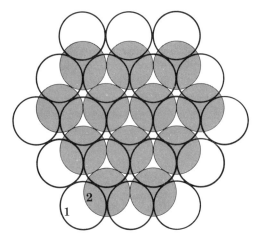

Fig. 4.29. Hexagonal close packing of spheres seen in plan

4.6 Cubic close packing

Study fig. 4.29 and note that the spheres of the third layer need not be placed in the hollows which are directly above the spheres of the lowest layer. The alternative positions of the spheres in the third layer are illustrated in fig. 4.31. If the packing sequence is continued, the spheres of the fourth layer will overlie the spheres of the first layer. The stacking pattern consists of repetitions of three layers and the form of the structure is a tetrahedron shown in plan in fig. 4.32 and as an oblique view in fig. 4.33. *Which element of symmetry exhibited by the tetrahedron determines its assigment to the cubic system?*

The tetrahedron possesses the four triad symmetry axes which is the characteristic feature of the cubic system. Since this packing arrangement exhibits four triad axes it is appropriate to call it **cubic close packing**.

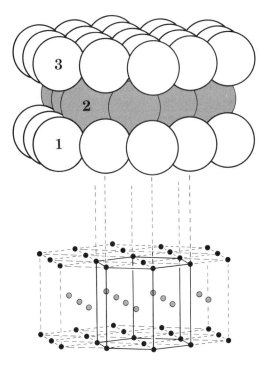

Fig. 4.30. Oblique view of hexagonal close packing and the hexagonal space lattice

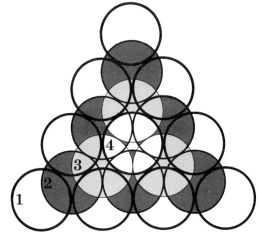

Fig. 4.31. Plan of the alternative packing of layers spheres

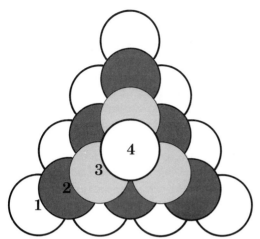

Fig. 4.32. Plan of the alternative packing of layers spheres

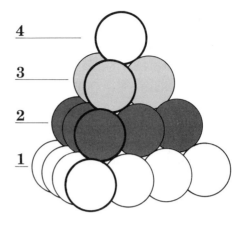

Fig. 4.33. The tetrahedron produced by cubic close packing

The native metals, copper, silver, gold, and lead occur with cubic close packing arrangements of the atoms.

The relationship of the tetrahedron to the cube was studied in section 3.8. Fig. 4.34 is an oblique drawing of the cube to show the close packing structure produced by the repetition of three layers of spheres. The relationship of the packing tetrahedron to the cube is shown by extending the packing in the [Ī10] direction. *Which of the cubic*

space lattices best describes the cubic close packing arrangement.

It can be seen from fig. 4.34 that the smallest cube in the cubic close packing structure consists of a cube with atoms at the corners and the centres of the cube faces. Consequently the **face-centred cubic space lattice** represents the repeat pattern of the cubic close packed structure. **This is a very important relationship between the symmetry of the cube and tetrahedron**, the **cubic**

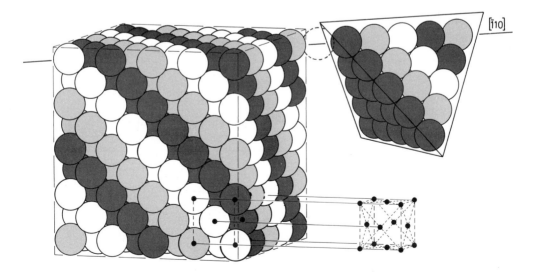

Fig. 4.34. Cube and tetrahedron produced by cubic close packing

close packing structure of spheres and the pattern represented by the cubic face-centred space lattice.

4.7 Body-centred cubic close packing

In both the hexagonal and the cubic close packing structures each sphere is in contact with twelve other spheres of equal size. In a third type of packing of spheres of equal size, each sphere is in contact with only eight other spheres as shown in fig. 4.35. This packing arrangement is represented by the body-centred space lattice and consequently it is called **body-centred cubic close packing**. Iron occurs in nature with a body-centred packing structure.

4.8 The relationship between crystal form and the space lattice

Three very important concepts can now be fully appreciated.

1. **The angular orientations of the crystal faces are determined by the layers of atoms in the crystal structure.**
2. **The manner in which the arrangement of the atoms is repeated in a crystal stucture can be described by a 3-dimensional space lattice composed of rows of equally spaced points.**

In crystal structures which are described by a primitive cubic space lattice, the planes parallel to the unit cell faces contain the closest arrangement of points

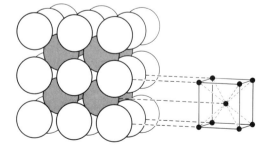

Fig. 4.35. Body-centred cubic close packing

which may represent atoms and it is likely that crystal faces will develop parallel to the {**100**} form. In crystal structures based on a body-centred cubic space lattice the diagonal planes have the closest arrangement of atoms and these may develop as crystal faces to produce the rhombdodecahedron form {**110**}. It can be seen from fig. 4.34 that in the cubic closest packing arrangement described by the face-centred cubic space lattice, the closely packed planes of points are parallel to the {**111**} form. Consequently crystals having this structural pattern are likely to develop faces or cleavages parallel to octahedron or tetrahedron forms. In chapter 7 the structures of the cube crystals of halite and fluorite are studied in order to understand why halite has cleavages parallel to the {**100**} form, while fluorite has cleavages parallel to the {**111**} form. Then in chapter 8 the similar structures of diamond and sphalerite are studied in order to understand why they crystallize with octahedron and tetrahedron forms respectively.

3. **The third important concept-the law of rational intercept ratios, is a direct consequence of the repetition of the crystal structure in three dimensions.** The orientation of layers of atoms in a crystal structure can be described by the appropriate space lattice in terms of repeat distances on the lattice rows. Consequently **the intercept ratios of a face determined by angular measurements will be simple ratios related to the repeat distances on the lattice rows which are parallel to the crystallographic axes** as shown in fig. 4.36. For example in the pyrite structure, well developed layers of atoms occur with indices of (210) and the structure will be studied in chapter 8. The law of rational intercept ratios is true even if the repeat distances are different in the three lattice rows and this will be

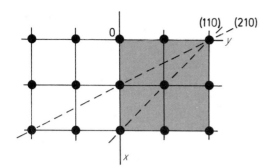

Fig. 4.36. The indices of crystal faces related to the space lattice points. The unit cell is shaded

considered when crystals from the other crystal systems are studied in the chapters following chapter 11.

4.9 Indexing the planes of points in the cubic space lattices

Fig. 4.37 shows plans of the unit cells of the three cubic space lattices projected on the plane containing the x and y reference axes. *Determine the intercept ratios and indices of the central planes of points which are parallel to the y and z crystallographic axes in the body-centred and face-centred cubic unit cells. Determine the intercept ratios and indices of the diagonal planes of points nearest the origin and indicated by the broken lines in the three space lattice unit cells shown in fig. 4.37.*

The intercept ratios of the front faces of the unit cells are x, ∞y, ∞z so that the indices are (100). In the body-centred and face-centred unit cells the intercept ratios of the central planes of points are $\frac{1}{2}x$, ∞y, ∞z, so that the Miller indices are (**200**). Note that the spacing of the (**200**) planes of points in the body-and face-centred lattices is one half the spacing of the (**100**) planes of points in the primitive unit cell.

In the primitive and body-centred unit cells the diagonal plane of points have indices of (**110**). In the face-centred unit cell the first diagonal plane of points from the origin has intercept ratios of $\frac{1}{2}x$, $\frac{1}{2}y$, and ∞z so that its indices are (**220**). Note that the spacing of the (**220**) planes of points in the face-centred lattice is one half the spacing of the (**110**) planes of points in the primitive and body centred space lattices.

In fig. 4.38 the unit cells of the three cubic space lattices are viewed along the diagonal plane (**110**) so that the planes of points parallel to the (**111**) plane can be seen. *What are the indices of the plane of points parallel to the (**111**) plane but nearest the origin in the three unit cells?*

In the primitive and face-centred cells the plane of points parallel to (**111**) and nearest the origin is in fact the (**111**) plane. However in the body-centred cubic cell the points occurring at the centres of the cells give rise to additional layers of points.

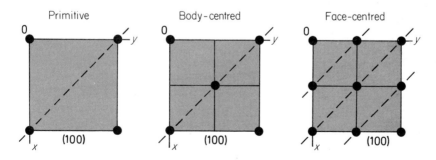

Fig. 4.37. Planes of points in the primitive, body-centred and face-centred cubic unit cells

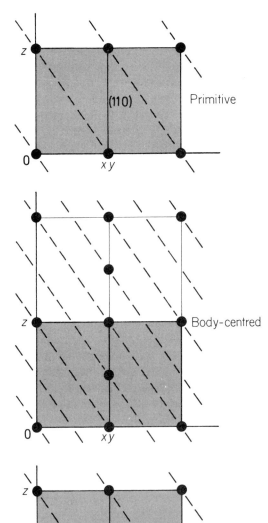

These have intercept ratios of $\frac{1}{2}x$, $\frac{1}{2}y$, $\frac{1}{2}z$ and indices of (**222**). Note that the spacing of the (**222**) planes of points in the body-centred lattice is one half the spacing of the (**111**) planes of points in the primitive and face-centred cubic lattices.

4.10 The ratios of the spacings of the planes of points in the cubic space lattices

Consider the ratio of the spacing of the (110) planes of points in the primitive cubic lattice in terms of the cell dimension *a*. Fig. 4.39 is a basal projection of the primitive unit cell. AB is the perpendicular distance between the (110) planes of points. AB is the hypotenuse of a right angle triangle with sides equal to $\frac{1}{2}a$. *What is the ratio of the spacing of the* (**110**) *planes in terms of the cell dimension* ***a***?

The ratio of the spacing of the (**110**) planes to the spacing of the (**100**) planes is given by the ratio AB/*a*. From the theorem of Pythagoras

$$AB^2 = (a/2)^2 + (a/2)^2 = a^2/2$$

therefore $AB = a/\sqrt{2}$

Fig. 4.38. The unit cells of the cubic space lattices viewed along the (110) planes to illustrate the spacing of the planes of points parallel to the (111) plane

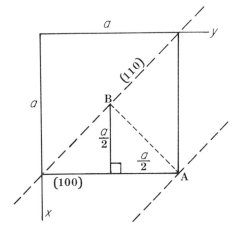

Fig. 4.39. The spacing of the planes of points parallel to (110) in the primitive cubic unit cell

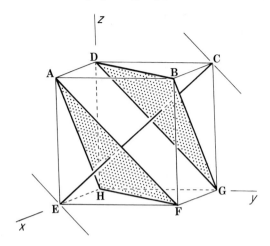

Fig. 4.40. The spacing of the planes of points parallel to the ($\bar{1}$11) plane in the primitive cubic unit cell

In fig. 4.40 it can be seen that the planes parallel to (111) in the cube pass through the points AFH and BGD, and that similar planes pass through the corner points E and C. The line CE is perpendicular to the ($\bar{1}$11) planes, and the distance CE is equal to three times the spacing of the (111) planes. *What is the spacing of the (111) planes in terms of the cube edge a?*

In the right-angle triangle ACE the base AC is double the spacing of the (110) planes and is $2a/\sqrt{2} = \sqrt{2}a$

Since
$$EC^2 = AE^2 + AC^2 = a^2 + (\sqrt{2}a)^2 = 3a^2$$
$$EC = \sqrt{3}a$$
and
$$EC/3 = a/\sqrt{3}$$

The ratios of the spacings (d) between the layers of atoms in the primitive cubic lattice in terms of the repeat distance (a) are therefore

$$d_{(100)} : d_{(110)} : d_{(111)} = a : a/\sqrt{2} : a/\sqrt{3}$$
$$= 1 : 0.707 : 0.577$$

It can be seen from fig. 4.38 that in the body-centred and face-centred cubic lattices the (200) planes have a spacing of $a/2$. It was also discovered in section 4.9 that in the

face-centred cubic lattice there are (220) planes with one half the spacing of the (110) planes, while in the body-centred lattice there are (222) planes with half the spacing of the (111) planes. *What are the ratios of the spacings of the planes of points which are parallel to the (100), (110), and (111) planes in the three cubic lattices?*

The ratios of the spacings of the principal planes of points in the three cubic lattices are tabulated below.

Simple cubic
$$d_{(100)} : d_{(110)} : d_{(111)} = 1 : 1/\sqrt{2} : 1/\sqrt{3}$$
$$= 1 : 0.707 : 0.577$$
Body-centred cubic
$$d_{(100)} : d_{(110)} : d_{(111)} = \tfrac{1}{2} : 1/\sqrt{2} : 1/2\sqrt{3}$$
$$= 1 : 1.414 : 0.577$$
Face-centred cubic
$$d_{(100)} : d_{(110)} : d_{(111)} = \tfrac{1}{2} : 1/2\sqrt{2} : 1/\sqrt{3}$$
$$= 1 : 0.707 : 1.154$$

The use of X-rays allows the determination of the spacings of the planes of atoms in crystals and this will be considered in chapter 6. Then in chapter 7 and 8 the ratios of spacing shown above will be used to determine which space lattice type is indicated by the X-ray measurements obtained from the cubic crystals halite, fluorite, diamond, and sphalerite.

The packing arrangements which have been considered in this chapter consist of atoms all of the same kind. Before continuing the study of the packing arrangements of atoms in crystals it is necessary to consider the manner in which atoms of different kinds are bonded together by chemical forces. This is subject of the following chapter.

4.11 Questions for recall and self assessment

1. What was the nature of the fundamental building unit of crystals proposed by Haüy?

2. What is the most fundamental concept of the hypotheses proposed by Haüy and Huygens to explain the regularity of crystal forms?
3. How can a 2-dimensional pattern be represented?
4. What is the shape of the simplest unit of a 2-dimensional pattern?
5. How many kinds of 2-dimensional patterns of points are possible and what are the shapes of the unit cells called?
6. What is the important feature of the 2-dimensional patterns which are indicated by the unit cells other than the parallelogram?
7. What is a space lattice?
8. How many points are associated with the unit cell in a primitive space lattice?
9. How many kinds of space lattice are possible?
10. What are the orientations of the crystallographic axes x, y and z?
11. Which letters are used to indicate the angles between the 3 crystallographic axes?
12. Which letters are used to indicate the repeat distances along the rows of points in a space lattice?
13. List the names of the five primitive space lattices which are described by reference to three axes?
14. List the relative values of the angles between the rows of points in the primitive space lattices.
15. List the relative values of the repeat distances of the points in the primitive spaces lattices.
16. List the axial symmetry of the lines of points in the primitive space lattices.
17. Which element of symmetry is characteristic of the cubic system?
18. Why is the multiple point, face-centred cubic space lattice used in preference to a primitive lattice based on the same array of points?
19. What are orientations of the crystallographic axes used to described hexagonal crystals?
20. Which symmetry element is characteristic of the trigonal system?
21. What is the characteristic symmetry of a structure composed of repetitions of two layers of closely packed spheres?
22. What is the characteristic symmetry of the form built up by the repetition of three layers of closely packed spheres?
23. The layers of spheres in the cubic close packing structure are parallel to which cubic form?
24. Which space lattice describes the arrangement of the spheres in the cubic close packing structure?
25. In crystal structures based on the body-centred cubic space lattice which cubic form is most likely to develop?
26. What are the indices of the central plane of points parallel to the y and z axes in a face-centred cubic unit cell?
27. In the face-centred cubic unit cell, what are the indices of the plane of points closest to the origin which intersects the x and y axes and is parallel to z?
28. In the face-centred cubic unit cell what are the indices of the plane of points which intersects all three axes and is closest to, but does not intersect the origin?
29. What is the ratio of the spacing of the (220) planes of points in the face-centred cubic lattice in terms of the unit cell dimension a?
30. In forms such as the pyritohedron (210), why are the intercept ratios simple rational numbers?

5 *Crystal Chemistry*

A mineral possesses a characteristic chemical composition which may range within certain limits. The characteristics of the atoms of the elements occurring in a mineral and the manner in which they are bonded together, determines the nature of the crystal structure and also the other physical properties such as density, hardness, cleavage, and melting point. The principal objectives of this chapter are listed below.

1. To understand the nature of ionic, covalent, and metallic types of bonds that occur in minerals.
2. To consider the chemical classification of minerals.
3. To study the reasoning involved in the estimate of the distance between sodium and chlorine ions along the edge of a halite crystal, in preparation for a study of the determination of the structure of halite by X-ray methods in chapter 7.

Contents of the sections

5.1 The structure of the atom

Most substances are compounds formed by the combination of two or more simple substances called the **elements** which cannot be decomposed by chemical reactions. Iron, lead, copper, aluminium, carbon, and sulphur are some of the most familiar elements which occur in an uncombined state and clearly these substances have very different and distinctive properties.

In 1805 an English schoolteacher, **John Dalton**, suggested that the laws of chemical combination could be explained if each element was made up of a particular kind of minute, indestructible, identical particles called **atoms**. Chemical combination of elements involves the union of atoms in simple ratios to form similar pairs or groups of atoms which are known as **molecules**. Dalton presented his views as the Atomic Theory of Matter, and this work was one of the most brilliant pieces of speculative reasoning in the whole field of science.

After the discovery of radioactivity in 1896, great advances were made in the study of matter and it was discovered that atoms were built up of even smaller particles of several kinds. However, a very simple model of the atom is adequate for this introductory study of crystal structures. Atoms consist largely of space. The mass of an atom is concentrated in the **nucleus** which is built up of two types of particle of almost equal mass. Some particles carry a unit positive electrostatic charge and these are called **protons**, while the other particles have no electrostatic charge and are known as **neutrons**. **Electrons** have a mass of about 1/1850 of the proton and carry a unit negative charge. Each electron moves round the nucleus at great speed and at great distances relative to the radii of the nucleus. The number of orbital electrons is equal to the number of protons in the nucleus so that the electrostatic charges are balanced.

5.2 The periodic table of the elements

Long before the structure of the atom was known, attempts had been made to classify the elements into groups on the basis of similar physical and chemical properties. The first satisfactory arrangement of the elements was made in 1869 by the Russian chemist, **Dmitri Mendeleeff**, who stated that the properties of the elements were periodic functions of their atomic mass. The mass of an atom is conveniently represented by a number which represents its mass relative to a selected atom. Dalton initially assigned a value of 1 to the hydrogen atom but later the oxygen atom with a value of 16 was used as a standard. Since 1961 the common atom of carbon which contains 6 protons and 6 neutrons has been used as a standard having exactly 12 atomic mass units.

Mendeleeff demonstrated that when the elements are arranged in the order of increasing **atomic mass**, similar properties recur periodically, and Mendeleeff constructed a chart in which the elements were placed in horizontal lines so that the similar properties occurred in vertical columns known as **groups** or **subgroups** as in fig. 5.1. Since the properties recur periodically the horizontal rows are known as periods. However, there were several minor discrepancies in Mendeleeff's chart and it was clear that the atomic mass itself was not the primary factor which determined the properties of the different elements. The position of an element on the periodic chart is determined by the number of protons forming part of the nucleus of the atom and this is known as the **atomic number** of the element. However an element may have different numbers of neutrons in its nucleus so that the element may occur with more than one characteristic atomic mass. For example the hydrogen atom has one proton but it may occur with one, two or three neutrons. Atoms of the same element which differ in their atomic masses are called **isotopes** of the element. It was the occurrence of isotopes that caused the discrepancies in Mendeleeff's chart.

In passing from one element to the next on the periodic chart, the atomic number increases by one, and in order to maintain a balanced electrostatic charge another electron carrying a unit negative charge is added to the structure. The arrangement of the elements with their atomic numbers is shown in the periodic chart fig. 5.1. The symbols and names of the elements and their relative atomic masses up to atomic number 92, uranium, are given in table 5.2.

For example group 1A consists of the **alkali metals** lithium (Li), sodium (Na), potassium (K), rubidium (Rb) and caesium (Cs). The alkali metals are all soft, and react violently with water. The melting points of these metals are low and range from 180°C for lithium to 280°C for caesium. The elements of group VII B, fluorine (F), chlorine (Cl), bromine (Br), and iodine (I) are also highly reactive. These elements known as

Periodic chart after Bohr

GROUPS	I	II	III	IV	V	VI	VII	O
1st period	H (1)							He (2)
2nd period	Li (3) 1.0	Be (4) 1.5	B (5) 2.0	C (6) 2.5	N (7) 3.0	O (8) 3.5	F (9) 4.0	Ne (10)
3rd period	Na (11) 0.9	Mg (12) 1.2	Al (13) 1.5	Si (14) 1.8	P (15) 2.1	S (16) 2.5	Cl (17) 3.0	Ar (18)

TRANSITION SERIES

GROUPS	IA	IIA	IIIA	IVA	VA	VIA	VIIA	VIII			IB	IIB	IIIB	IVB	VB	VIB	VIIB	O
4th. period	K (19) 0.8	Ca (20) 1.0	Sc (21) 1.3	Ti (22) 1.5	V (23) 1.6	Cr (24) 1.6	Mn (25) 1.5	Fe (26) 1.8	Co (27) 1.9	Ni (28) 1.9	Cu (29) 1.9	Zn (30) 1.6	Ga (31) 1.6	Ge (32) 1.8	As (33) 2.0	Se (34) 2.4	Br (35) 2.8	Kr (36)
5th. period	Rb (37) 0.8	Sr (38) 1.0	Y (39) 1.2	Zr (40) 1.4	Nb (41) 1.6	Mo (42) 1.8	Tc (43) 1.9	Ru (44) 2.2	Rh (45) 2.2	Pd (46) 2.2	Ag (47) 1.9	Cd (48) 1.7	In (49) 1.7	Sn (50) 1.8	Sb (51) 1.9	Te (52) 2.1	I (53) 2.5	Xe (54)
6th. period	Cs (55) 0.7	Ba (56) 0.9	57-71 1.0-1.2	Hf (72) 1.3	Ta (73) 1.5	W (74) 1.7	Re (75) 1.9	Os (76) 2.2	Ir (77) 2.2	Pt (78) 2.2	Au (79) 2.4	Hg (80) 1.9	Tl (81) 1.8	Pb (82) 1.9	Bi (83) 1.9	Po (84) 2.0	At (85) 2.2	Rn (86)
7th. period	Fr (87) 0.7	Ra (88) 0.9	89-103 1.3-1.4	104														

Fig. 5.1. Periodic chart after Bohr. The atomic number is shown above the symbol of the element and the names of the elements are listed in table 5.2. Some important elements are indicated by bold type. The numbers below the element symbol are electronegativity values determined by Pauling which are discussed in section 5.8

the **halogens**, have unpleasant smells and are irritant, corrosive poisons. In contrast group 0 consists of the **inert gases**, helium (He), neon (Ne), argon (A), krypton (Kr) and xenon (Xe). These gases are all unreactive (inert) and they do not form compounds of conventional type. Clearly the periodic recurrence of similar chemical properties cannot be due to the atomic number of the elements because this increases by one throughout the table. It is necessary therefore to consider the arrangement of the electrons in the atoms of the elements.

5.3 Electron orbitals

Material when heated to high temperature emits radiation which can be passed through

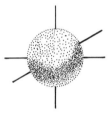

Fig. 5.2. Representation of the probability that the two electrons in the *s* suborbit will occupy an approximately spherical orbital

a spectrograph and be resolved into its component wavelengths. It is found that generally the spectra are not continuous but consist of a series of lines representing discrete wavelengths. In 1913 **Bohr** suggested that an electron in orbit about an atomic nucleus could occur only with particular values of angular momentum, and consequently only at particular energy levels which remained constant as long as the electron does not change its orbit. Bohr postulated that when an electron moves from one energy level to an orbit with a lower energy level, a particular **quantum of radiation** is emitted and this explains the discrete wavelengths observed in the spectrographs. The electrons of an atom occur in certain permitted **orbital zones** or **energy levels**. The energy levels are sometimes called **quantum shells** and are usually referred to by the numbers **1** to **7** or the letters **K, L, M, N,** etc., beginning with the orbital zone nearest the nucleus. The quantum shells are equivalent to the periods on fig. 5.1.

The first satisfactory suggestion concerning the number of electrons that could occur in the various orbital zones around an atomic nucleus was advanced independently by **Bury** and by **Bohr** in 1921. They proposed that the maximum number of electrons that could occur in each orbital zone shell would

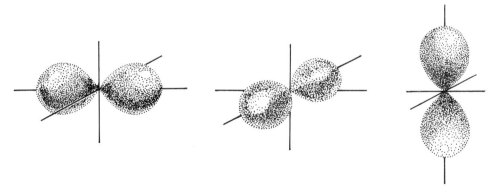

Fig. 5.3. Representation of the probable distribution of the pairs of electrons in the three orbitals of the *p* suborbit

be given by the formula $2n^2$ in which n represents the number of the quantum shell. Therefore the theoretical maximum number of electrons in the quantum shells must be 2, 8, 18, 32, 50, etc.

Within each quantum shell the electrons are distributed in **suborbits (subshells)** which are denoted by the letters s, p, d, and f.

The term **orbital** is used to describe the region in which an electron is most probably to be found. Two electrons may occur within the s suborbit and there is the highest probability that they occupy approximately spherical orbitals as represented by fig. 5.2. The shapes of the p, d, and f suborbitals are increasingly complex and are strongly dependent on direction. There are three p suborbitals and these can be represented by three mutually perpendicular dumb-bell shaped lobes as shown in fig. 5.3.

Each orbit can accommodate only two electrons and these must differ in a manner termed **spin**. This may be visualized as a clockwise and anticlockwise spin of the electrons about an axis through their centres. Repulsion between electrons of opposite spin is not as great as that between electrons with the same spin. Thus pairing between electrons of opposite spin has a marked stabilizing effect on the orbit.

There are five possible d suborbitals and seven possible f suborbitals each containing two electrons. The distribution of the electron pairs in the subshells is shown in table 5.1 below.

Table 5.1.

Quantum shell number	Maximum number of electrons in the quantum shells (orbital zones)	Distribution of the electrons in subshells (suborbitals)				Total number of electrons in each shell
n	$2n^2$	$s.$	$p.$	$d.$	$f.$	
1.	2.	2.				2.
2.	8.	2.	6.			8.
3.	18.	2.	6.	10.		18.
4.	32.	2.	6.	10.	14.	32.
5.	50.	2.	6.	10.	14.	32.
6.	72.	2.	6.	10.		18.
7.	98.	2.	6.			8.

On the whole the energy of the orbital electrons is greater in each successive shell but the relative energy of the electrons in the various subshells is shown in fig. 5.4. It can be seen from the diagram that the energy of the electrons in the d subshell is higher than the energy of the electrons in the s subshell of the succeeding quantum

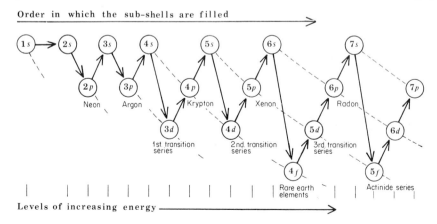

Order in which the sub-shells are filled

Levels of increasing energy

Fig. 5.4. The relative energy levels of the orbital electrons. The arrows indicate the order in which the subshells (suborbits) are filled with electrons

shell. For instance the energy of the electrons in subshell 3*d* is greater than 4*s*, and the energy of the electrons in 4*d* is greater than in 5*s*. Because of this, no *d* subshell receives electrons until the *s* subshell of the next quantum shell is filled. Consequently there are never more than eight electrons in the outermost shell of the structure of an atom because the *s* and *d* can hold only two and six electrons respectively.

5.4 The sequence of elements in the periodic table

The distribution of the electrons in the quantum shells of the first 92 elements are listed in table 5.2. The first element, hydrogen (1), has only one electron but the second element, helium (2), has two electrons and these completely fill the first shell because it can contain only two electrons. Lithium (3) has three protons in its nucleus and a third electron enters the second shell of the structure. In the succeeding elements beryllium (4), boron (5), carbon (6), nitrogen (7), oxygen (8), fluorine (9), and neon (10), the electrons are added to the second shell. The atom of the gas **neon**, has eight electrons in the second shell and this now contains the maximum number of electrons.

Sodium (11) has one electron in the third shell and additional electrons are added to this shell in the elements from magnesium (12) and argon (18). **Argon** has eight electrons in the third shell. A maximum of eighteen electrons could be contained in the third shell but because the energy of the electrons in the 3*d* subshell is greater than the energy of the electrons in the 4*s* subshell as shown in fig. 5.4, the 4*s* subshell must be filled before electrons can enter the 3*d* subshell to complete the third shell. Potassium (19) has one electron in the 4*s* subshell and calcium (20) has two. This fills the 4*s* subshell and the following ten additional electrons go into 3*d* and complete the third

shell. The series of elements in which the process of filling the next to the outermost shell occurs, is known as a **transition series**. The ten elements following calcium from scandium (21) to zinc (30) are known as the **first transition series**.

In zinc (30) the third shell and 4*s* are complete and then additional electrons enter the fourth subshell 4*p* until there is a total of eight electrons in the fourth shell in the gas, krypton (36). Before any more electrons can be added to the fourth shell, two electrons must enter the 5*s* subshell, and this happens in the elements rubidium (37) and strontium (38). The next ten electrons enter the 4*d* subshell in the elements from yttrium (39) to cadmium (48) which are known as the **second transition series** of elements. However this does not complete the fourth shell which can contain a maximum of thirty-two electrons. Six electrons enter the 5*p* subshell and two electrons enter the 6*s* subshell before the fourteen electrons enter the 4*f* subshell to complete the fourth shell. The elements in which the 4*f* subshell is filled have atomic numbers from 57 to 71 and are known as the **rare earth elements**. The rare earth elements form a transition series within a transition series because in the succeeding elements with atomic numbers from 72 to 80 additional electrons occur in the 5*d* subshell.

In the next six elements additional electrons occur in the 6*p* subshell so that the gas radon has eight electrons in its outermost shell. The atoms of the elements francium and radium, contain one and two electrons respectively in the 7*s* subshell. The remaining elements with atomic numbers from 89 to 100 have additional electrons in the 5*f* subshell and they are known as the **actinide series**. These atoms with complex structures sometimes emit particles which may be single electrons known as β particles or groups of two protons and two neutrons similar to the nuclei of helium but known as α particles. When α particles break away, the atomic number changes and consequently it becomes an atom of another element, two places lower in the periodic table. This disintegration of an atomic structure is known as **radioactivity** and the most familiar occurrence is the gradual breakdown of uranium to form lead as a stable end product.

5.5 Valency and the chemical composition of minerals

Minerals are compounds which have characteristic chemical compositions. Long before the electronic structure of the atom was understood, chemists had described the compositions of compounds by assigning formulas which indicated the proportions of the elements present. The **valence** of an element was originally described as the number of valence bonds formed by an atom of an element with other atoms.

If the weight percentages of the elements which compose a mineral have been determined by one of the methods of chemical analysis, the proportions of the atoms in the mineral can be calculated easily. For example, **halite** consists of approximately 39.4% Na (sodium) and 60.6% Cl (chlorine) by weight. Since Na and Cl have different atomic weights different numbers of atoms of each element would be required in order to make up the same approximate weight, and therefore it is necessary to calculate the proportions of Na and Cl atoms that make up the weight percentages which are characteristic of halite crystal. The calculation of the atomic proportions is shown below.

	% weight	Atomic wt.		Atomic proportions
Na	39.4 ÷ 22.99		=	1.713 ≃ 1
Cl	60.6 ÷ 35.457		=	1.709 ≃ 1

This indicates that halite consists of approx-

imately equal numbers of Na and Cl atoms and it can be represented by the formula NaCl. The atoms of Na and Cl each have unit valence.

The mineral **fluorite** is composed of approximately 51.3% Ca and 48.7% F by weight and the atomic weights are 40.08 and 18.99 respectively. *What are the atomic proportions and valences of calcium and fluorine in fluorite?*

	% weight	Atomic wt.		Atomic proportions
Ca	51.3 ÷ 40.08		=	1.2799 ≃ 1
F	48.7 ÷ 18.99		=	2.5645 ≃ 2

In fluorite there are twice as many F atoms as there are Ca atoms so that the composition may be represented by CaF_2. The Ca atom as a valence of two compared with the unit valence of the F atom.

In the mineral **chalcopyrite** three elements are present.

	Copper (Cu)	Iron (Fe)	Sulphur (S)
Weight %	34.88	30.05	34.50
Atomic weight	63.546	55.847	32.06
Combining proportion	0.548	0.538	1.076
Combining ratio	1.0	0.982	1.964
	1	1	2

The composition of chalcopyrite can be represented by the formula $CuFeS_2$.

In many minerals the positions of some of the atoms of a characteristic element may be taken by another element. For example in the mineral **sphalerite** ZnS, some of the positions of zinc (Zn) atoms may be occupied by iron. The Fe atoms are said to substitute for the Zn atoms. **Substitution** of this kind in sphalerite is represented in the formula of the composition in this manner (ZnFe) S. The brackets indicate that Fe may substitute for the characteristic element Zn which is placed first in the formula and this

Table 5.2 The elements

Table 5.2. The distribution of the electrons in the quantum shells of the first 92 elements.

Atomic number	Symbol	Name	Relative atomic mass	Addition of electrons to quantum shells						Ionic radii in ångstroms			
1	H	Hydrogen	1.008	**1**									
2	He	Helium	4.00260	**2**									
3	Li	Lithium	6.941	2	**1**					Li^+	0.68		
4	Be	Beryllium	9.01218	2	**2**					Be^+	0.35		
5	B	Boron	10.81	2	**3**					B^+	0.23		
6	C	Carbon	12.011	2	**4**					C^{+4}	0.16		
7	N	Nitrogen	14.0067	2	**5**					N^{+5}	0.13		
8	O	Oxygen	15.9994	2	**6**					O^{-2}	1.40		
9	F	Fluorine	18.9984	2	**7**					F^-	1.36		
10	**Ne**	**Neon**	20.179	2	**8**								
11	Na	Sodium	22.9898	2	8	**1**				Na^+	0.97		
12	Mg	Magnesium	24.305	2	8	**2**				Mg^{+2}	0.66		
13	Al	Aluminium	26.9815	2	8	**3**				Al^{+3}	0.51		
14	Si	Silicon	28.086	2	8	**4**				Si^{+4}	0.42		
15	P	Phosphorus	30.9738	2	8	**5**				P^{+5}	0.35		
16	S	Sulphur	32.06	2	8	**6**				S^{+6}	0.30	S^{-2}	1.84
17	Cl	Chlorine	35.453	2	8	**7**				Cl^-	1.81		
18	**Ar**	**Argon**	39.948	2	8	**8**							
19	K	Potassium	39.102	2	8	8	**1**			K^+	1.33		
20	Ca	Calcium	40.08	2	8	8	**2**			Ca^{+2}	0.99		
21	Sc	Scandium	44.9559	2	8	9	**2**			Sc^{+3}	0.81		
22	Ti	Titanium	47.90	2	8	10	**2**			Ti^{+4}	0.68	Ti^{+3}	0.76
23	V	Vanadium	50.9414	2	8	11	**2**			V^{+5}	0.59	V^{+3}	0.74
24	Cr	Chromium	51.996	2	8	13	1			Cr^{+6}	0.52	Cr^{+3}	0.63
25	Mn	Manganese	54.9380	2	8	13	**2**			Mn^{+4}	0.60	Mn^{+2}	0.80
26	Fe	Iron	55.847	2	8	**14**	2			Fe^{+3}	0.64	Fe^{+2}	0.74
27	Co	Cobalt	58.9332	2	8	**15**	2			Co^{+3}	0.63	Co^{+2}	0.72
28	Ni	Nickel	58.71	2	8	**16**	2			Ni^{+2}	0.69		
29	Cu	Copper	63.546	2	8	**18**	1			Cu^{+2}	0.72	Cu^+	0.96
30	Zn	Zinc	63.37	2	8	18	**2**			Zn^{+2}	0.74		
31	Ga	Gallium	69.72	2	8	18	**3**			Ga^{+2}	0.62		
32	Ge	Germanium	72.59	2	8	18	**4**			Ge^{+4}	0.53		
33	As	Arsenic	74.9216	2	8	18	**5**			As^{+5}	0.46	As^{+3}	0.58
34	Se	Selenium	78.96	2	8	18	**6**			Se^{+6}	0.42	Se^{+4}	0.50
35	Br	Bromine	79.904	2	8	18	**7**			Br^-	1.95		
36	**Kr**	**Krypton**	83.80	2	8	18	**8**						
37	Rb	Rubidium	85.4678	2	8	18	8	**1**		Rb^+	1.47		
38	Sr	Strontium	87.62	2	8	18	8	**2**		Sr^{+2}	1.12		
39	Y	Yttrium	88.9059	2	8	18	9	**2**		Y^{+3}	0.92		
40	Zr	Zirconium	91.22	2	8	18	**10**	2		Zr^{+4}	0.79		
41	Nb	Niobium	92.9064	2	8	18	**12**	1		Nb^{+5}	0.69		
42	Mo	Molybdenum	95.94	2	8	18	**13**	1		Mo^{+6}	0.62	Mo^{+4}	0.70
43	Tc	Technetium	98.9062	2	8	18	**14**	1					
44	Ru	Ruthonium	101.07	2	8	18	**15**	1					
45	Rh	Rhodium	102.9055	2	8	18	**16**	1					
46	Pd	Palladium	106.4	2	8	18	**18**						
47	Ag	Silver	107.868	2	8	18	18	**1**		Ag^+	1.26		
48	Cd	Cadmium	112.40	2	8	18	18	**2**		Cd^{+2}	0.97		

First Transition Series (elements 21–30)

Second transition series (elements 39–48)

Table 5.2 The elements

49	In	Indium	114.82	2 8 18 18 **3**	In⁺³ 0.81	

No.	Sym	Name	Mass	Config	Ion 1	Ion 2
49	In	Indium	114.82	2 8 18 18 **3**	In^{+3} 0.81	
50	Sn	Tin	118.69	2 8 18 18 **4**	Sn^{+4} 0.71	Sn^{+2} 0.93
51	Sb	Antimony	121.75	2 8 18 18 **5**	Sb^{+5} 0.62	Sb^{+3} 0.76
52	Te	Tellurium	127.60	2 8 18 18 **6**	Te^{+6} 0.56	Te^{+4} 0.70
53	I	Iodine	126.9045	2 8 18 18 **7**	I^- 2.16	
54	**Xe**	**Xenon**	131.30	2 8 18 18 **8**		
55	Cs	Cesium	132.9055	2 8 18 18 **8 1**	Cs^+ 1.67	
56	Ba	Barium	137.34	2 8 18 18 **8 2**	Ba^{+2} 1.34	
57	La	Lanthanum	138.9055	2 8 18 18 **9 2**	La^{+3} 1.14	
58	Ce	Cerium	140.12	2 8 18 **20** 8 2		
59	Pr	Praseodymium	140.0977	2 8 18 **21** 8 2		
60	Nd	Neodymium	144.24	2 8 18 **22** 8 2		
61	Pm	Promethium	(147)	2 8 18 **23** 8 2		
62	Sm	Samarium	150.4	2 8 18 **24** 8 2		
63	Eu	Europium	151.96	2 8 18 **25** 8 2		
64	Gd	Gadolinium	157.25	2 8 18 25 **9 2**		
65	Tb	Terbium	158.9254	2 8 18 **27** 8 2		
66	Dy	Dysprosium	162.50	2 8 18 **28** 8 2		
67	Ho	Holmium	164.9303	2 8 18 **29** 8 2		
68	Er	Erbium	167.26	2 8 18 **30** 8 2		
69	Tm	Thulium	168.9342	2 8 18 **31** 8 2		
70	Yb	Ytterbium	173.04	2 8 18 **32** 8 2		
71	Lu	Lutetium	174.97	2 8 18 32 **9 2**	Lu^{+3} 0.85	
72	Hf	Hafnium	178.49	2 8 18 32 **10 2**	Hf^{+4} 0.78	
73	Ta	Tantalum	180.9479	2 8 18 32 **11 2**	Ta^{+5} 0.68	
74	W	Tungsten	183.85	2 8 18 32 **12 2**	W^{+6} 0.62	W^{+4} 0.70
75	Re	Rhenium	186.2	2 8 18 32 **13 2**		
76	Os	Osmium	190.2	2 8 18 32 **14 2**		
77	Ir	Iridium	192.2	2 8 18 32 **15 2**		
78	Pt	Platinum	195.09	2 8 18 32 **16 2**	Pt^{+2} 0.80	
79	Au	Gold	196.9665	2 8 18 32 **18 1**	Au^+ 1.37	
80	Hg	Mercury	200.59	2 8 18 32 18 **2**	Hg^{+2} 1.10	
81	Tl	Thallium	204.37	2 8 18 32 18 **3**		
82	Pb	Lead	207.2	2 8 18 32 18 **4**	Pb^{+4} 0.84	Pb^{+2} 1.20
83	Bi	Bismuth	208.9806	2 8 18 32 18 **5**	Bi^{+3} 0.96	
84	Po	Polonium	210	2 8 18 32 18 **6**		
85	At	Astatine	210	2 8 18 32 18 **7**		
86	**Rn**	**Radon**	222	2 8 18 32 18 **8**		
87	Fr	Francium	223	2 8 18 32 18 **8 1**		
88	Ra	Radium	226.0254	2 8 18 32 18 **8 2**		
89	Ac	Actinium	277	2 8 18 32 18 **9 2**		
90	Th	Thorium	232.0381	2 8 18 32 **19** 9 2	Th^{+4} 1.02	
91	Pa	Protactinium	231.0359	2 8 18 32 **20** 9 2		
92	U	Uranium	238.029	2 8 18 32 **21** 9 2	U^{+4} 0.97	

(Elements 57–71 are bracketed and labeled vertically: *Rare earth elements*)

The relative atomic masses and ionic radii are according to Ahrens, L. H., *Geochim. et Cosmochim, Acta*, **2**, 168 (1952). The whole numbers are the mass numbers of the most stable isotope.

is the explanation of the range of colour shown in plate 11.

5.6 The electrovalent or ionic bond

Cavendish discovered that the electric conductivity of water is greatly increased when common salt (NaCl) is dissolved in the water. **Arrhenius** proposed in 1887 that in solution the sodium chloride dissociates into electrostatically changed atomic particles Na^+ and Cl^- which are called **ions**. When electrodes are put into a sodium chloride solution the sodium ions are attracted towards the **cathode** while the chlorine ions move towards the **anode**. The movement of the ions in opposite directions in the solution provides the mechanism for the conduction of electricity in the solution. Positively charged ions such as Na^+ are called **cations** while negatively charged ions like Cl^- are called **anions**.

It was observed that each horizontal sequence of elements shown on the periodic chart ends with a gas. These gases are helium, neon, argon, krypton, xenon, and radon, and because they do not react readily they are known as the **inert gases**. With the exception of helium which has two electrons, all the inert gases have eight electrons in their outermost shell. This observation led to the idea that the most stable structure for an atom is one in which the outer shell of electons is completely filled with eight electrons.

J. J. Thompson suggested that electrons of atoms of different elements might be transferred from one atom to another so that the atoms of both elements could achieve the stable eight electron outer shell. Sodium (Na) has only one electron in the third shell but chlorine (Cl) has seven. When the elements combine to form sodium chloride vapour, each sodium atom loses one electron while each chlorine atom gains an extra electron. The sodium atom now has eight electrons in the second shell but it has

an unbalanced unit positive electrostatic charge on the nucleus so it forms the **cation** Na^+. The chlorine atom has eight electrons in the third shell but the extra electron carries an unbalanced unit negative electrostatic charge so it forms the **anion** Cl^-. In the sodium chloride vapour, pairs of sodium and chlorine ions are held together by the electrostatic attraction of the opposite charges. Because the combining force is the electrostatic attraction of oppositely charged ions, this kind of bond is called the **electrovalent** or **ionic bond**. The valency electrons are localized on particular atoms and the **electrovalency** of the atom is equal to the number of electrons it must gain or lose in order to attain a stable electron configuration by electrovalence bonding. The electrovalency of sodium is +1 while chlorine has unit negative electrovalency −1. The theory of the ionic bond was developed by **W. Kossel** in 1916.

The role of the electrons in the outermost shells in the sodium and chlorine ions and in the sodium chloride vapour molecules may be represented by a diagram introduced by the American chemist **G. N. Lewis** in which the electrons of the outermost shell are represented by dots.

$$Na \cdot + \overset{\cdots}{\underset{\cdots}{Cl}} : \rightarrow Na^+ + : \overset{\cdots}{\underset{\cdots}{Cl}} : {}^-$$

In aqueous solutions containing sodium and chlorine the two elements occur as free ions which may collide but they remain dissociated. If the volume of the solvent is decreased below a certain critical value the mutual attraction of the unlike electrical charges of sodium and chlorine ions may be greater than the disruptive forces arising from the collisions. Under these conditions the sodium and chlorine ions may become organized in a three-dimensional structure which gives rise to the mineral **halite** (rock salt). The chemical characteristics of the halite crystal, such as its saltiness, are those of the solution and are quite unlike the properties of the elements since sodium occurs as a metal and chlorine is a green, acid gas. It can be inferred therefore that the **halite crystal is made up of ions of sodium and chlorine and that the structure is characterized by ionic bonding.**

5.7 The covalent bond

An unattached chlorine atom is very reactive and it readily combines with another chlorine atom to form the chlorine gas molecule Cl_2. Bonding of this kind cannot be explained by the transfer of electrons and the attraction of oppositely charged ions. G. N. Lewis suggested in 1916 that atoms can also attain the electronic structure of an inert gas by sharing pairs of electrons and the shared pair of electrons is known as a **covalent bond**. The valency electrons involved in the covalent bond occupy molecular orbits which extend over two atomic nuclei so that they are effectively localized between a pair of atoms. Lewis illustrated the role of the electrons in the outer shells of the chlorine gas molecule with the diagram below.

$$:\overset{..}{\underset{..}{Cl}} + \overset{..}{Cl} \rightarrow :\overset{..}{\underset{..}{Cl}} : \overset{..}{\underset{..}{Cl}} :$$

Since the Cl atom shares only one electron it is said to have a **covalency** of one. What are the covalencies of C and H in the molecule of the gas methane CH_4?

$$\overset{.}{\underset{.}{C}} + 4H \rightarrow H : \overset{\overset{..}{H}}{\underset{\underset{..}{H}}{C}} : H$$

In the methane gas molecule H shows a covalence of one while C has a covalence of four (tetracovalent).

Most molecules are held by covalent bonding and their characteristic feature is that these substances do not conduct electricity when molten or in solution. **Diamond** is composed entirely of carbon atoms. The carbon atom has four electrons in its outermost shell and in the diamond structure each carbon atom is able to attain the eight

electron configuration by sharing a pair of electrons with four adjacent carbon atoms. The hardness of diamond is due to the strength of the covalent bonding of the carbon atoms. In most naturally occurring minerals there are two or more kinds of bonds and these may be intermediate in character between the ionic and covalent bonds.

5.8 Electronegativity

Although it has been convenient up to this point to regard atoms and ions as rigid spheres, when ions approach each other they interact so that there may be a change in size (**dilation**) or a change in shape (**deformation**) of the electron cloud of the atom or ion. As a result of the interaction, the positive charges on the nucleus and the negative charges on the electrons lose their spherical symmetry and are distributed so that one part of the ion may have an excess positive charge and another part will have an excess negative charge. This is known as **polarization** of the ion. The larger negative ions (anions) are the more easily polarized because the outer electrons are less strongly bound to the nucleus. Cations of small size and high charge suffer least polarization and are said to have a high polarization power because they cause the greatest amount of polarization on the adjacent anions. Consequently strong polarization is to be expected in structures composed of small highly charged cations of strong polarization power and large weakly charged anions of large polarizability.

In polarized bonds the valency electrons tend to be concentrated in the space between the cation and anion so that some covalent characteristics are introduced into the ionic bonding. Polarization will tend to decrease the equilibrium distance between cations and anions in a structure. In view of the unequal sharing of bonding electrons, attempts have been made to assign to each atom a number which measures the tendency of the bonding electrons to be drawn towards that atom. **Electronegativity** is the name given to the index which represents electron-attracting tendency.

Pauling calculated electronegativity values from bond strengths measured by heats of formation. Pauling's scale of electronegativity values are shown below the element symbols in the periodic chart, fig. 5.1, and they range in value from 0.7 for caesium (Cs) to 4.0 for fluorine (F). The electronegativity values may be used to predict whether the bonding in a compound is predominantly ionic or covalent. In general the bond between the atoms of two elements will be more ionic if the electronegativity values are very different. The compound caesium fluoride (CsF) shows a maximum difference of electronegativities (4.0 − 0.7 = 3.3) and therefore it would form bonds which are almost entirely ionic. This is so because the element F is strongly electronegative and attracts an electron into its structure while the weakly electronegative element Cs offers little resistance to the transfer of its single electron. Atoms with the same electronegativity such as Cl in the Cl_2 molecule and C in diamond will form covalent bonds because the atoms have equal attraction for the electrons.

5.9 Bonding in metals

Unlike ionic and covalent compounds, metals conduct electricity in the solid state, and they are also ductile. To account for these main characteristics of metals it was postulated that a metal can be regarded as a regular array of positively charged ions (cations) with an equal number of negative charges on mobile electrons so that overall electrostatic neutrality is preserved. The electrons involved in the bonding are not associated with particular atoms and are said to be delocalized. The electrical conductivity of metals is explained by the presence of mobile electrons, and the ductility is explained by the ease with which ions of like charge are able to slip over one another.

5.10 Van der Waals forces

In the inert gases the outer electronic shells are completely filled and the electron density distribution is spherically symmetrical about the nucleus, so that the atom is nonpolar over a period of time. However at any particular instant the electrons may not be spherically symmetrical so that the atom may be polarized. Although when averaged over a period of time the dipole moment which varies in magnitude and direction will be zero, maximum attraction will occur when atoms are closely packed together and these are known as the Van der Waals forces. These relatively weak intermolecular forces are important in organic molecular crystals but they are present in all other crystalline solids.

5.11 Ionic Radii

Since atoms and ions consist of a tiny, densely charged nuclei surrounded by a space containing a cloud of orbiting electrons, the size of an ionic particle can be described only in terms of its interaction with other ions. Between a pair of oppositely charged ions there exists a long range attractive electrostatic force which according to Coulomb's law is directly proportional to the product of their charges but inversely proportional to the square of the distance between their centres, as shown in fig. 5.5. When the oppositely charged ions approach close to each other the clouds of electrons and also the positively charged nuclei set up repulsive forces, of short range. The resultant force curve is shown by the heavy line in fig. 5.5 and the equilibrium separation of the ions occurs where the force is zero.

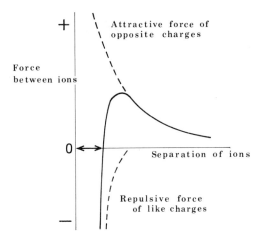

Fig. 5.5. The equilibrium separation of ions
shown by the arrows

Measurements of the interatomic distance may be obtained from the interplaner spacing indicated by X-ray studies which will be described in chapters 6 and 7. If the radius of one atom is assumed, the radii of the other may be calculated. Estimates of ionic radii of some common ions are given in table 5.2. The ionic radii apply only to ionic bonded crystal structures and strictly only for those in which the cations are surrounded by six anions. It can be seen from table 5.2 that the ionic radii of the cations decreases as the charge on the ion increases and this is brought out by fig. 5.6. Note that the common anions tend to be relatively large.

5.12 What is the distance between the Na^+ and Cl^- ions on the edge of a halite crystal cube?

Estimates of the distance between the ions on the edges of a crystal of halite had been made long before the discovery of X-rays. The reasoning used is set out below.

The **molecular weight** of a pure substance is a relative number taking 12 as the mass of

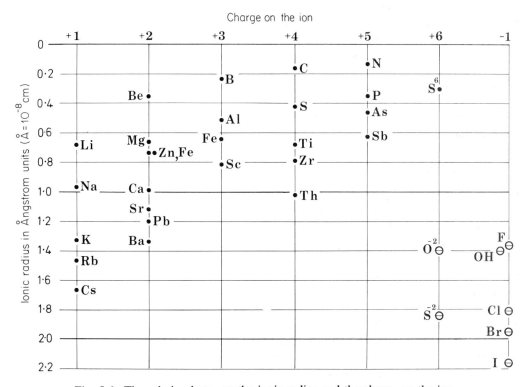

Fig. 5.6. The relation between the ionic radius and the charge on the ion

the carbon 12 atom as a reference. NaCl has a molecular weight of 58.44.

The gram molecule or **mole** of a substance was originally defined as the quantity having a mass in grams equal to its molecular weight. A mole of NaCl consists of $22.989 + 35.453 = 58.44$ grams.

The number of molecules in a mole is the same for all substances. This physical constant known as **Avogadro's number** is 0.6022×10^{24} molecules per mole. Because the isotopes of an element have different atomic masses, Avogadro's number is now more accurately defined as the number of carbon 12 atoms in exactly 12 g of carbon 12. Consequently a **mole** of a substance is defined as Avogadro's number of molecules of the substance.

Since Na and Cl occur as **ions** there are 1.2044×10^{24} ions in a crystal of halite which weighs 58.44 grams. The number of ions along a halite cube edge is approximately the cube root of the number of ions

$$(1.2044 \times 10^{24})^{-3} = 1.064 \times 10^8 \text{ ions.}$$

The density of halite is 2.16 g/cm³ therefore a halite crystal weighing 58.44 grams will occupy a volume of $58.44/2.16 = 27.06$ cm³ approximately, and the side of the cube will be approximately 3 cm.

We now have an estimate of the number of ions (1.064×10^8) occurring along a particular crystal edge of known length, (3 cm), and consequently an estimate of the distance occupied by an ion can be obtained.

$$3/1.064 \times 10^8 \text{ cm} = 2.819 \times 10^{-8} \text{ cm}$$

Consequently the distance between the centres of two ions on the edge of a halite cube crystal is estimated to be 2.819 Å (1 ångstrom Å = 10^{-8} cm) or 2.819×10^{-10} m. The metre is the basic unit of length used in the Systeme International d'Unites (SI), but the angstrom unit is used here for convenience when referring to crystal structure determinations.

5.13 Relative bond strengths

Ions of sodium also combine with ions of the other halogens to form crystal structures similar to that of halite. The relative strengths of the ionic bonds in these crystals can be estimated from the melting temperature of the mineral, and these are given below.

Mineral	Radius of anion	Interionic distance	Melting temperature
NaF	1.36	2.33 Å	980 °C
NaCl	1.81	2.78 Å	800 °C
NaBr	1.95	2.92 Å	755 °C
NaLi	2.16	3.13 Å	650 °C

It can be seen that in these ionic crystal structures there is an inverse correlation between the bond strength as indicated by the melting temperature and the interionic distance; the greater the interionic distance, the weaker the bond.

A number of oxides including MgO, FeO, MnO, CaO, and BaO form crystal structures similar to that of halite, but both the ions forming the oxide crystals are divalent. In the crystal of BaO the interionic distance is 2.77 Å and is closely comparable with the interionic distance of 2.189 Å in halite. The melting temperature of BaO is 1920 °C, more than double the melting temperature of halite which is about 800 °C. *Does this observation give any indication of the relative strengths of the bonds set up between monovalent and divalent ions?* It is apparent that the strength of the bond depends directly on the total charge involved between the ions; the greater the charge the greater the bond strength.

The size of the ion depends on the state of ionization. For example in sphalerite (ZnS), sulphur occurs as a divalent anion, S^{-2}, which has an ionic radius of 1.84 Å. In contrast, in the sulphate mineral barytes, $BaSO_4$, the sulphur occurs as a hexavalent cation coordinated with the 4 oxygen anions, to form a group called a **radical**, $(SO_4)^{-2}$. The sulphur provides 6 electrons for the four oxygens so that the radical has a divalent negative charge which is balanced by the presence of the Ba^{+2} ion in barytes. The size of the hexavalent sulphur ion S^{+6} is only 0.30 Å, because the excess positive charge on the nucleus attracts the remaining orbital electrons.

5.14 The classification of minerals

The chemical composition of a mineral is of fundamental importance and minerals are classified into main **classes** on the basis of their composition.

Some elements such as gold (**Au**), silver (**Ag**), copper (**Cu**), and sulphur (**S**) occur uncombined with the atoms of other elements, and they are classified as **native elements**. Diamond consists entirely of carbon atoms and it is grouped with the native elements.

Apart from the class of native elements, minerals are classified according to the character of the negative ion (**anion**) or **anion group** which is combined with positive ions. Some of the main classes of minerals are indicated below.

Sulphides— Sulphur in a divalent form S^{-2} occurs as the anion, e.g. galena PbS, sphalerite ZnS.

Halides— The characteristic anions are the halogens F, Cl, Br, e.g. Halite NaCl, fluorite CaF_2.

Oxides— Oxygen occurs as the anion, O^{-2}, e.g. corundum Al_2O_3.

Carbonates— The anion is the carbonate radical $(CO_3)^{-2}$, e.g. calcite $CaCO_3$.

Sulphates— The anion is the sulphate group (radical) $(SO_4)^{-2}$ in which sulphur has a valency of +6, e.g. barytes $BaSO_4$, gypsum $CaSO_4, 2H_2O$.

Silicates— In the most abundant group of rock forming minerals, silicate groups of the kind $(SiO_4)^{-4}$ form the anion.

A more complete classification of minerals is given at the beginning of Appendix D which consists of brief notes on the characteristics of common minerals and important ore minerals. Appendix E consists of simplified tables which will aid the identification of the selected minerals in handspecimen, on the basis of the physical properties which can be determined easily.

Simple qualitative chemical tests are useful for the identification of some minerals. The most common test is the use of cold, dilute hydrochloric acid (**HCl**) to identify some carbonate minerals by the effervescence produced when carbon dioxide gas is released. Usually **acid solubility tests** are carried out by placing some mineral powder in a test tube so that it may be heated gently if necessary. A **blowpipe** may be used to blow a pinhole jet of air through a burner flame to produce a small flame which may reach a temperature as high as 1500 °C. Some minerals such as quartz are **infusible** even in the hottest part of flame. The presence of certain elements produce characteristic **flame colorations**. For example lithium gives a crimson colour and sodium produces an intense yellow. There are a series of tests which can be made by holding a mineral grain in the oxidizing or reducing parts of flames but these are often difficult to interpret because of the complexities of the composition of minerals and due to the presence of impurities or contaminants.

It was recognized in the eighteenth century that the physical properties of crystals depend on their chemical composition. In addition to possessing a characteristic composition or compositional range, a mineral is characterized by particular physical and chemical properties such as its hardness, cleavage, melting point, coefficient of thermal expansion, electrical, and thermal conductivity. In general strong bonding results in greater hardness, higher melting point, and lower coefficient of thermal expansion.

However it was not until X-rays were applied to the study of the structure of minerals that the relationships between the chemical composition, internal structure and external crystal form were fully understood. The nature of X-rays are considered in chapter 6 and the main concepts of crystal structure analysis are introduced in chapters 7 and 8 with reference to the determination of the structures of some common minerals which belong to the cubic system.

5.15 Questions for recall and self-assessment

1. What was the nature of the hypothesis proposed by John Dalton in 1805?
2. What are the names and electrical characteristics of the three main components of an atom?
3. What was the nature of the hypothesis proposed by Dmitri Mendeleeff in 1869?
4. What are the names of the two features which form the basis of Mendeleeff's classification?
5. What is the name given to the things that caused the discrepancies in Mendeleeff's scheme?
6. What is the name of the quantity which is more appropriate than the property used by Mendeleeff for the arrangement of the elements in a sequence?
7. What explanation was proposed by Bohr in 1913 to explain the observation that material at high temperatures emits radiations which have particular wavelengths?

8. How many electrons can occur in the two inner suborbitals s and p, and which suborbital contains the electrons at the next higher energy level?
9. What was concluded about the electron structures of the inert gases?
10. What is the nature of the atoms of Na and Cl when common salt is dissolved in water?
11. What is the probable nature of the bonding which gives rise to the NaCl molecule in the gaseous state?
12. What is the nature of the bonding of the halite crystal?
13. What is the nature of the bonding in the chlorine gas molecule Cl_2?
14. What may be deduced from the electronegativity values of the elements in a compound?
15. What is the nature of the bonding in metals which are characterized by ductility and good conductivity of electricity.
16. What is the relationship between the strengths of ionic bonds and the magnitude of the ionic radii of the monovalent anions that may occur in similar structures?
17. What is the relationship between the strength of a bond and the valency of the ions that may occur in similar crystal structures?
18. What is the basis of the chemical classification of minerals other than the native elements?
19. Pyrite is composed of approximately 46.5% Fe and 53.5% S by weight. What is the composition of pyrite?
20. What is the probable nature of the **Fe–S** and **S–S** bonds in the pyrite structure? (refer to fig. 5.1.)

6 *The Nature of X-rays*

Crystal structures are determined by the use of X-rays. However there is no simple lens available which is capable of forming an image of the crystal structure from the X-ray beams. The atomic arrangements of the crystal structures have to be deduced from observations of the X-rays diffracted from the planes of atoms in the mineral using the principles of physical optics. The principal objectives of this chapter are listed below.

1. To understand the nature of X-rays and their emission from an atom as the result of electrons moving from one energy state to another. This extends the comcepts concerning the atomic structures of the elements which were introduced in chapter 5.

2. To understand the nature of the absorption of X-rays by an atom and the excitation of secondary, fluorescent X-rays.

3. To understand the diffraction of X-rays by layers of atoms and the importance of the Bragg equation $n\lambda = 2d \sin \theta$, for the determination of the size and shape of the unit cell of the crystal.

4. To consider the interference of X-rays scattered from a unit cell which consists of layers of different kinds of atoms, and to understand how the measurement of the intensities of the diffracted beams leads to the determination of the arrangement of the atoms in the crystal.

Contents of the sections

6.1 The wave-like nature of X-rays

The presence of an object is recognized by the reflection and refraction effects of light which were introduced in section 1.8. Another effect of light known as **diffraction** gives rise to the halos about the moon or other bright light sources when these are viewed through fog. Fog consists of small dispersed water particles. Diffraction halos are also produced when light is seen through textiles which consist of regular arrangements of fine threads.

Two main conditions are required for the production of **diffraction effects**. Clearly the first condition is that there must be **a regularity in the material** which gives rise to the diffraction effect, and this condition is met in the examples given above by the regularity of the size of the water particles in fog, and by the regularity of the pattern of threads in a cloth. The second condition is that there must be **a regularity in the light** and this is accounted for by regarding light as a wave motion.

After the discovery of **X-rays** by Roentgen in 1895 there was considerable controversy concerning the possibility that X-rays could be regarded as a wave motion similar to the light waves but with wavelengths about one thousandth of the wavelengths of light. In 1912 **Laue** considered the possibility of producing a diffraction effect with X-rays and sought a regular arrangement of particles in which the separation of the particles would be of the same order of magnitude as the postulated wavelengths of X-rays. Laue postulated that a regular arrangement of particles with an appropriate spacing might be found in crystals, and he suggested an experiment in which a fine beam of X-rays was allowed to pass through a crystal and fall on a photographic plate. When the film was developed

the large spot made by the main beam was surrounded by a pattern of spots which was related to the symmetry of the crystal. For example fig. 6.1 is a Laue pattern produced when an X-ray beam passes perpendicularly through a section of a quartz crystal parallel to the side face. This direction in the quartz crystal is a diad axis of symmetry and this is revealed by the pattern of X-ray beams.

Laue's experiment confirmed the wave-like nature of X-rays and also demonstrated that X-rays provide a method of studying the internal structure of crystals. The Laue method gives useful information concerning the internal symmetry of crystals but because of the geometrical arrangement, the patterns always show a centre of symmetry whether or not one is present in the crystal. Consequently only eleven of the thirty-two symmetry classes can be distinguished in Laue X-ray photographs.

6.2 The energy quantum

It is possible to cause a material to emit radiant energy by heating it or by subjecting it to an electric discharge. If these radiations are analysed in a way similar to the separation of white light into its component colours considered in section 1.11, a pattern of lines representing particular wavelengths is

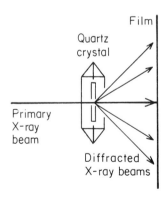

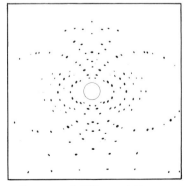

Pattern of diffracted X - rays

Fig. 6.1. A Laue X-ray diffraction pattern

found. An instrument for separating radiation of different wavelengths is called a spectograph if a photographic plate is used to detect the radiation, or a spectrometer if an electronic detector is used. The pattern of lines is different for each element and since it originates within the atom the pattern is known as the **atomic spectrum**.

Planck when attempting to explain the emission of radiations with particular energies from a hot body, was led to the original concept that the energy was emitted from vibrating particles in bursts rather than as a continuous flow. Each burst of radiation corresponds to a definite amount of energy known as a **quantum**. Planck assumed that transitions between allowed energies could occur, and that the energy quantum emitted when an oscillator moved to a lower allowed energy state was proportional to the frequency of the emitted radiation.

$$\Delta \varepsilon = v \cdot h$$

where $\Delta \varepsilon$ = Energy
v = Frequency of radiation.
h = Proportionality constant, known as Planck's constant.

Bohr applied Planck's quantum theory to the orbits of electrons in the hydrogen atom and assumed that:

1. The electron revolves about the nucleus in a particular allowed orbit and does not radiate energy.
2. Radiation energy is emitted or absorbed when the electron moves from one allowed orbit into another allowed orbit.
3. In any transition of the electrons between two allowed orbits the energy difference is a quantum of energy related to the frequency by Planck's constant.

Consequently radiation with a particular wavelength is emitted and it forms a line in the atomic spectrum. Electrons which absorb energy from heat or an electric discharge, jump into higher energy levels and when they spontaneously resume their original lower energy level, the absorbed energy is emitted as discrete wavelengths. The analysis of atomic spectra led to the interpretation that electron orbits can be grouped into the seven quantum shells (1 to 7) and the four subshells ($s, p, d,$ and f) which have been considered in sections 5.3 and 5.4.

6.3 The generation of X-rays

X-rays are generated when electrons, travelling at high speed, collide with atoms of an obstacle known as the target. Any X-ray tube therefore must contain a source of electrons, a high accelerating voltage and a metal target. The type of X-ray tube in general use is known as a filament tube. It consists of an evacuated glass tube which insulates the electron source from the target, fig. 6.2. The source of electrons is a tungsten filament heated by a filament current which can be varied between for example 4 and 36 milliamps, this determines the temperature and the number of electrons it can emit per second. The filament also acts as the high voltage cathode with a high negative potential up to 60,000 volts. The metal target acts as the anode and is usually kept at ground potential. Since 98% of the kinetic energy of the electrons is converted to heat in the target, the latter must be water cooled to prevent it melting. The water connections serve to ground the target. The high voltage applied across the tube accelerates the electrons towards the target. Usually the filament is surrounded by a metal cup at the same high potential as the filament. This repels the electrons and tends to focus them into a narrow line on the target known as the line focus. X-rays are emitted from the line focus in all directions but are allowed to escape from the tube through 4 windows in the tube housing. These windows must be vacuum tight but

X-ray tube

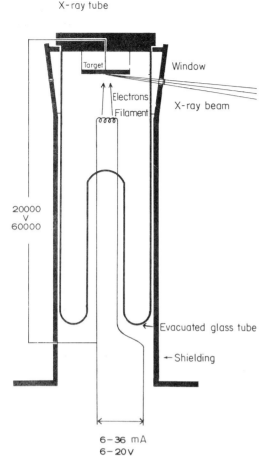

Target

Window

Electrons

X-ray beam

Filament

20000
V
60000

Evacuated glass tube

← Shielding

6−36 mA
6−20V

Fig. 6.2. A section of a filament X-ray tube

must be thin so that they are highly transparent to X-rays. They are usually made of beryllium or mica.

6.4 The continuous spectrum

An electron from the filament, accelerated by a difference in potential of *V* volts, has the energy = *eV*, where *e* is the charge on the electron. When an electron enters the electric field surrounding the nucleus of an atom in the target, the electron loses some kinetic energy. This results in the emission of a quantum of X-radiation, the energy of which is inversely proportional to its

wavelength in accordance with the Einstein equation.

$$E = hc/\lambda$$

where *c* = the speed of light,
and   *h* = the proportionality constant
            known Planck's constant
and   λ = wavelength.

If the electron is instantaneously stopped by a direct collision, the energy of the radiation is equal to the energy of the electron.

$$eV = hc/\lambda \quad \text{and} \quad \lambda = hc/eV$$

Normally the energy of the electron is lost in a stepwise fashion by a series of collisions, and gives rise to a number of X-ray quanta whose wave-lengths are longer than that produced by a direct collision. It can be seen from the equation $\Delta E = hc/\lambda$, that when the change in energy is small, the wavelength is large. Consequently a stream of electrons striking the target produces a continuous

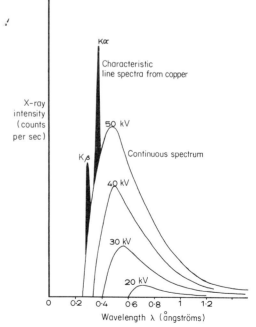

Fig. 6.3. The wavelengths of X-rays produced by a filament tube at different voltages

spectrum of wavelengths with a short wave length limit related to the maximum electron energy. This is known as **the continuous spectrum** or white radiation. The curve of intensity against wavelength is a smooth curve which begins abruptly at a short wavelength limit (max. quantum energy), rises to a maximum (average quantum energy) and declines towards infinity and minimum quantum energy as shown in fig. 6.3.

An increase in the voltage across the X-ray tube increases the speed of the bombarding electrons and these arrive at the target with greater energy, so both the short wave limit and the intensity maximum of the continuous spectrum shift to shorter wavelengths. The total intensity of the X-radiation is proportional to the atomic number of the target material so this is always made from a heavy metal, usually copper.

6.5 The characteristic line spectra

If the voltage across an X-ray tube is raised to a critical level, there is a marked increase in the intensity of X-rays of certain wavelengths. Since the range in wavelength is narrow and the wavelengths are characteristic of the target metal, these are known as the **characteristic line spectra** and they are illustrated in fig. 6.3.

The lines occur in groups or series called the **K,L,M** and **N** series in order of increasing wavelength. In general each series has the same appearance in all elements but the lines shift to shorter wavelengths as the atomic number increases. As the voltage across the tube is raised the characteristic lines from a particular element appear in the order, first, the **M** series in 5 sections; then the **L** series in 3 sections; and finally all the lines of the **K** series simultaneously. Only the **K** lines are useful in diffraction because the longer wavelength lines are too easily absorbed.

The characteristic line spectra are produced when the bombarding electrons have sufficient energy to penetrate to the interior of the atoms in the target and dislodge an electron from one of the quantum shells. The quantum shells of copper are listed below starting with the inner shell.

1st or **K** shell has 2 electrons.
2nd or **L** shell has 8 electrons in three energy levels.
3rd or **M** shell has 18 electrons in five energy levels.
4th or N shell has 1 electron.

The outer electrons have the highest energies and consequently they are most easily removed. The **K** electrons are the most difficult to displace because they have the lowest energy level and are most closely held to the nucleus. Only the **K** lines are important in diffraction studies and the formation of these lines is illustrated in fig. 6.4.

If an electron is displaced from the **K** shell, then it is most likely that the vacancy will be filled by the movement of an electron from the next outer shell rather than a more distant shell. The vacancy created in the second shell would be filled from the next outer shell and so on, so that the atom returns to its normal energy state in a series of steps. When an electron moves into an inner shell it moves to a lower energy level, and this transition is accompanied by the emission of energy in the form of **X-radiation**, the wavelength of which is precisely fixed by the difference in energy between the shells. The X-rays produced when electrons move in to fill a vacancy in the **K** shell form the **K** series of lines. The eight electrons in the L shell are distributed over three energy sublevels, two in L_1 and in L_2 and four in L_3. Transition does not occur from the L_1 subshell. The L_2 and L_3 subshells represent two slightly different energy levels. When an electron moves from the L_3 subshell to the **K** shell it gives rise to X-rays with a characteristic wavelength which are known as $K\alpha_1$. X-rays with a slightly longer

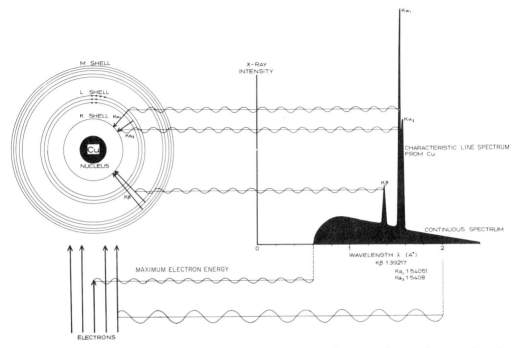

Fig. 6.4. The generation of X-rays of the line spectrum of copper due to the transfer of electrons into the K shell, and the generation of the continuous spectrum due to complete or partial electron collisions

wavelength known as $K\alpha_2$ are produced when an electron moves from the lower energy L_2 subshell into the **K** shell. The **Kα** line is actually a doublet and because there are four electrons in L_3 and only two electrons in L_2, the $K\alpha_1$ is about twice as intense as $K\alpha_2$. The $K\alpha_1$ and $K\alpha_2$ lines are rarely resolved in routine X-ray measurements.

When a **K** vacancy is filled by the movement of an electron from the **M** shell the energy change gives rise to X-rays which are called **Kβ**, and since the drop in energy between the **M** and **K** shells is greater than the drop in energy between L and K, the **Kβ** X-rays have a shorter wavelength than **Kα**, as shown in fig. 6.4. Actually the **Kβ** line is also a doublet because the electrons move in from two of the five **M** energy sublevels. The relative intensities of the $K\alpha_1$, $K\alpha_2$, and **Kβ** lines are about $10:5:2$. A third doublet called **Kγ** is formed when electrons move

into the **K** shell from two of the five **N** subshells, but the line is so weak that it does not produce detectable effects. The wavelengths of the copper **K** lines are $K\beta = 1.39217$ Å, $K\alpha_1 = 1.54050$ Å, $K\alpha_2 = 1.54434$ Å.

6.6 Absorption, and fluorescent X-ray radiation

When the X-rays encounter matter they are partly transmitted and partly absorbed. Experiment shows that the fractional decrease in the intensity I as the X-rays pass through an homogeneous substance is proportional to the distance x tranversed. In differential form this is expressed as $-dI/I = \mu dx \cdot \mu$ is known as the **linear absorption coefficient** and is dependent on the substance, its density ϱ and the wavelength of the X-rays. It is more useful to consider the mass traversed rather than the thickness. The quantity μ/ρ is called the **mass absorption coefficient**, and it is a constant for the absorption of X-rays of a particular wavelength. The mass absorption coefficient possesses the useful property that it is independent of the physical and chemical state of the material. For example μ is different in water, steam and in a mixture of oxygen and hydrogen, but μ/ρ is the same in all three.

Now consider the transmission of X-rays of different wavelengths through a thin sheet of nickel, **Ni**. The variation in the absorption of the X-rays with different wavelengths is shown by plotting the values of the mass absorption coefficient against wavelength in fig. 6.5. The longer wavelength X-rays are greatly absorbed but as the wavelength decreases so does the absorption and the mass absorption coefficient. This is because the X-rays with shorter wavelengths have higher energies and are more successful in passing through the material.

The steady decrease in the mass absorption coefficient stops abruptly and there is a sudden marked increase in the amount of absorption. This is known as the **absorption edge**, and for nickel it occurs at a wavelength of 1.49 Å. The reason for the sudden increase in the absorption is that the energy of the X-ray quanta at this particular wavelength is just sufficient to displace electrons from the **K** shell of the atoms of the absorber, in this case nickel. Most of the energy of the incident X-rays is used in displacing the **K** shell electrons so most of the incident X-rays of this wavelength are absorbed. X-rays of shorter wavelength have the energy to displace **K** shell electrons but as the energy increases, the X-ray quanta find it easier to pass through the absorber so that the mass absorption coefficient decreases steeply for X-rays with the shorter wavelengths as shown in fig. 6.5.

Consider the nature of the absorption process. The scattering of the incident X-rays by atoms takes place in all directions and since the energy does not appear in the primary beam, it can be said to have been absorbed as far as the primary beam is concerned. Two scattering processes occur simultaneously. Tightly bonded electrons are set in oscillation and radiate X-rays of the same wavelength. This is called **coherent scattering**. When an X-ray quanta collides

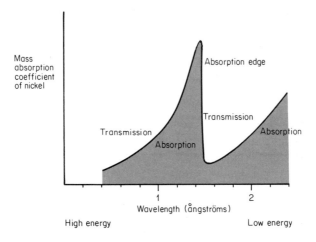

Fig. 6.5. The relative values of the mass absorption coefficient of nickel at different X-ray wavelengths

with a loosely held or free electron, some of the energy may be used in displacing the electron while the X-ray continues on a slightly modified path with less energy and consequently with a slightly longer wavelength. Since the phase of these scattered X-rays bears no relation to the phase of the incident beam it is described as **incoherent scattering** or Compton radiation. The energy absorbed from the incident beam causes electronic transitions within the absorbing atoms. If the wavelength of the incident X-ray is shorter than the wavelength of the absorption edge of the absorber, then the incident X-rays have sufficient energy to displace the **K** shell electrons of the absorber. Consequently the incident beam is heavily absorbed, but when the electron vacancies in the **K** shell of the absorber are filled by the falling in of electrons from the **L** and **M** shells, the characteristic **K** line spectrum of the absorber is produced. This secondary X-ray radiation is known as **fluorescent radiation**. It is characteristic of the absorber but its wavelength is always longer than of the primary beam because it has less energy.

6.7 The use of filters to produce monochromatic X-rays

The **K** shell electrons of a substance can be displaced either by bombardment by other electrons, or by being excited by X-rays which have wavelengths shorter than the wavelength of the absorption edge of the substance. In both cases the **K** lines will be produced as the atoms return to their normal state. X-ray excitation differs from electron excitation in that no continuous spectrum is produced.

Consider the passage of X-rays produced by a **Cu** target through a thin sheet of nickel. The absorption edge of nickel occurs at a wavelength of 1.49 Å and this lies between the wavelengths of the **Cu Kβ** and **Cu Kα** X-rays as shown in fig. 6.6. Consequently

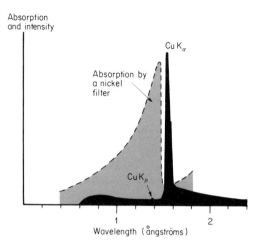

Fig. 6.6. The use of a nickel filter to reduce the intensity of the K*β* X-rays from copper

the **Cu Kβ** wavelengths are strongly absorbed by the nickel while most of the **Cu Kα** X-rays pass through the nickel because they have insufficient energy to displace the **K** electrons of nickel. If the radiation from a copper X-ray tube, is passed through a strip of nickel (called a **filter**), of a particular thickness, the intensity of the **Cu Kβ** line can be greatly reduced relative to the **Cu Kα** line. It is usual to use a filter of a thickness which changes the ratio of the intensities **Kα₁** to **Kβ** from 5:1 to 600:1. Under these conditions the intensity of **Kα₁** is reduced by about half, but the **Kβ** line is completely lost in the background and produces no detectable effect.

Study the absorption edges of the 4 elements below **Cu** in the periodic table, shown in fig. 6.7. Ni absorbs **Cu Kβ** but transmits Cu Kα so it can be used as a filter. The next element **Co** absorbs **Cu Kα** strongly and as a result it produces **Co** fluorescent radiation. This is not very troublesome because cobalt is not very abundant. The next element is iron and this also absorbs **Cu Kα** X-rays heavily and emits the characteristic **K** line spectra of iron as fluorescent radiation. Iron is one of the most abundant elements and when it is present in a sample being studied

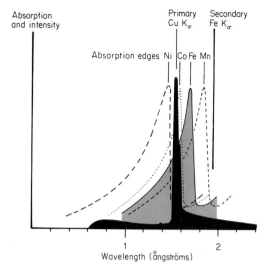

Fig. 6.7. The generation of secondary, fluorescent X-rays from iron due to the absorption of primary X-rays from copper

with **Cu** radiation, the fluorescent radiation produces a marked background effect which spoils the peak to background ratio. Generally three elements fluoresce strongly with each target element. With camera techniques the only way that this can be dealt with is by using an X-ray tube with a different target material. For example iron bearing samples may be studied with primary X-rays

derived from a tube containing a cobalt target.

The first 10 questions in section 6.13 refer to the preceding sections.

6.8 The scattering of X-rays by electrons and atoms

An X-ray beam can be regarded as an electromagnetic wave characterized by an electric field whose strength varies sinusoidally with time at any point in the beam. Since an electric field exerts a force on a charged particle such as an electron, the oscillating electric field of an X-ray beam will set any electron it encounters into oscillatory motion about its mean position. An electron which has been set into oscillation by an X-ray beam is continuously accelerating and decelerating during its motion, and therefore it emits an electromagnetic wave.

An electron lying in the path of this beam of monochromatic X-rays vibrates with the frequency of the incident beam, periodically absorbing energy and emitting it as X-radiation of the same frequency. The wavelength of the incident beam is unmodified by the interaction, but the secondary rays are radiated in all directions and we can

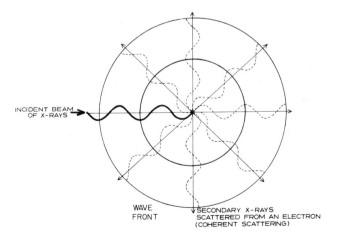

Fig. 6.8. Spherical wave fronts of X-rays scattered coherently by an electron

represent wave fronts in two dimensions by circles around the source of secondary X-rays as in fig. 6.8. This effect is called **coherent scattering** because the scattered waves have the same wavelength as the incident beam.

Although rays are scattered in all directions by an electron, the intensity of the diffracted beam depends on the angle of scattering, and it is greatest in the forward or backward directions, and least in directions perpendicular to the incident X-ray beam. When an X-ray beam encounters an atom, each tightly bound electron in the atom scatters part of the incident beam coherently. The nucleus because of its relatively high mass does not oscillate greatly and therefore it makes a negligible contribution to the coherently scattered radiation. A quantity called the **atomic scattering factor** is used to describe the efficiency of scatter-

ing of an atom in the direction of the incident beam, and it is the ratio of the amplitudes of the waves scattered by the atom and by one electron. In an idealized situation in which all the electrons occurred at a point, the atomic scattering factor would have the value of the atomic number Z. In directions other than that of the incident beam, phase differences occur between the waves scattered by the individual electrons so that the waves interfere and the atomic scattering factor decreases. The loosely bonded electrons scatter part of the incident X-ray at a slightly increased wavelength and this modified scattering is called **incoherent** or Compton **scattering**.

6.9 The scattering of X-rays by layers of atoms: The Bragg equation

A crystal is an orderly arrangement of atoms

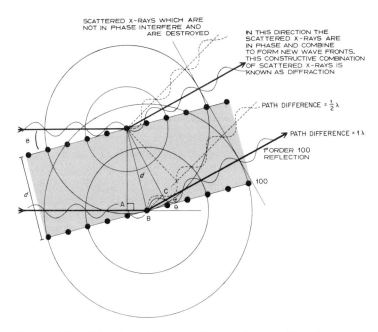

Fig. 6.9. The diffraction of X-rays by layers of atoms. The diagram represents the generation of the first order diffraction beam (reflection) which is produced when the path difference of the X-rays scattered from the adjacent layers of atoms is equal to one wavelength

and all the atoms in the path of a monochromatic beam of X-rays scatter the X-rays simultaneously and in all directions. In fig. 6.9 a crystal structure is represented by two rows of atoms and the wave fronts of the X-rays scattered from atoms perpendicularly opposite each other are represented by circles. Consider the two scattered rays represented by dashed lines which are travelling in the same direction. In this particular direction the two scattered waves are exactly opposite in phase and the oscillations perpendicular to the wave direction oppose each other, with the result that the waves cancel each other completely. In most directions the scattered waves tend to interfere and partly cancel out. However in certain specific directions the scattered X-rays are in phase and combine to form new wave fronts. Consider the waves scattered in the direction represented by the black lines in fig. 6.9. In this direction the scattered waves are exactly in phase and the oscillations perpendicular to the wave direction are intensified so that the beam intensity increases. This cooperative scattering which results in the building up of a beam of secondary X-rays in a particular direction is known as **diffraction**. Note in fig. 6.9 that **the scattered waves are enhanced in the direction along which the wave scattered from the lower layer of atoms has travelled further than the wave scattered by the upper layer and that the increased distance is equal to one wavelength.**

A few months after the original Laue experiments had been published, **W. L. Bragg discovered that the phenomena of diffraction could be pictured as a reflection** of the incident beam from sets of planes of atoms. Consider a beam of parallel monochromatic X-rays penetrating layers of atoms which have a spacing d as in fig. 6.9. The angle of incidence of the X-ray beam is θ and each layer of atoms will scatter a portion of the primary beam simultaneously. In the directions in which the scattered

beams are in phase they will combine to produce a diffracted beam of X-rays. The scattered X-rays will be in phase if the distance travelled by the waves scattered from the upper layer differs from the distance travelled by the waves scattered from the lower layer by a whole number n of wavelengths λ. Fig. 6.9 represents the situation in which diffraction occurs when the path difference $AB + BC = $ one wavelength. It can be seen that

$$AB + BC = 2d \sin \theta$$
$$\therefore \quad n\lambda = 2d \sin \theta$$

where n is a whole number of wavelengths.

This is the Bragg equation which gives the angle of incidence θ at which planes of atoms of spacing d, will diffract X-rays of wavelength λ. The diffracted beam which is formed when the phase difference is one wavelength is called the **first order reflection**, and this is illustrated in fig. 6.9.

6.10 Higher order X-ray reflections

Fig. 6.10 shows the layers of atoms in a different angular position and here the difference in the distances travelled by the X-rays scattered from the two layers is equal to 2 wavelengths. This diffracted beam of X-rays is known as the **2nd order reflection**. Note that when $n = 2$, sin θ is doubled.

Now consider fig. 6.11 which shows the layers of atoms in the angular position in which the difference in the distances travelled by the X-rays diffracted from the two layers of atoms is equal to 3 wavelengths. This diffracted beam is known as the **3rd order reflection**. Consequently for a fixed wavelength λ, a set of planes of atoms with spacing d will give a number of diffracted beams of X-rays when the layers are rotated through positions at which the path difference between the waves scattered from the adjacent layers is **1, 2, 3** or more wavelengths. This is the reason for the lines of spots on the Laue photograph shown in

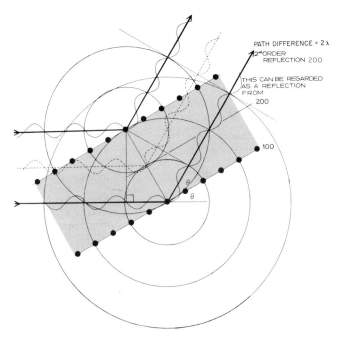

Fig. 6.10. The second order X-ray reflection produced when
the path difference equals two wavelengths

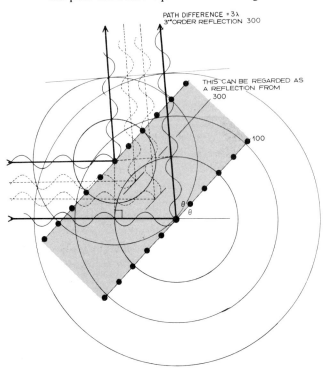

Fig. 6.11. The third order X-ray reflection produced when
the path difference equals three wavelengths

fig. 6.1, and the pattern of lines is determined by the orientation of the planes of atoms. Because θ cannot exceed 90°, and sin 90° = 1, then $\lambda/2d$ cannot exceed unity. A crystal has a large number of sets of planes with a wide range in the values of d, but only those planes for which d is equal to, or greater than $\lambda 2$ can diffract the X-ray beam.

6.11 Scattering from a unit cell

Consider a crystal structure consisting of layers of different kinds of atoms as illustrated in fig. 6.12. ABCD represents a unit cell of the crystal structure. Assume that Bragg's law is satisfied for the layers of atoms parallel to AB and CD with a spacing d so that a diffracted beam is produced in the direction AE. The path difference of the rays scattered from the layers AB and CD is FC + CG = $2d \sin \theta = n\lambda$.

X-rays will also be diffracted from the layers of atoms of a different kind which may occur at K between the layers AB and CD. *How is the reflection AE from the AB and CD planes of atoms affected by the beam of X-rays diffracted from the parallel layers of atoms through point K?*

The phase difference between the X-rays diffracted from the AB and K planes of atoms is given by the path difference JK + KL.

Since the ratio $\dfrac{JK + KL}{FC + CG} = \dfrac{AK}{AC}$

and FC + CG = $n\lambda$, and AC = d

then $\dfrac{JK + KL}{n\lambda} = \dfrac{AK}{d}$

∴ JK + KL = $\dfrac{n\lambda\,AK}{d}$

This **phase difference** is given in terms of the wavelength. If $n = 1$ and AK = $d/2$, then the phase difference will be $\frac{1}{2}\lambda$ and destructive interference will occur. Almost com-

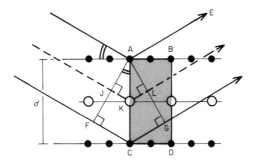

Fig. 6.12. The diffraction of X-rays from layers of different atoms

plete cancellation will occur if the X-rays diffracted from the two different layers are of approximately equal intensity. This situation might arise if the atomic numbers of the two kinds of atoms were close so that they have similar atomic scattering factors, and the atoms are present in approximately equal numbers. Cancellation might also occur if there were twice as many of one kind of atom which has an atomic number about one half of that of the other kind of atom. *What is likely to happen if the layers of atoms have almost equal scattering power but* $n = 2$ *and* **AK** $= d/2$?

This is the position of the second order X-ray reflection from the layers of the same kind of atoms because the path difference will be 2λ. Consequently the X-ray reflections from the adjacent layers of different atoms will have a path difference of λ and so enhancement will occur.

6.12 Phase differences expressed in terms of angular measure

In the preceding sections the difference in phase between sinusoidal X-ray waves scattered from two layers of atoms has been described in terms of the path difference measured in wavelengths. Clearly it is necessary to specify the wavelength if measurements are to be made. Phase differences are usually expressed in angular measure because this is independent of the wavelength.

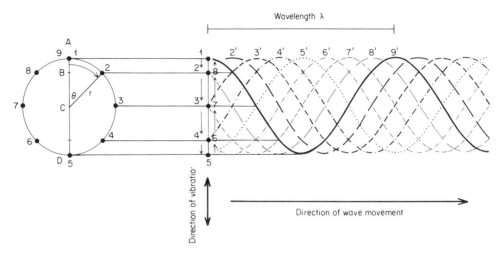

Fig. 6.13. Simple harmonic motion

The **reference circle** is a method of representation which can be used for all wave systems whatever the nature of the disturbance that is being propagated. When a point moves with uniform velocity in a circle, the projection of the position of the point on a diameter AD in fig. 6.13 may be used to represent the oscillation (vibration) in a direction perpendicular to the direction of wave propagation.

If the point on the circle moves with a uniform angular velocity ψ in a circular path, and zero time is taken when the point is in position **1** of fig. 6.13, at a later time t the point will have moved through an angle $\theta = \psi t$. In the same time the projection of the point will have moved from **A** to **B**, and its displacement from the centre **CB** = $r \cos \theta = r \cos \psi t$ where r is the radius of the circle.

If position **3** in fig. 6.13 had been taken as zero time, the displacement of the projection of the point would have been given by $r \sin \psi t$. Motion which can be defined by equations of the kind **CB** $= r \cos \psi t$ or **CB** $= r \sin \psi t$ depending on the zero point which is selected, is called **simple harmonic motion**.

In fig. 6.13 the positions **1** and **3** on the reference circle differ in phase by 90°. Positions **1** and **9′** on the wave are one wavelength apart and can be regarded as differing in phase by one wavelength or by 360° or 2π radians on the reference circle.

Two waves which differ in path length by one wavelength are said to differ in phase by 2π radians. Therefore if the difference in path length p is regarded as a fraction of the wavelength, p/λ, the corresponding angular phase difference ϕ can be regarded as a fraction of 2π. Consequently $\phi = 2\pi p/\lambda$. Study fig. 6.12 again and consider the difference in phase of the waves diffracted from the atoms **A** and **K**. It was found that the path difference **JK** + **KL** = $n\lambda$**AK**/d. Therefore the corresponding angular phase difference is $\phi = 2\pi n\lambda$**AK**/$d\lambda = 2\pi n$**AK**/d. Note that the wavelength does not remain in the expression of the phase difference in angular measure.

The waves scattered from layers of different atoms differ in phase and also in amplitude (the magnitude of the vibration) if the atomic scattering factors of the atoms are different as shown in fig. 6.14. In order to find the resultant wave diffracted from a unit cell it is necessary to add waves which differ in phase and amplitude. **Fourier** established that any strictly periodic pheno-

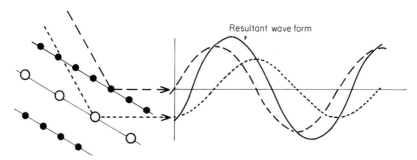

Fig. 6.14. The superposition of waves of different phase and amplitude

menon can be regarded as being built up by the principle of superposition from a succession of constituent waves of simple harmonic type. If one wave is regarded as having zero phase then a second wave can be represented by the equation

$$\delta = a \cos (\psi t + \phi)$$

when δ = displacement of the vibration
  a = amplitude (r of the reference circle)
  ψ = angular velocity
  t = time
  ϕ is the phase when $t = 0$

The displacement of the resultant wave at any particular time t is given by the summation of the displacements of all the constituent waves at time t as shown in fig. 6.14. The resultant wave has a shape similar to the constituent waves but it has different amplitude and phase. The intensity of the diffracted beam is proportional to the square of the amplitude of the resultant X-ray beam.

Since there is no lens capable of forming an image of the crystal structure from the diffracted beams of X-rays, all that can be done is to measure the angular positions and intensities of the diffracted beams produced from an incident beam of X-rays of known wavelength. The study of the diffraction angles leads to the determination of the *d* spacings and ultimately the shape of the crystal lattice, while the study of the intensi-

ties of the diffracted beams leads to the determination of the arrangement of the atoms in the crystal. The methods used by W. H. and W. L. Bragg when they first determined the structures of simple cubic crystals will be studied in chapters 7 and 8. The powder method used for the routine identification of minerals is described in appendix C.

6.13 Questions for recall and self assessment

1. What are the two conditions which are necessary for the appearance of diffraction effects such as halos around bright light sources?
2. Why did Laue suggest an experiment in which X-rays were passed through a crystal?
3. Which letters are used to label the main quantum shells in which the electrons are distributed about an atomic nucleus?
4. When a beam of electrons strikes a metal target such as copper, which electrons in the copper atom are displaced only by the highest energy electrons from the beam, and what is the reason for this observation?
5. What is the explanation for the line spectra?
6. What are the relative values of the wavelengths of the lines from different shells?

7. What is the nature and origin of the α and β lines in the **K** spectrum?

8. What is the reason for the continuous spectrum which is produced by an X-ray filament tube.

9. What is the meaning of absorption edge?

10. Why is **Ni** used as a filter for primary X-rays from **Cu**?

11. What is the meaning of coherent scattering?

12. What is the reason for the diffraction of a beam of X-rays by a crystal structure?

13. Which features are related by the Bragg equation?

14. What is the meaning of 3rd order reflection.

15. What is the atomic scattering factor?

16. In the position of a 1st order diffraction beam from the planes parallel to the top and bottom of a unit cell, what would be the effect of a (002) layer of different atoms in similar numbers and with similar scattering power.

17. What would be the effect in the situation considered in question 16 if there were twice as many atoms of the kind which outline the unit cell, as the kind that occur on the (002) plane?

18. What is the definition of simple harmonic motion?

19. What is the angular equivalent of a phase difference of one wavelength?

20. What is the main advantage of the angular measurement of phase difference?

7 *The Structures of Halite and Fluorite*

The vast increase in knowledge in recent decades and the interest generated by spectacular developments such as moon exploration, has resulted often in a neglect of some of the earlier and most fundamental advances in the Earth Sciences. One of the most important fundamental advances occurred within a few years of Laue's discovery of the diffraction of X-rays by crystals in 1912. During this period W. H. Bragg and particularly W. L. Bragg determined the structures of a number of simple cubic crystals and initiated the development of the X-ray analysis of crystalline materials. The importance of these initial studies was recognized by the award of the Nobel Prize for physics to Sir Lawrence Bragg in 1915. It is interesting and very rewarding to study the simple steps in the reasoning used in the first determinations of the structural arrangements of the cubic crystals. The principal objectives of this chapter are listed below.

1. To study the reasoning involved in the determination of the structures of halite and fluorite which are characterized by ionic bonding.
2. To appreciate that crystal structures are not composed of discrete molecules but consist of continuous three-dimensional arrays of atoms with repeated patterns.
3. To appreciate that a mineral may be described by a chemical composition which represents the approximate pro-

portions of the elements present, but that there are two important constraints on the geometrical packing of the atomic particles;
   (a) the relative sizes of the different atomic particles, and
   (b) the electrostatic charges associated with the atoms of the different elements.
4. To understand why cleavages develop parallel to the {100} form in halite and the {111} form in fluorite.

Contents of the sections

119

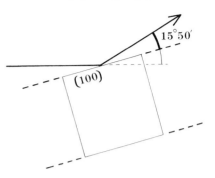

Fig. 7.1. The angular orientation of the 1st order diffracted X-ray beam from the planes parallel to (**100**) in halite when the primary beam is Cu Kα

7.1 The ratios of the spacings of the (100), (110), and (111) planes of ions in the halite crystal

The crystals of halite have cube forms with very prominent cleavage planes parallel to the cube faces as illustrated in fig. 1.19. Clearly in halite the ions must have an arrangement which can be described by one of the three cubic spaces lattices. The indexing of the planes of points in the cubic space lattices was considered in section 4.9 and the ratios of the spacings of the (**100**), (**110**), and (**111**) planes of points in the cubic space lattices were determined in section 4.10. It is now appropriate to consider the use of X-rays to determine the actual ratios of the spacings of the (**100**), (**110**), (**111**) planes of ions in halite so that the cubic space lattice most suitable for the description of the halite crystal structure can be selected.

 The structure of halite was the first crystal structure to be determined by X-ray methods, and at the time nothing was known about the arrangement of the ions of **Na⁺** and **Cl⁻** in the crystal, nor was the wavelength of X-rays known (Bragg, W. L., *Proc. Roy. Soc.*, 1913, **A89**, 468). The structure was first deduced from Laue photographs, but Sir Lawrence Bragg soon demonstrated that crystal structures could be determined more effectively by an instrument similar to the diffractometer which is described in appendix C.

 A crystal of halite was arranged in a diffractometer so that the angles between

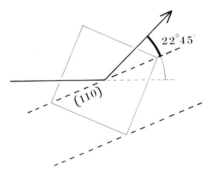

Fig. 7.2. The angular orientation of the 1st order diffraction beam from the planes parallel to (**110**) in halite

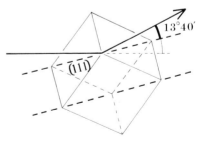

Fig. 7.3. The angular orientation of the 1st order diffraction beam from the planes parallel to (**111**) in halite

the primary X-ray beam and the X-ray beams diffracted from the layers of ions lying parallel to the (**100**) face of the cube could be measured in turn. The position of the first recorded beam diffracted from the (**100**) layers of ions in halite occurs at a θ angle of **15°50′** when a **Cu Kα** primary X-ray beam is used, and this is represented in fig. 7.1. The second and third recorded X-ray reflections from the (**100**) planes are observed at angles of **33°08′** and **55°** respectively.

The halite crystal was then reoriented so that the X-ray beams diffracted from the (**110**) planes of ions could be recorded as shown in fig. 7.2. The first and second recorded X-ray reflections are observed at angles of **22°45′** and **50°38′** respectively.

The first, second, and third recorded X-ray reflections from the layers of atoms parallel to the (**111**) planes of halite are observed at angles of **13°40′**, **28°14′** and **45°11′** respectively, and the position of the first recorded reflection is represented in fig. 7.3.

Consider the first recorded reflections from the (**100**), (**110**) and (**111**) planes of ions in halite. It can be seen from the Bragg equation,

$$n\lambda = 2d \sin \theta, \text{ that } d = n\lambda/2 \sin \theta.$$

If *n* = 1 and the wavelength λ is constant, then the spacings *d* between the planes of ions parallel to (**100**), (**110**), and (**111**) will be proportional to the 1/sin θ values. The calculation of the ratios of the spacings of the (**100**), (**110**), and (**111**) planes of ions in halite is shown in tabulated form below.

	(100)	(110)	(111)
θ of first observed reflection	15°50′	22°45′	13°40′
sin θ	0.2729	0.3867	0.2362
1/sin θ	3.688	2.596	4.241
Ratios of the *d* spacings	1	0.704	1.15

These are the ratios of the *d* spacings determined from observations of the first recorded X-ray reflections from the halite crystal. *Which of the three cubic space lattices studied in section 4.10 best represents the arrangement of the planes of ions in the halite crystal structure?*

7.2 Indexing the X-ray diffraction beams from halite

It can be seen that the ratios of the spacings calculated from the angles of X-ray beams diffracted from the (**100**), (**110**), and (**111**) planes of ions in halite, correspond to the ratios of the spacings of the (**100**), (**110**), and (**111**) planes of points occurring in the face-centred cubic space lattice. **The simple procedure involved in the determination of the cubic lattice most appropriate for the description of the halite structure illustrates the first step in the study of a crystal structure; the determination of the shape and size of the unit cell from the angular positions of the diffraction beams.**

It is the spacing between the adjacent planes of atoms which produces the first observed X-ray reflection from the (**100**) face of the halite cube. The planes of points in the cubic space lattices were indexed in section 4.9. *What are the indices of the first observed X-ray reflection from the (100) layers of ions in halite?*

The X-ray reflection observed at the lowest θ angle occurs when there is one wavelength phase difference between the X-rays scattered from the adjacent layers of ions parallel to the (**100**) plane and these were indexed (**200**) in section 4.9, because the centre plane has intercept ratios of 1/2*x*, ∞*y* and ∞*z*. **When referring to the X-ray reflection the indices 200 are given without brackets.** The angular position of the **200** reflection is shown in fig. 7.4 and it can be seen that a path difference equivalent to two wavelengths, occurs between the X-rays diffracted from the alternate layers of ions

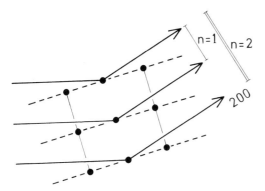

Fig. 7.4. The angular position of the **200** reflection from halite

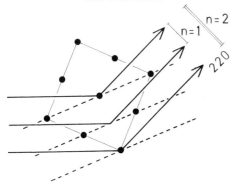

Fig. 7.5. The angular position of the **220** reflection from halite

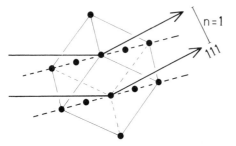

Fig. 7.6. The **111** reflection from halite

which have the spacing of the face-centred unit cell. Consequently the **200** reflection can also be regarded as a second order reflection from the unit cell planes.

The second observed X-ray reflection from the (100) planes in halite will occur when the path difference between the X-rays diffracted from adjacent (100) planes is equivalent to two wavelengths. This reflec-

tion can be regarded as a 1st order reflection from an imaginary plane, 1/4 of the repeat distance from the zero plane so that its intercept ratios would be $1/4x$, ∞y, ∞z and the indices would be **400**. In a later section the second observed reflection from the (100) planes of halite will be referred to by the indices **400**.

Study the planes of atoms which are parallel to the (110) plane of the face-centred unit cell shown in fig. 7.5. *What are the indices of the planes of atoms from which the first observed X-ray reflection from the (110) planes is recorded?*

The intercepts of the layer of atoms in between the (110) plane and the origin of the face-centred unit cell are $\frac{1}{2}x$, $\frac{1}{2}y$, ∞z; and therefore the indices of this layer of atoms are (220) as explained is section 4.9. Consequently the first observed reflection may be indexed **220** as in fig. 7.5, and it may be regarded as a second order reflection from the (110) planes of the face-centred unit cell. *What are the indices of the first observed reflection from the (111) planes in the face-centred cubic lattice represented in fig. 7.6?*

Since the points at the face centres fall within the (111) plane of the unit cell, the first observed reflection from these planes is indexed **111**.

7.3 The size and content of the unit cell of halite

The second major step in the determination of a crystal structure, is to calculate the number of atoms per unit cell from the shape and size of the unit cell, the density and the chemical composition. The chemical composition of halite was calculated in section 5.5 and it was shown to be **NaCl**. An estimate of the distance between the centres of the **Na⁺** and **Cl⁻** ions on the halite cube edge was studied in section 5.12. and it was found to be approximately **2.819 Å**.

The gram molecule or **mole** of a substance is the mass in grams equal to its molecular weight, and therefore the gram molecule of

NaCl = 22.989 + 35.463 = 58.44   grams.
The density of halite is 2.16 grams/cm³.
Therefore a halite crystal weighing
58.44 grams will occupy a volume of 58.45/2.16 cm³.

The number of molecules in a mole of any substance is given by **Avogadro's number** and is **6.022 × 10²³** as explained in section 5.12. In section 5.12. it was calculated that the distance between the (100) planes of ions in halite was 2.819 Å and therefore the edge of the face-centred unit cell will be approximately **5.64 Å**. *How many molecules of NaCl are there in a cube of halite which has an edge of 5.64 Å?*

Since there are 6.022 × 10²³ molecules in 27.06 cm³ of halite, the number of molecules in a face-centred unit cube with an edge 5.64 Å in length is given by:

$$\frac{\text{Volume of unit cell} \times \text{Avogadro's number}}{\text{Volume occupied by a gram molecule}}$$

$$= \frac{(5.64 \times 10^{-8})^3 \times 6.022 \times 10^{23}}{27.06} = 3.992$$

Therefore the nearest whole number of NaCl molecules in a unit cell of halite is four. The contents of the unit cell is usually represented by **Z** and for halite **Z = 4**. The spacing between the adjacent layers of ions forming the cube face was shown to be approximately 2.819 Å in section 5.12. This calculated spacing may be used to determine the wavelength of the X-rays from the Bragg equation. However, halite is not suitable for accurate measurements because the crystals frequently contain defects. Calcite has a more constant crystal structure and it has been used as a standard for comparison. If the wavelength of the main beam (Kα) from a copper X-ray tube is 1.5418 Å *what are the d spacings of the planes of atoms lying parallel to* (**100**), (**110**), *and* (**111**) *in halite?*

The spacings between the planes of atoms may be determined from the θ angles of the first observed reflections shown in figs. 7.4–7.6 and the relationship $d = \lambda/2 \sin \theta$. In halite,

$d_{(200)} = 2.819$ Å, $d_{(220)} = 1.994$ Å,
$$d_{(111)} = 3.258 \text{ Å}.$$

Consequently the edge of the face-centred unit cell of halite is approximately 5.64 Å.

7.4 The postulated arrangement of the ions in the unit cell of halite

The third major step in the analysis of a crystal structure is to postulate the positions of the ions of the appropriate number of formula units in the unit cell pattern indicated by the angular orientations of the X-ray reflections. The procedure is essentially one of trial and error. A structure is postulated and the probable intensities of the diffracted beams are calculated and then compared with the observed intensities. If there is good agreement the structure determination is accepted.

With regard to halite the next problem is: *How can the equivalent of* **4** *molecules of NaCl be accommodated in the unit cell of a face-centred cubic lattice?* If Cl⁻ ions were placed at the lattice points of a face-centred cubic space lattice, *how many Cl⁻ ions can be assigned to a particular unit cell?*

Consider an arrangement in which Cl⁻ ions occur at the corners and face centres of a unit cell as shown in fig. 7.7. Each of the ions at a face centre is shared between two adjacent unit cubes and therefore one half of a face-centred ion can be assigned to a particular unit cube as illustrated by the hemisphere in fig. 7.7. There are six Cl⁻ ions at the face centres, and the effective contribution to the unit cell is three whole Cl⁻ ions. One of the corner ions is shown as 1/8th of a sphere because each of the corner ions is shared by eight adjacent unit cubes. Consequently if the Cl⁻ ions are placed at the corners and face-centres of a unit cell there will be 14 Cl⁻ ions present, but only 4 Cl⁻ ions can be assigned wholly to a particular face-centred unit cube.

How can the **Na⁺** *ions be arranged within*

the face-centred cubic cell based on the **Cl⁻** ions, so that **4 Na⁺** *ions can be assigned to the unit cell, while the maximum symmetry is maintained and the charges on the ions are balanced?*

If the **Na⁺** ions are placed half way between the **Cl⁻** ions in the directions of the cube edges, there will be twelve **Na⁺** ions at the centres of the edges, and also one **Na⁺** ion at the centre of the cube as shown in fig. 7.8. One of the **Na⁺** ions at the centre of an edge is shown as a quarter sphere in fig. 7.8 because it is shared by four adjacent unit cubes. Therefore only $\frac{1}{4}$ of each of the **Na⁺** ions at the centres of the edges can be assigned to the unit cube, and the edge ions contribute 3 whole **Na⁺** ions to the unit cell.

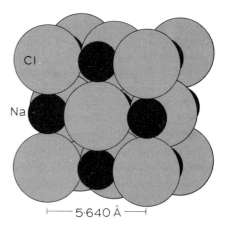

Fig. 7.9. The packing arrangement of **Cl⁻** and **Na⁺** ions in a face-centred unit cell of halite, in which the **Cl⁻** ions occupy the lattice positions

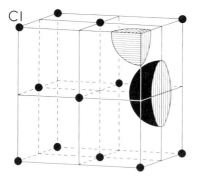

Fig. 7.7. **Cl⁻** ions placed at the lattice points of the cubic face-centred unit cell

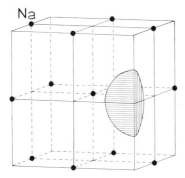

Fig. 7.8. The postulated positions of the **Na⁺** ions in relation to face-centred unit cell in which **Cl⁻** ions occur at the lattice points

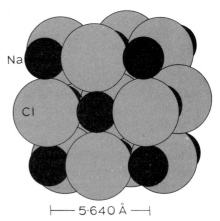

Fig. 7.10. The packing arrangement of the **Na⁺** and **Cl⁻** ions in the face centred unit cell of halite, in which the **Na⁺** ions occupy the lattice positions

Consequently with the **Na⁺** at the centre, **4 Na⁺** ions are associated with **4 Cl⁻** ions for each unit cell, and the requirement that the unit cell content $Z = 4$ is satisfied. *If only the Na⁺ ions are considered as a continuous array, what kind of space lattice is formed?*

The **Na** ions also form a face-centred cubic space lattice which is displaced along the cube edge by an amount equal to one half of the repeat distance. The postulated structure of halite consists of two interpenetrating face-centred cubic space lattices, one composed of Cl⁻ anions and the other composed of Na⁺ cations. The repeat distance along the cube edge is 5.64 Å and the ionic radii of Cl⁻ and Na⁺ are 1.81 Å and 0.97 Å respectively. The packing arrangement of a unit cell in which Cl⁻ ions occupy the corners and face-centres is shown in fig. 7.9. The alternative unit cell in which Na⁺ cations occur at the corners and face-centres is shown in fig. 7.10.

It is now necessary to study the observed diffraction pattern in terms of the postulated structure of the halite crystal.

7.5 The intensities of the X-ray beams diffracted from halite

The intensity of an X-ray reflection depends in the first place on the mass of the atoms responsible for the diffraction. The heavier the atom, the greater the number of electrons which contribute scattered X-rays. When the planes of atoms are identical and evenly spaced the intensities of the diffracted beams decrease with the increase in the order of the reflection. The relative intensities of the 1st, 2nd, 3rd, 4th, and 5th order reflections are in the approximate

ratios 100:20:5:3:1. The intensities of the diffracted beams are given as relative values using as a reference the strongest reflection which is given an intensity value of 100. A marked departure from this regular diminution of intensities with increasing order of reflection indicates that the planes of atoms are not identical. *What are the characteristics of the relative intensities of the reflections of increasing order, observed from the (100), (110), and (111) planes of ions in halite which are shown in the table below?*

Reflection	Intensity	Reflection	Intensity	Reflection	Intensity
1st	200	100	220	50	111	9
2nd	400	20	440	6	222	33
3rd	600	5	660	1	333	1
4th	800	1			444	3

The intensities of the X-ray reflections from the (100) planes of atoms decrease progressively with increasing order. This suggests that the (100) planes contain similar arrangements of Na⁺ and Cl⁻ ions as is shown in figs. 7.9 and 7.10. The progressive decrease in the intensities of the higher order reflections from the (110) planes also indicates that these planes have similar arrangements of Na⁺ and Cl⁻ ions as shown in fig. 7.11.

The relative intensities of the reflections of increasing order from the (111) planes alternate in magnitude. When one recalls that one of the primary reasons for the variation in the intensity of scattered X-rays is the mass of the atom, *what is the most likely explanation of the alteration of the relative intensities of the X-ray reflections from the (111) planes?*

The alternating intensities of the reflections from the (111) planes suggests that the alternating layers of ions parallel to (111) in halite are composed entirely of ions of chlorine or sodium as shown in fig. 7.12. The direction [22$\bar{1}$] within the (111) planes is

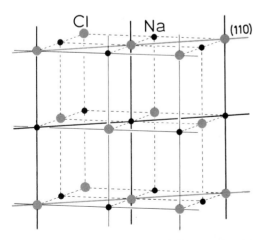

Fig. 7.11. The arrangement of the equal numbers of Na⁺ and Cl⁻ ions in the planes parallel to the {110} form in halite

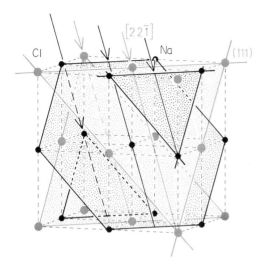

Fig. 7.12. The **[22Ī]** direction in the planes
of ions parallel to **(111)** in halite

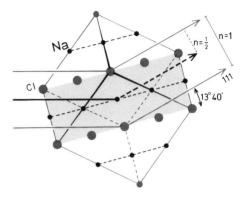

Fig. 7.13. Projection of the positions of
the ions on a plane perpendicular to the
[22Ī] direction in halite, and the position of
the 111 reflection

indicated in fig. 7.12, and fig. 7.13 is a
projection of the position of the ions on a
plane perpendicular to the [22Ī] direction.

The first observed X-ray reflection from
the **(111)** planes is the first order reflection
from the layers of **Cl⁻** ions as shown in fig.
7.13. It is also the first order reflection from
the layers of **Na⁺** ions, but since the path
difference between the X-rays diffracted
from the adjacent layers of **Na⁺** and **Cl⁻**
ions is equivalent to $\frac{1}{2}$ **wavelength**, partial
cancellation will occur and the observed
intensity of the **111** reflection is low.

Fig. 7.14 represents the position of the
second order reflection **222**. The X-rays
diffracted from the adjacent layers of **Cl⁻**
ions and also from the adjacent layers of
Na⁺ ions differ in phase by **two wavelengths**.
The X-rays diffracted from the layers of Cl⁻
ions will be **one wavelength** out of phase
with the X-rays diffracted from the layers of
Na⁺ ions, and so the two diffracted beams
will combine to produce a second order
reflection, **222**, of high intensity.

At the position of the third order reflection **333** the X-ray beams diffracted from the
layers of **Cl⁻** and **Na⁺** ions will be out of
phase by **one and a half wavelengths** and

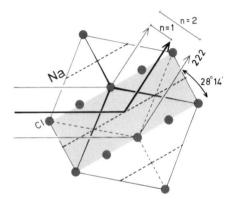

Fig. 7.14. The position of the **222** reflection from halite

they will again partially cancel out and result
in a low intensity reflection. At the position
of the fourth order reflection **444** the X-rays
reflected from the adjacent layers of **Cl⁻**
and **Na⁺** ions will be out of phase by **two
wavelengths** and they will build up to produce a reflection of greater intensity than
the 333 reflection. Since the observed intensities of the X-ray beams diffracted from
halite can be accounted for by the postulated arrangement of the **Na** and **Cl** ions, it
can be concluded that the postulated structure of halite is correct.

7.6 The X-rays diffracted from sylvite

It is important to note that a crystal structure should not be studied in isolation. The mineral **sylvite, KCl,** has a structure like that of halite, but the two ions **K** and **Cl** are very similar in mass, having atomic numbers of 19 and 17 respectively. The relative intensities of the observed X-ray reflections from sylvite are shown in the table below.

Reflec-tion	Inten-sity	Reflec-tion	Inten-sity	Reflec-tion	Inten-sity
1st	200	100	220	59	222	23
2nd	400	8	440	1	444	1
3rd	600	6				

The X-rays scattered from the adjacent (111) layers of **K** and **Cl** ions almost completely cancel each other out and no **111** reflection may be recorded. Consequently the ratios of the *d* spacings calculated from the first observed reflections from the (**100**), (**110**) and (**111**) planes of sylvite have the ratios $1 : 1/\sqrt{2} : 1/\sqrt{3}$, and one might deduce incorrectly that sylvite has a primitive cubic lattice.

7.7 Coordination number and Pauling's rules

In nearly all minerals the chemical formula unit is not distinguishable as a molecule. Crystals consist of continuous three-dimensional structures. In an ionic structure such as halite, each cation is surrounded by anions. The number of anions that can fit around each cation is known as the **coordination number** of the cation. The coordination number of the Na^+ cation at the centre of the unit cell of the halite structure shown in fig. 7.9 is 6 because each Na^+ cation is surrounded by six Cl^- anions.

The principles which determine the structure of complex inorganic ionic crystals were established by **Linus Pauling** in 1929. **The first of Pauling's rules** states that in an ionic crystal structure the cation-anion distance is the sum of the radii of the two ions, and **the coordination number is determined by the ratio of the radius of the cation to that of the anion**. In halite the ionic radii of Na^+ and Cl^- are 0.97 Å and 1.81 Å respectively so that the radius ratio Na^+/Cl^- is 0.54. When the radius ratio is in the range 0.41 to 0.73, the cation tends to occur in 6-fold coordination.

The ranges of the radius ratios which give rise to the different coordination numbers are shown in the table below. The structural arrangements of the anions at the different coordination numbers are indicated in the table and are illustrated in figs. 7.15 to 7.19.

Radius ratio	Co-ordination number	Arrangement of the anions
0.15–0.22	3	Corners of a triangle, Fig. 7.15
0.22–0.41	4	Corners of a tetrahedron, Fig. 7.16
0.41–0.73	6	Corners of an octahedron, Fig. 7.17
0.73–1.0	8	Corners of a cube, Fig. 7.18
1	12	Mid-points on cube edges, Fig. 7.19

Paulings's second rule states that in a stable coordination structure the total strength of the valency bonds which reach a cation from the anions surrounding it, is equal to the charge on the cation. This is known as the **electrostatic valency principle**. The relative strength of any bond in an ionic structure may be determined by dividing the total charge on an ion by the number of nearest neighbours to which it is linked. For example in halite each Na^+ cation which has a single charge is surrounded by six Cl^- anions as nearest neighbours, and therefore the relative bond strength is 1/6. This is called the electrostatic valency of the cation.

128

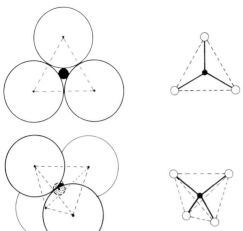

Fig. 7.15. Triangular coordination group

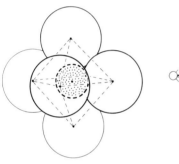

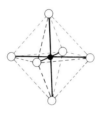

Fig. 7.16. Tetrahedron coordination group

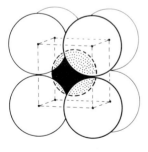

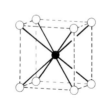

Fig. 7.17. Octahedron coordination group

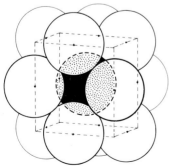

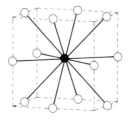

Fig. 7.18. Cube coordination group

Fig. 7.19. Coordination group
in which ions occur at the mid
points of the cube edges

The **Cl⁻** anions are also in six-fold coordination with **Na⁺** cations so that the electrostatic valency of the **Cl⁻** anion is also 1/6. Crystals in which all the bonds are equal in strength are said to be **isodesmic**.

Thus the chemical law of valency is satisfied by balancing the total positive and negative charges and not by grouping individual positive and negative charged ions in the crystal structure. The potential energy of an ionic crystal structure is mainly electrostatic and the stable configuration of the ions is the one which has the minimum potential energy.

The number of cations that surround an anion is also given as a coordination number so that it is said that in the halite structure the **Na⁺** and **Cl⁻** ions are in 6.6 coordination. All crystals which have equal numbers of cations and anions in 6-fold coordination have structures similar to halite. The minerals sylvite (**KCl**), periclase (**MgO**), and galena (**PbS**) have the same structure type as halite and are said to be **isostructural** or **isotypous**.

7.8 Why do cleavage planes develop in halite parallel to the {100} form?

The strength of the bonds may be significantly different across the different planes in the crystal structure. Fig. 7.20 represents the halite structure when viewed along the [22$\bar{1}$] direction which is parallel to the (**111**) planes. It can be seen that the ions have a cubic close packing arrangement in which the ions of the fourth layer are directly above those of the first layer. The alternate {111} planes consist entirely of **Na⁺** cations or **Cl⁻** anions. Consequently the ionic bonding is very strong between the adjacent {111} planes so that the corners do not fall off halite cubes. The ionic bonding is also relatively strong in directions perpendicular to the {110} planes because within these planes the **Na⁺** and **Cl⁻** ions occur in alter-

nate rows along the ⟨110⟩ directions as shown in fig. 7.11.

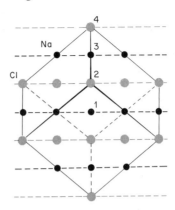

Fig. 7.20. The positions of the ions in halite projected on a plane perpendicular to the [22$\bar{1}$] direction which is parallel to (**111**), to show the stacking of the layers of ions

In contrast the {**100**} planes are composed of equal numbers of **Na⁺** and **Cl⁻** ions arranged alternately so that the repulsive forces between the like ions in adjacent layers are strong. Consequently the bonding forces between the {**100**} planes are relatively weak and this is the reason why the cleavages in halite develop parallel to the {**100**} form.

7.9 The postulated structure of fluorite

The crystals of fluorite have well developed cube forms but in contrast to halite, the corners of fluorite crystals frequently break away because of the presence of well developed cleavages parallel to the octahedron form {111} as shown in fig. 7.21. The chemical composition of fluorite was considered in section 5.5 and it was found to be **CaF₂**.

X-ray measurements similar to those made in the study of halite, show that fluorite also has a face-centred cubic unit cell with the cube edge equal to 5.4626 Å. The density of fluorite is 3.18 grams/cm³.

Fig. 7.21. Fluorite crystals showing the well developed cleavages parallel to the {111} form

How many molecules of CaF₂ can be assigned to the unit cell of fluorite? The calculations are similar to those used to calculate the contents of the unit cell of halite in section 7.3.

The volume occupied by a gram molecule of **CaF₂** = 78.08/3.18 = 24.55 cm₃ therefore the number of molecules **Z** in a face-centred unit cell is equal to the size of the unit cell ÷ the volume occupied by the gram molecule × Avogadro's number.

In fluorite,

$$\mathbf{Z} = \frac{(5.4626 \times 10^{-8})^3 \times 6.022 \times 10^{23}}{24.55} = 3.99$$

Consequently the equivalent of four molecules of **CaF₂** can be assigned to the unit cell of fluorite. If it is assumed that the **Ca⁺²** ions occupy the face-centred cubic lattice positions, then four **Ca⁺²** ions can be assigned to each unit cell. The composition

of fluorite indicated that there must be twice as many **F⁻** anions in the unit cell as there are **Ca⁺²** ions. *Where can the eight **F⁻** ions be positioned within the unit cell of fluorite in order to provide maximum symmetry?*

If the face-centred unit cell is divided into eight smaller cubes as shown in fig. 7.22, the eight **F⁻** anions must occupy the centres of the smaller cubes in order to obtain the highest degree of symmetry. Note that the **F⁻** ions occur on the triad axes of the unit cell.

The structure of fluorite may be represented by projecting the positions of the atoms on the base of the unit cell as in fig. 7.23. The vertical heights of the atoms are indicated as fractions of the repeat distance in the vertical direction. The **Ca⁺²** ions at the face centres are labelled ½. When an atom occurs at the base of the unit cell (**0**) with another atom in the corresponding

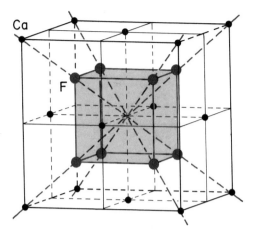

Fig. 7.22. The face-centred unit cell of the fluorite structure with Ca^{+2} ions at the lattice points. The F^- ions lie on the triad axes at the centres of the eight small cubes

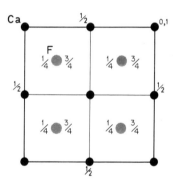

Fig. 7.23. The positions of the Ca^{+2} and F^- ions in the fluorite structure projected on the base of the unit cell. The heights of the ions above the base are indicated as fractions of the repeat distance in the vertical direction. Ions which occur in the base (0) and top (1) planes of the unit cell are not labelled

7.10 The intensities of the X-ray reflections from fluorite

The relative intensities of the X-ray reflections produced by fluorite are listed in the table below and they have been indexed on the basis of a face-centred unit cell represented by fig. 7.24.

Reflection	Intensity	Reflection	Intensity	Reflection	Intensity
200	1	220	100	111	94
400	12	440	5	222	1
600	1				

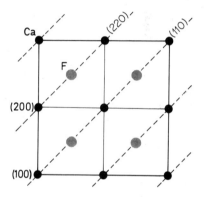

Fig. 7.24. The vertical planes of ions in the unit cell of fluorite

The next task is to consider whether the observed intensities of the X-ray reflections produced by fluorite are consistent with the postulated structure which is represented by the unit cell shown in fig. 7.22. *Is the fact that the 220 reflection has the maximum intensity, consistent with the postulated structure of fluorite?*

The F^- ions lie within the layers of Ca^{+2} ions which are parallel to (110) and therefore the F^- and Ca^{+2} ions combine to give a 220 reflection of maximum intensity from the (110) planes. *Is an extremely weak 200 reflection likely to be produced by the postulated unit cell, particularly when one con-*

position at the top (1), as is the case with the atoms at the corners of the unit cell, it is the convention not to indicate the numbers. Within the unit cell of fluorite the F anions occur at heights of $\frac{1}{4}$ and $\frac{3}{4}$ of the repeat distance of the unit cell.

siders that the atomic numbers of **Ca** *and* **F** *are 20 and 9 respectively?*

Fig. 7.25 represents the X-rays diffracted from the (**100**) planes of fluorite. The first observed reflection from the layers of atoms lying parallel to (**100**) in the face-centred unit cell would be indexed **200** and the X-rays scattered from the adjacent layers of Ca^{+2} ions would differ in phase by one wavelength as shown in the lower part of fig. 7.25. In the postulated structure the layers of F^- ions occur in between the (**200**) layers of Ca^{+2} ions and therefore in the direction of the first observed reflection, the X-rays scattered from the layers of F^- anions would differ in phase from the X-rays scattered by the Ca^{+2} ions by $\frac{1}{2}$ wavelength and destructive interference would occur as represented by the dashed lines in the lower part of Fig. 7.25. Since the reflecting power of two F^- ions (atomic number **9**) is approximately equal to the reflecting power of one Ca^{+2} ion (atomic number **20**), the X-rays scattered from the alternate layers of **Ca** and **F** ions almost cancel each other so that the **200** reflection is extremely weak and is sometimes not recorded. *Is the presence of the* **400** *reflection consistent with the postulated structure of fluorite?*

In the direction of the reflection indexed **400**, represented in the upper part of fig. 7.25, the X-rays scattered from the adjacent

layers of Ca^{+2} ions differ in phase by two wavelengths, while the X-rays scattered from the layers of F^- ions in between differ in phase from the Ca^{+2} reflections by one wavelength. Consequently constructive interference occurs and the **400** reflection is relatively strong. Similarly the **600** reflection from fluorite is weak while the **800** reflection is relatively strong.

Fig. 7.26 is a diagram showing the positions of the Ca^{+2} and F^- ions when projected on a plane perpendicular to the [**22Ī**] **direction which is parallel to the (111)** plane. The spacing of the layers composed of F^- ions is one half the spacing of the layers composed entirely of Ca^{+2} ions, but note the position of the layers of F^- anions.

In the direction of the **111** reflection, the X-rays scattered from the adjacent layers of Ca^{+2} ions will differ in phase by **one wavelength**. Since the spacing between the layers of F^- ions is $\frac{1}{2}$ the spacing of the layers of Ca^{+2} ions, the X-rays scattered from the adjacent layers of F^- ions will differ in phase by $\frac{1}{2}$ **wavelength**, and therefore destructive interference will eliminate the rays from the

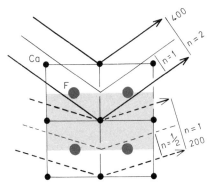

Fig. 7.25. The angular orientations of the X-ray reflections from the planes parallel to (**100**) in fluorite

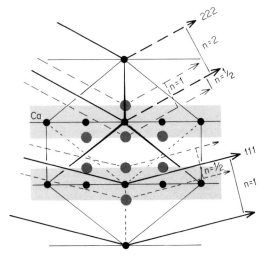

Fig. 7.26. The positions of the ions of the fluorite structure projected on a plane perpendicular to the [**22Ī**] direction, to show the angular orientations of the X-rays diffracted from the planes of ions parallel to (**111**).

F⁻ ions which are represented by the dashed lines in the lower part of fig. 7.26. Consequently the **111** reflection is due entirely to diffraction from the layers of Ca^{+2} ions and this is represented by the black lines in the lower part of fig. 7.26. *Is the absence of the* **222** *reflection consistent with the postulated structure of fluorite?*

In the direction of the **222** reflection the X-rays from adjacent layers of Ca^{+2} ions will differ in phase by **two wavelengths**, while the X-rays diffracted from the layers of F⁻ ions will differ in phase by **one wavelength** as shown in the upper part of fig. 7.26. However, the **222** diffraction beams from the separate layers of Ca^{+2} and F⁻ ions will differ in phase by $\frac{1}{2}$ **wavelength** and destructive interference will occur. Consequently the reflection from the layers of Ca^{+2} ions (atomic number **20**) will be cancelled by the reflection from the layers containing twice as many F⁻ ions (atomic number **9**) and no **222** reflection is observed.

Since the intensities of the X-ray reflections can be explained by the postulated arrangement of the Ca^{+2} and F⁻ ions in a face-centred unit cell, it is reasonable to conclude that the postulated structure of fluorite is correct.

7.11 The coordination of the ions in the fluorite crystal structure

The radii of the Ca^{+2} and F⁻ ions are 0.99 Å and 1.36 Å respectively, so that the radius ratio is 0.72. This would permit a structure in which there were equal numbers of cations and anions in 6-fold coordination as in the halite structure. However there are only half the number of divalent Ca^{+2} cations so that the packing arrangement of the ions in the unit cell can be represented by fig. 7.27. This is a good illustration of Pauling's second rule which states that in any stable crystal structure the total number of ions must be such that the positive and negative charges are balanced. *What are the coor-*

dination numbers of the **F⁻** *and* Ca^{+2} *ions in fluorite.*

The F^{+} anions in fluorite have four Ca^{+2} cations as closest neighbours, and the F⁻ anion is said to be in 4-fold coordination with the Ca^{+2} cations which occupy the apices of tetrahedra as shown in fig. 7.28. The Ca^{+2} **cation** at the centre of the base of the unit cell shown in fig. 7.28 is in 8-fold coordination with respect to the F⁻ anions which are arranged at the corners of a cube. The mineral fluorite (CaF_2) is said to have **8:4 coordination**.

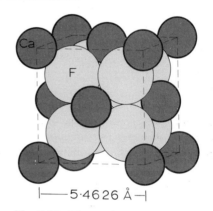

Fig. 7.27. The packing arrangement of the Ca^{+2} and F⁻ ions in the structure of fluoride, CaF_2

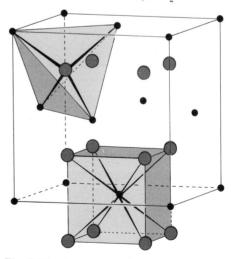

Fig. 7.28. The coordination groups around the Ca^{+2} and F⁻ ions in the fluorite structure

7.12 Why do the corners fall off fluorite crystals?

The perfect {111} cleavage planes in cube crystals of fluorite and the octahedra form of the cleavage fragment of fluorite are shown in plate 15 and fig. 7.21. The arrangement of the Ca^{+2} and F^- ions when viewed along one of the {111} planes is shown in fig. 7.26. It can be seen that the layers of Ca^{+2} ions are flanked on each side by a layer of F^- ions. The bonding between these three layer units is very weak, because the mutual repulsion of the F^- ions largely counteracts the attractive forces between the Ca^{+2} and F^- ions. Consequently in fluorite, perfect cleavages develop very easily on the {111} planes.

Good accounts of the early development of X-ray crystal analysis will be found in the following books.

Bragg, W. H., *An Introduction to Crystal Analysis*, G. Bell & Sons Ltd., London, 1926.

Bragg, W. L. *The Development of X-ray Analysis*, G. Bell & Sons Ltd., London, 1975.

7.13 Questions for recall and self assessment

1. What is the first step in the study of a crystal structure?
2. What are the indices of the first observed X-ray reflection from the planes of atoms parallel to the cube faces in a face-centred lattice?
3. In which way is the content of the unit cell of a crystal determined?
4. What is the nature of the third major step in the determination of a crystal?
5. What is the significance of the progressive decrease in the intensities of the 1st, 2nd, 3rd, and 4th order reflections from the planes parallel to (100) in halite?
6. What is the significance of the alternation of the intensities of the 1st, 2nd, 3rd, and 4th order reflections from planes of ions which are parallel to (111) in halite.
7. What is the significance of comparing the X-ray diffraction patterns of **halite NaCl** and **sylvite KCl**?
8. What is the meaning of coordination number?
9. What is the first of Pauling's rules?
10. What is the form of the group which has a coordination number of 6?
11. What is the electrostatic valency principle?
12. Why is the structure of halite said to be isodesmic?
13. What are the names given to minerals such as **halite NaCl, sylvite KCl,** and **galena PbS** which have similar structures.
14. The X-ray reflections indicate that fluorite has a face-centred space lattice, and the radius ratio, Ca/F = 0.72 suggests a 6:6 coordination. Why is the fluorite structure different from the structure of halite?
15. In which way can the position of the ions in fluorite be indicated on a plan parallel to the base of the unit cell?
16. What is the coordination of fluorite, CaF_2?
17. Which kind of close packing is represented by the face-centred space lattice?
18. What is the packing arrangement of the Na^+ and Cl^- ions in halite?
19. What is the nature of the bonding in the ⟨111⟩ directions of form in halite?
20. What is the consequence of the bonding in the ⟨111⟩ directions of form in fluorite?

8 The Structures of Diamond, Sphalerite, and Pyrite

In this chapter the X-ray analysis of crystals is extended by considering the determination of the structures of three more cubic minerals made by Sir W. L. Bragg and his team in the first few years of development of this subject. The principal objectives are listed below.

1. To study the analysis of the diamond structure and introduce two additional symmetry operations, the screw axis and the glide plane.
2. To understand the reasons for the octahedron form of the diamond crystal and the presence of {111} planes which are capable of cleavage.
3. To recall and apply the simple principles involved in the determination of the structure of a cubic crystal, by working out the structure of sphalerite from a sequence of observations and questions.
4. To understand why sphalerite crystals usually develop without a symmetry centre.
5. To study the pyrite structure in order to further appreciate the location of atoms by parameters within the unit cell and to understand the reasons for the development of the pyritohedron form.

Contents of the sections

8.1 The structure of diamond

The carbon atom has four electrons in its outermost shell and in the diamond structure each carbon atom is able to attain the eight electron configuration by sharing a pair of electrons with four adjacent carbon atoms which occur at the corners of an enclosing tetrahedral shaped structural unit represented in fig. 8.1. The covalent bonding between the carbon atoms is repeated to form a continuous structure within which

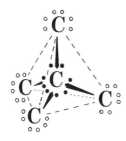

Fig. 8.1. The tetra-
hedral arrangement of
the covalently bonded
carbon atoms in diamond

the energy of the covalent bonds is concentrated in the vicinity of the shared electrons,
and this gives rise to the exceptional hardness of diamond.

Diamond crystals frequently occur with
octahedral forms as illustrated in fig. 1.4.
Conspicuous growth steps can be seen on
some of the faces of the diamond crystal.
The structure of diamond was determined
by W. H. and W. L. Bragg (*Proc. Roy.
Soc.,* 1913, **A89,** 277).

X-ray diffraction measurements show that
the structure of diamond is based on a
face-centred cubic space lattice with the
edge of the unit cube 3.5667 Å in length.
Since the density of diamond is 3.5 grams/
cm³ and the atomic weight of carbon is
12.001, *how many* **C** *atoms must be assigned
to the unit cell of diamond?*

The calculation of the unit cell content
explained in sections 7.3 and 7.9 indicates
that the equivalent of eight **C** atoms are
associated with the unit cell of diamond. If
C atoms occur at the face-centred lattice
points, these contribute the equivalent of
four **C** atoms to the unit cell and therefore it
is necessary to find sites for the other four **C**

atoms. Clues concerning the positions of the
C atoms within the unit cell of diamond may
be obtained from a comparison of the relative intensities (**I**) of the main X-ray reflections from diamond and fluorite which are
shown in the table below.

Explanations of the absence of the 200, 600,
and 222 reflections from fluorite were considered in section 7.10. A comparison of the
intensities shown in the tables suggests that
the four **C** atoms within the unit cell of
diamond occupy similar positions to the F⁻
ions in the face-centred unit cell of fluorite.
The **F⁻** ions occur at the centres of the eight
small cubes within the face-centred unit cell
and in four-fold coordination with **Ca⁺** ions.
In the diamond structure only four **C** atoms
are available to occupy sites at the centres of
small cubes within the unit cell, but a symmetrical arrangement is produced if the four
C atoms are positioned in four tetrahedrally
related small cubes as shown in the projection fig. 8.2. The **C** atoms which occur at the
centres of four of the small cubes are labelled $\frac{1}{4}$ and $\frac{3}{4}$.

A face-centred unit cell of the diamond
structure is shown in black in fig. 8.3 and the
four **C** atoms which occur at the centres of
small cubes are shown as grey circles with
bonds to the **C** atoms at the face-centred
lattice points. It can be seen that the **C**
atoms in the internal positions also form a
face-centred cubic space lattice shown in
grey in fig. 8.3. The two unit cells shown in
fig. 8.3 are related by displacement along
the cube diagonal $[\bar{1}\bar{1}\bar{1}]$ for a distance equal
to $\frac{1}{4}$ of the diagonal of the unit cell.

The reason that the 200 and 222 reflections are absent from diamond is that the
atoms are all of the same kind whereas in

Fluorite						Diamond					
Reflection	I	Reflection	I	Reflection	I	Reflection	I	Reflection	I	Reflection	I
200	1	220	100	111	94	200	0	220	25	111	100
400	12	440	5	222	1	400	8			222	0
600	1					600	0				

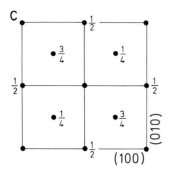

Fig. 8.2. Projection on the zero plane parallel to (**001**) to show the positions of the **C** atoms in the unit cell of diamond

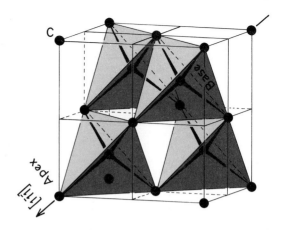

Fig. 8.4. The stacking of the tetrahedral coordination units in the diamond structure

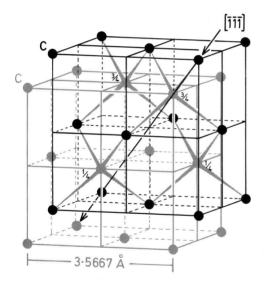

Fig. 8.3. Oblique view of the unit cell of diamond shown in black. The **C** atoms occurring within the unit cell are indicated by bond lines to the adjacent **C** atoms. The internal **C** atoms also belong to a face-centred lattice which is displaced along the cube diagonals. The unit cell illustrated is displaced in the [$\bar{1}\bar{1}\bar{1}$] direction by a distance equal to one quarter of the cube diagonal

fluorite the reflections from the Ca^+ ions are not completely cancelled by the reflections from twice as many F^- ions occurring on planes in between the Ca^+ ions.

The structure of diamond consists of covalently bonded tetrahedral units of the kind represented in fig. 8.1. The coordination tetrahedron enclosing each **C** atom is oriented so that its faces are parallel to the {**111**} form and in fig. 8.4 they are shown by bond lines and shading. The tetrahedra are stacked in a pyramid-like arrangement which can be visualized most easily by turning fig. 8.4 to the reading position indicated by the labels 'apex' and 'base'. In this position the apices of the coordination tetrahedra point in the [$1\bar{1}\bar{1}$] direction. Since the diamond structure consists of **C** atoms only, a similar stacking arrangement occurs in which the apices of the coordination tetrahedra point in the opposite direction [$\bar{1}11$]. Similar stacking arrangements occur parallel to all four triad axes of symmetry. Consequently although the structure of diamond appears to have tetrahedral symmetry, because the atoms are all of the same kind the crystals of diamond are cubic holosymmetric and tend to occur as octahedra as illustrated in fig. 8.5.

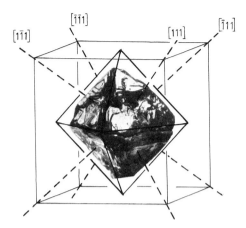

Fig. 8.5. The holosymmetric symmetry of
the diamond crystal

8.2 Screw axes of symmetry in the diamond structure

Diamond crystals such as the octahedral form illustrated in fig. 8.5 possess the highest degree of cubic symmetry and exhibit three tetrad axes of symmetry. *Does the internal structure of diamond illustrated in figs. 8.2 and 8.3 possess tetrad axes of symmetry?*

It is clear from figs. 8.2 and 8.3 that since carbon atoms occur at the centres of only four of the eight small cubes within the unit cell, the structure of diamond does not possess tetrad symmetry axes. However, the positions of the **C** atoms in diamond can be related by **tetrad screw axes** as shown in fig. 8.6. On a screw axis, rotation is combined with simultaneous translation parallel to the screw axis.

(1) Starting with the atom at the centre of the base of the unit cell in position (0), the **first operation** consisting of upward translation equal to $\frac{1}{4}$ of the repeat distance combined with anticlockwise rotation of 90°, places the atom at the centre of the small cube in the position labelled $\frac{1}{4}$ on the left side of the screw axis.

(2) The **second operation** brings the atom

to the centre of the front face of the unit cell in the position labelled $\frac{1}{2}$.

(3) The **third operation** brings the atom to the centre of one of the small cubes in position $\frac{3}{4}$.

(4) The **fourth operation** brings the atom to the centre of the top face of the unit cell to the position (1) corresponding to the initial position (0).

It can be seen from fig. 8.6 that the operation of a screw axis is similar to a spiral staircase, and since in this particular situation, four steps were required to bring the atom back to a position corresponding to the initial position, this axis is described as a **tetrad screw axis**. Another screw axis is shown by the dashed line in fig. 8.6. *What is the direction of rotation associated with upward translation on this second screw axis?*

On the screw axis shown by the dashed line in fig. 8.6 the upward translation is combined with simultaneous rotation in a clockwise direction. The positions of four vertical tetrad screw axes in the unit cell of diamond are shown in fig. 8.7. Rotation in a clockwise direction occurs on two of the tetrad screw axes, while rotation in an anticlockwise direction occurs on the other two screw axes.

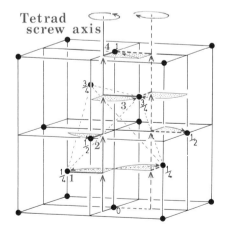

Fig. 8.6. The tetrad screw axis in the
diamond structure

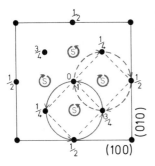

Fig. 8.7. Projection on the basal plane to show the position of the tetrad screw axes in the diamond structure

8.3 Glide planes in the diamond structure

The diamond structure does not possess reflection planes of symmetry parallel to the cube faces, but the position of the atoms may be related by an element of symmetry called a **glide plane** illustrated in fig. 8.8. The positions of the atoms on either side of a glide plane may be brought into coincidence by **reflection combined with simultaneous translation parallel to the glide plane**. In the diamond structure there are glide planes parallel to the cube faces and situated midway between the planes of atoms as illustrated in fig. 8.8. The atom on the base labelled **0** may be brought into position of the atom labelled $\frac{1}{4}$ by simultaneous reflection across the glide plane and translation in the glide direction. The same operation also results in the movement of an atom from position $\frac{1}{4}$ to the position $\frac{1}{2}$ at the face centre, and so on. The positions of the vertical glide planes on the basal projection of the unit cell of diamond are shown by the dashed lines in fig. 8.9.

The group of symmetry elements consisting of **reflection planes, rotation axes,** and **inversion combined with rotation** exhibited by a crystal is known as a **point group** because it refers to symmetry about the origin of the crystallographic axes. Each of

the 32 crystal classes is characterized by a point group as explained in section 2.8. In order to describe the symmetry of a crystal structure it may be necessary to introduce **screw axes** and **glide planes** as well as rotation axes and reflection planes. The group of symmetry operations used to define the crystal structure is called a **space group**, because it is not related to a single point but extends through space.

The **translations** associated with **glide planes** and **screw axes** are of atomic dimensions and they are not apparent in the

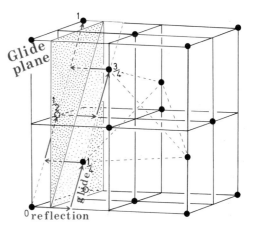

Fig. 8.8. The operation of a glide plane in the diamond structure

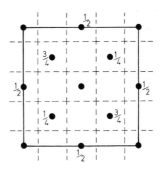

Fig. 8.9. The position of vertical glide planes shown by dashed lines on a basal projection of the diamond structure

external symmmetry of the crystal. Consequently the crystals of fluorite and diamond
possess the same point group but they have
very different space groups. There are **230**

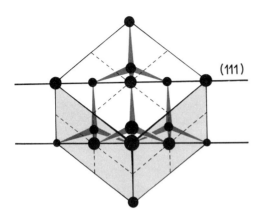

Fig. 8.10. The unit cell of diamond viewed
along the [22$\bar{1}$] direction which is parallel to
the (111) plane. The carbon atoms more
distance from the observer are represented by
smaller circles. The tetrahedra are indicated
by the bond lines so that the stacking arrangement in the direction perpendicular to the
(111) plane can be seen

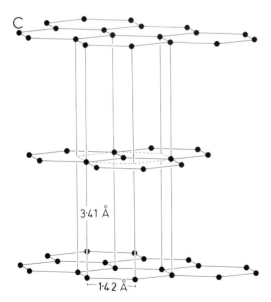

Fig. 8.11. The structure of graphite

space groups distributed amongst the **32
point groups** and this subject will be considered again in chapter 9.

8.4 Why can diamond crystals be cleaved parallel to the {111} form?

Fig. 8.10 is a view of the unit cell of diamond
along the direction [22$\bar{1}$] which is parallel to
the (111) plane, to show the stacking of the
tetrahedra. It can be seen that there is a
greater distance between the atoms at the
apices and those in the base layer, than
there is between the atoms within the base
layer. This greater spacing is associated with
slightly weaker bonding in the directions
perpendicular to the (111) plane, so that
diamond can be cleaved on the planes
parallel to the {111} form.

8.5 Polymorphism: The structure of graphite

The mineral **graphite** also consists entirely
of carbon atoms, but whereas diamond
forms colourless cubic crystals and is the
hardest of all minerals, graphite is a soft
black mineral shown in fig. 1.12, and its
crystals have hexagonal symmetry. Chemical substances which occur as two or more
physically distinct minerals are said to be
polymorphous. In graphite the carbon atoms
occur in layers composed of hexagonal rings
of atoms so that each atom has three neighbours, fig. 8. 11. The distance between the
centres of the carbon atoms in the layers is
1.42 Å which is much less than the distance
of 1.54 Å between the atoms in the diamond
structure. The layers of atoms in graphite
are separated by a relatively great distance
of 3.41 Å, and alternate atoms lie directly
above atoms in the adjacent layer.

The reason for the softness of graphite is
that the bonding between the layers of
atoms is very weak, whereas the atoms
within the layers are drawn together more
closely than in the diamond structure. Be-

cause the layers can slide easily over each other graphite is an excellent lubricant, and graphite crystals exhibit a perfect cleavage parallel to the layers.

8.6 Observations and questions concerning the structure of sphalerite

In order to recall and practice each step in the determination of a simple crystal structure, this section consists of observations and questions concerning the structure of sphalerite. The answers have been arranged in section 8.7 which gives the explanations of the steps in the determination of the structure of sphalerite. The structure of sphalerite was determined by W. L. Bragg, (*Proc. Roy. Ass.*, 1913, **A89**, 468).

Q1. A group of crystals of sphalerite is shown in fig. 8.12. *What is the characteristic form shown by individual crystals of sphalerite in this group?*

Q2. *What are the elements of symmetry of the form exhibited by the sphalerite crystals?*

Q3. The θ angles at which the first X-ray reflections are recorded from the (**100**), (**110**), and (**111**) planes of the sphalerite crystal when the primary X-ray beam is Cu K_α ($\lambda = 1.5418$ Å) are shown in the table below. *What are the ratios of the spacings of the layers of atoms lying parallel to (**100**), (**110**), and (**111**)?*

1st reflection from (**100**) θ = 16°34′

1st reflection from (**110**) θ = 23°45′

1st reflection from (**111**) θ = 14°17′

Q4. *Which of the cubic space lattices described in section 4.10 is characterized by d spacings in the ratios calculated from the information given above?*

Fig. 8.12. A group of sphalerite crystals showing the characteristic crystal forms. (×4)

Q5. *What is the length of the edges of the unit cell of sphalerite?*

Q6. *What are the indices of the first observed reflections from the* (**100**), (**110**), *and* (**111**) *planes from this kind of space lattice?*

Q7. Pure sphalerite is composed of approximately 67% **Zn** and 33% **S** by weight. *What are the atomic proportions of* **Zn** *and* **S** *in sphalerite?*

Q8. The density of sphalerite is 4.096 grams/cm³. *How many formula units are associated with each of the unit cells of sphalerite?* The calculation of the unit cell content was explained in section 7.3.

Q9. The intensities of the main X-ray reflections from sphalerite are shown in the table below, and for comparison the intensities of the reflections from halite, fluorite and diamond have been repeated. *Which two characteristics of the reflections suggest that the structure of sphalerite may be similar to the structures of fluorite and diamond?*

	Reflection	I	Reflection	I	Reflection	I
Sphalerite	200	10	220	51	111	100
	400	6	440	3	222	2
Halite	200	100	220	50	111	9
(Section	400	20	440	6	222	33
7.5)	600	5	660	1	333	1
	800	1			444	3
Fluorite	200	0	220	100	111	94
(Section	400	12	440	5	222	1
7.10)	600	1				
Diamond	200	0	220	25	111	100
(Section	400	8			222	0
8.1)						

Q10. *Considering the nature and size of the unit cell of sphalerite and the cell contents* **Z**, *what would be an appropriate arrangement of the* **S** *atoms within the unit cell if the* **Zn** *atoms are placed at the lattice points?*

Draw a plan of the unit cell of **sphalerite** as a projection on the basal plane and indicate the positions of the atoms above the base as fractions of the unit cell height.

Make an accurate drawing similar to fig. 8.3 to show an oblique view of the unit cell of sphalerite.

Q11. *What is the coordination number of each* **Zn** *atom with respect to the neighbouring* **S** *atoms in the sphalerite structure, and what is the relationship between the form of the coordination unit and the unit cell?*

Q12. *What is the coordination number of* **S** *and the relationship between the* **S** *centred coordination groups and the unit cell?*

Q13. *Do the* **Zn** *and* **S** *coordination groups have the same stacking arrangement in the direction of a triad axis?*

Q14. *What is the likely explanation for the characteristic form of the sphalerite crystals.*

8.7 Explanations of the steps in the determination of the structure of sphalerite

A1. The sphalerite crystals illustrated in fig. 8.12 occur as the tetrahedron

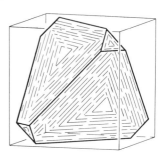

Fig. 8.13. The tetrahedron form of sphalerite crystals. The large faces are composed of the (1$\bar{1}$1) tetrahedron form but the corners are replaced by small faces of the alternative (111) tetrahedron

form. In fig. 8.13 the corners of the $\{1\bar{1}1\}$ tetrahedron are absent due to the presence of faces of the alternative $\{111\}$ tetrahedron form.

A2. The tetrahedron has the **four triad axes** which are characteristic of all the symmetry classes in the cubic system. In addition there are **three diad axes which are also tetrad axes of rotary inversion** described in section 2.7, and these are used as the crystallographic axes. There are **six planes of reflection symmetry** which are parallel to the $\{110\}$ form but there is **no symmetry centre**.

A3. The *d* spacings may be calculated from the Bragg equation $d = \lambda/2 \sin \theta$ and for sphalerite the *d* spacings of the (100), (110), and (111) planes are respectively $d_{(100)} = \textbf{2.703 Å}$, $d_{(110)} = \textbf{1.914 Å}$ and $d_{(111)} = \textbf{3.125 Å}$. The ratios of the *d* spacings are

$$d_{(100)} : d_{(110)} : d_{(111)} = \textbf{1} : \textbf{0.7080} : \textbf{1.1561}$$

A4. Since the crystals of sphalerite exhibit the symmetry of one of the classes of the cubic system, the crystal structure may be described by one of the three cubic space lattices. It will be found in section 4.10 that the ratios of the *d* spacings given above are characteristic of the **face-centred cubic space lattice**.

A5. Since the structure of sphalerite is based a face-centred cubic space lattice the edge of the unit cell is double the spacing between the planes of atoms lying parallel to (**100**) and is **5.406 Å**.

A6. It will be recalled from section 4.9 that the indices of the first observed reflections from the (**100**), (**110**), and (**111**) planes of a face-centred cubic space lattice are **200, 220** and **111** respectively.

A7. Pure sphalerite is composed of approximately 67% **Zn** and 33% **S** by weight and when these values are divided by the respective atomic weights which are 65.38 and 30.066, it is found that the proportions of the **Zn** and **S** atoms are equal. Consequently the composition of sphalerite can be represented by the formula **ZnS**.

A8. The calculation of the unit cell contents **Z** was considered in section 7.3. The volume occupied by a gram molecule of sphalerite is 97.446/4.096 = 23.79 cm^3. Therefore Z = Volume of unit cell × Avogadro's number ÷ Volume occupied by the gram molecule

$$= \frac{(5.406 \times 10^{-8})^3 \times 6.022 \times 10^{23}}{23.79} = 3.999$$

Consequently **Z = 4**.

A9. Comparison of the X-ray reflections from sphalerite, halite, fluorite and diamond which are tabulated in the preceding section, suggests that the **first characteristic** is that the **200** reflection from sphalerite is abnormally low compared with the reflection from halite. It is significant that the **200** reflections are also extremely weak or absent from fluorite and diamond.

The reflections from the (**220**) and (**440**) planes in all four minerals are not greatly different. The **second characteristic** feature is that the **111** reflections are strong from sphalerite, fluorite and diamond, while the **222** reflections are very weak or absent. In contrast the **111** reflection from halite is weak while the **222** reflection is relatively strong.

A10. It has been determined that the crystal structure of sphalerite is based on a face-centred cubic space lattice and that the unit cell content is 4 molecules of **ZnS**. A comparison of the intensities of the X-ray reflections suggests that the structure of sphalerite, **ZnS**, is similar to the structures of fluorite,

CaF$_2$, and diamond rather than to halite, NaCl. Consequently if the Zn atoms are placed at the face-centred cubic space lattice points, these contribute the equivalent of four Zn atoms to the unit cell, and it is necessary to position four S atoms within the unit cell. The most appropriate positions for the four S atoms are at the centres of four of the smaller cubes, with a tetrahedral arrangement as in the

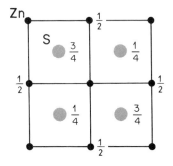

Fig. 8.14. Projection of the unit cell of sphalerite on the basal plane

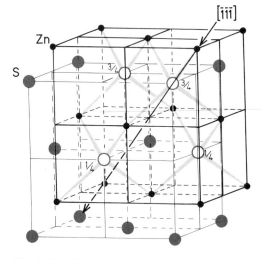

Fig. 8.15. Oblique diagram of the structure of sphalerite showing the interpenetrating face-centred, cubic space lattices composed of Zn atoms (black) and S atoms (grey). The unit cell based on the S atoms is displaced in the [$\bar{1}\bar{1}\bar{1}$] direction in this illustration

diamond structure. The postulated structure of sphalerite is illustrated by the plan fig. 8.14, and the oblique view fig. 8.15.

The Zn atoms occupy the face-centred lattice points and the four S atoms within the unit cell shown in fig. 8.15 are indicated by bond lines to Zn atoms. It can be seen that the S atoms also occur in face-centred cubic lattice positions, but this space lattice is displaced along the triad axes by one quarter of the repeat distance. The unit cell shown in fig. 8.15 is displaced in the [$\bar{1}\bar{1}\bar{1}$] direction.

The observation that the 200 reflection from sphalerite is relatively low is due to the presence of planes of atoms of one kind in between the (200) planes of atoms of the other kind as explained in sections 7.10 and 8.1. However the reflections from the planes of Zn atoms (atomic number 30) are not completely cancelled by the reflections from the planes of S atoms (atomic number 16), so that complete destructive interference does not occur in sphalerite as it does in diamond.

A11. Each Zn atom is in 4-fold coordination with S atoms and the triad axes of the (ZnS$_4$) tetrahedra are parallel to the triad axes of the unit cell, with the planes of S atoms parallel to the {111} tetrahedron form as shown in fig. 8.16.

A12. Each S atom is in 4-fold coordination with Zn atoms and it can be seen in fig. 8.16 that the faces of the (SZn$_4$) tetrahedra are parallel to the {1$\bar{1}$1} tetrahedron form.

A13. The apices of the (ZnS$_4$) and (SZn$_4$) tetrahedra are shown in fig. 8.16 and it can be seen that they point in opposite directions, [$\bar{1}\bar{1}\bar{1}$] and [111] respectively in the illustration, with the bases parallel to the (111) plane. The (ZnS$_4$) and

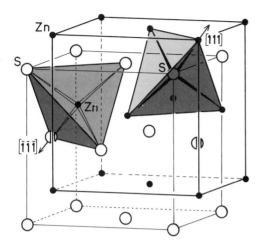

Fig. 8.16. The apices of the two kinds of coordination tetrahedra, (**ZnS₄**) and (**SZn₄**) in the sphalerite structure point in opposite directions

(**SZn₄**) coordination tetrahedra are stacked in opposite directions along all four triad axes. In plate 20 a model of the sphalerite structure is arranged with a triad axes vertical to show the S-centred and **Zn**-centred coordination groups pointing upwards and downwards respectively.

A14. A model representing the expanded structure of sphalerite is shown in plate 19 and is arranged so that the nature of the layers of atoms parallel to the ($\bar{1}11$) plane can be seen clearly. The distinctive feature of the layers of atoms parallel to the ($\bar{1}11$) plane is that the **Zn** atoms occur towards one side of the layer whereas the S atoms are prominent on the other side. This is a consequence of the stacking arrangement of the (**ZnS₄**) and (**SZn₄**) tetrahedra as illustrated in plate 20.

Since the opposite sides of the {111} layers of atoms have very different chemical characteristics it is likely that during crystallization one side will grow more easily than the other side. This may result in the growth of only

four of the faces of the {111} form so that the tetrahedron form without a symmetry centre develops. The (**111**) and ($\bar{1}\bar{1}\bar{1}$) faces cannot be distinguished by their X-ray reflections, but the fact that the faces have very different characteristics may be deduced from the postulated structure of sphalerite which is consistent with the observed intensities of the X-ray reflections.

The bonding between the **Zn** and **S** atoms in the sphalerite structure could be purely covalent, but the size of the unit cell which is determined by the effective radii of the atoms in the structure indicates that the bonding is partially ionic in character. The packing arrangement of the **Zn** and **S** atoms in sphalerite is shown in fig. 8.17.

The two interpenetrating face-centred cubic space lattices of the sphalerite structure are composed of different atoms, **Zn** and **S**, and consequently there are no tetrad screw axes nor glide planes parallel to the cube faces as in the diamond structure.

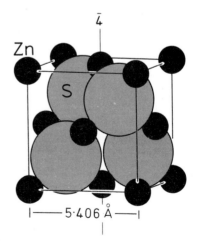

Fig. 8.17. The packing arrangement of the **Zn** and **S** atoms in sphalerite

There are reflection planes parallel to {110} in the sphalerite structure as in the diamond structure. Tetrad axes of rotary inversion parallel to the cube edges pass through each Zn and S atom as shown in fig. 8.17 and consequently the sphalerite structure has the symmetry of the tetrahedron form.

In the sphalerite structure **Fe** may substitute for **Zn** up to about 50 mole per cent of **FeS**. The iron rich varieties of sphalerite (**ZnFeS**) are black or dark in colour, while the pure varieties are nearly colourless or amber in colour with a distinctive resinous lustre as shown in plate 11.

8.8 The structure of pyrite

In the crystal structures of halite, diamond, and sphalerite the atoms occur at the points of two interpenetrating face-centred space lattices. However, just as in wall paper patterns only the selected points are at the lattice points, in the majority of crystals many of the atoms do not occur at the lattice points. In the fluorite structure the F^- ions occur at the centres of the eight small cubes which make up the face-centred unit cell. Consequently in the structures of most minerals it is necessary to describe the positions of some of the atoms in terms of parameters with respect to the lattice points.

Pyrite crystals frequently take the form of cubes characterized by conspicuous striations on the faces or the pyritohedron form {210} which was studied in sections 2.6 and 3.7. The cube and pyritohedron forms and a crystal showing a combination of these forms are illustrated in plate 6. Clearly pyrite crystals do not possess the high degree of symmetry which is characteristic of halite, fluorite or diamond. In the pyrite crystals the {100} planes are planes of reflection symmetry but the symmetry axes perpendicular to these planes are diad axes and not tetrad axes as in halite. The {110}

planes in pyrite are not planes of reflection symmetry.

Pure pyrite is composed of approximately 46.5% **Fe** and 53.45% **S** by weight and the atomic proportions may be determined using the atomic weights of Fe and S which are 55.85 and 32.066 respectively. It will be found that there are approximately twice as many S atoms in the pyrite structure and therefore the composition can be represented by **FeS₂**.

X-ray observations reveal that the unit cell of pyrite has an edge length of 5.412 Å. Pyrite has a density of 5.01 g/cm³ and it will be found that the unit cell content of pyrite is Z = 4. *Is it likely that pyrite (**FeS₂**) has the same structure as fluorite (**CaF₂**)?*

Since pyrite does not possess the same high degree of symmetry which is characteristic of fluorite, it must be assumed that the arrangement of the atoms in the pyrite structure is not as symmetrical as in the fluorite structure. The X-ray reflections obtained from pyrite are very much more complicated than the set of reflections from fluorite considered in section 7.10 and this confirms that the structures of the minerals are different.

In fluorite the F^- ions occur at the centres of the eight small cubes in the face-centred unit cell and consequently the **200** and **600** reflections are weak while the **400** and **800** reflections are relatively strong. In contrast the structure of pyrite produces a strong **200** reflection, with relatively weak **800** and **1000** reflections.

In fluorite the four triad axes intersect at the F^- ions which are at the small cube centres. If the atoms are displaced from the small cube centres then only one triad axis can be retained, but Sir W. L. Bragg realized that it is possible to have a system of non-intersecting triad axes in a cubic crystal and that this might explain the structure of pyrite, (Bragg, W. L., *Proc. Roy. Soc.*, 1913, A89, 468). Bragg postulated that in pyrite the **S** atom is displaced from the

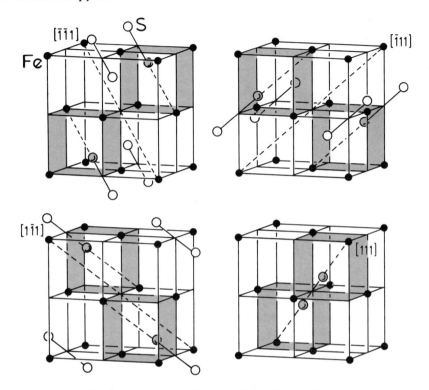

Fig. 8.18. The positions of the S atoms within the small cubes of the unit
cell of pyrite shown in separate diagrams. The S atoms which occur within
the unit cell are shaded

centre of the small cube along one of the triad axes so that the three fold symmetry is retained, but the other triad axes are provided by adjacent small cubes in the face-centred unit cell. The positions of the S atoms in the small cubes of the unit cell of pyrite are shown separately in fig. 8.18.

The distance of the centre of the S atom from the small cube centre measured along a triad axes has to be determined, and this was first instance of the determination of a parameter of an atom in a crystal structure. The expected intensities of the X-ray reflections from the layers of S and Fe atoms can be calculated from the known scattering efficiencies of Fe and S at different angles, and it is found that the observed intensities are best explained if the S atoms are displaced along the triad axis by a ratio of 0.228 of the cube diagonal.

The S atoms occur as 'dumbell like' pairs which are centred on the mid points of the edges of the face-centred unit cell in a similar manner to the arrangement of the ions in the halite structure. The axes of the pairs of S atoms are parallel to one of the four non-intersecting triad axes of the cube and the orientations of the pairs of S atoms are shown in the four separate diagrams in fig. 8.18.

8.9 The coordination of the Fe and S atoms in the pyrite structure

The pair of S atoms at the centre of the unit cell shown in fig. 8.19 is aligned along the [111] triad axis and the distance between the centres of the S atoms in a pair is comparable with the distance between covalently bonded S atoms. The central pair of S atoms

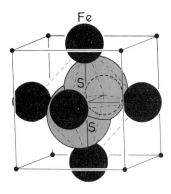

Fig. 8.19. The packing arrangement of pyrite showing the central pair of **S** atoms coordinated to six **Fe** atoms

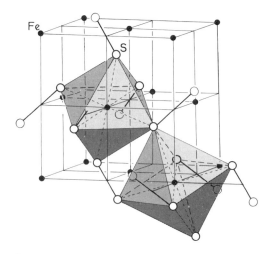

Fig. 8.20. The coordination of the **Fe** atom in the pyrite structure. Each **Fe** atom is surrounded by six **S** atoms in octahedral structual units which have four different orientations, two of which are illustrated

is coordinated to six **Fe** atoms which occur at the centres of the faces of the unit cell and the packing arrangement is illustrated in fig. 8.19. Each **S** atom is equidistant from three **Fe** atoms which form planar groups that are parallel to the (111) and $(\bar{1}\bar{1}\bar{1})$ planes in the unit cell shown in fig. 8.19. Each **Fe** atom is surrounded by six **S** atoms in octahedral structural units which have one of four orientations. Two (**FeS₆**) octahedral units with different orientations are illustrated in fig. 8.20.

8.10 Why does pyrite have a crystal face parallel to (021) but not parallel to (021)?

The pyritohedron form and the orientations of the (021) and (012) planes are shown in figs. 8.21 and 8.22. Fig. 8.23 is a plan of the

Fig. 8.21. The pyritohedron form. (×2)

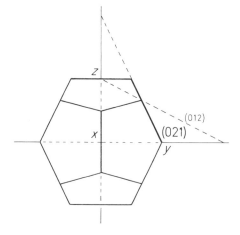

Fig. 8.22. The orientation of the (**021**) and (**012**) planes

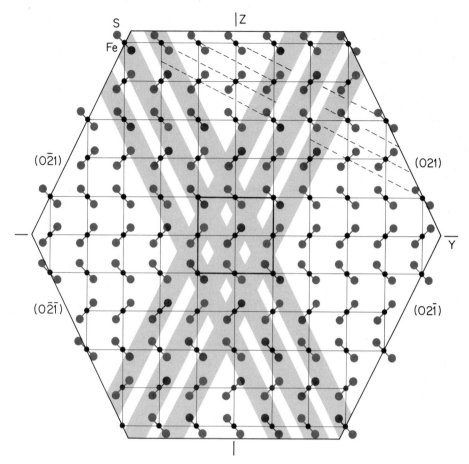

Fig. 8.23. A projection of the pyrite structure on the (**100**) plane to show the densely packed layers of atoms parallel to the (021) and (0$\bar{2}$1) planes. The layers of atoms are shown by the shaded bands. In contrast the (012) plane indicated by the dashed lines does not consist of distinct layers of atoms

pyrite structure in which the positions of the atoms have been projected onto the (100) plane. The Fe atoms are shown in black and the S atoms are in grey. A unit cell is outlined at the centre of the lattice.

It can be seen from the projection of the structure in fig. 8.23 that there are layers of densely packed atoms parallel to the (021) and (0$\bar{2}$1) planes. In contrast the (012) plane does not consist of a distinct layer of atoms so no crystal face develops parallel to this plane.

The intersection of the (021) and (02$\bar{1}$) layers of atoms gives rise to the striations parallel to the x axis which are observed on the (010) faces of pyrite cubes. The pyrite structure has a similar appearance when the positions of the atoms are projected on to the (001) plane but in this case the dense layers of atoms are seen to be parallel to the (210) and (2$\bar{1}$0) faces so that the striations on the (100) faces of the pyrite cube are parallel to the z crystallographic axis as shown in fig. 1.2.

8.11 Questions for recall and self assessment

1. How are the eight **C** atoms distributed in a unit-cell of the diamond structure?
2. What is the nature of the space lattice which describes the diamond structure?
3. What is the nature of the bonding and coordination of the **C** atoms in diamond?
4. How are the coordination groups arranged in the diamond structure?
5. Why do diamond crystals commonly exhibit the octahedron form?
6. Why is it possible to cleave diamond on planes parallel to {111}?
7. Which operations are described by a screw axis?
8. What is the nature of the axial symmetry of the diamond structure in directions parallel to the lattice rows?
9. Which operations are described by glide planes?
10. What is the meaning of space group?
11. What is the meaning of polymorphism?
12. What are the characteristics of the space lattice which describes the structure of sphalerite?
13. What are the forms and orientations of the coordination groups in sphalerite?
14. In sphalerite are the layers of atoms parallel to the planes of the {111} form similar on both sides?
15. What is the explanation of the common occurrence of the tetrahedron form exhibited by sphalerite crystals?

9 Space Groups

In the space lattice every point has three main characteristics, (1) every point is identical, (2) the surroundings of every point are identical with those of all other points, and (3) each point has an identical orientation. All fourteen kinds of space lattice possess the maximum symmetry of the appropriate crystal systems and this symmetry about a point, which is the origin of the reference axes, is described by the point group.

Bravais realized that the occurrence of crystals with lower symmetry must be due to the arrangement of the groups of atoms in the crystal. In chapter eight, two additional symmetry operations, the screw axis and the glide plane, were introduced and used to describe the symmetry of the atoms in particular crystal structures. The symmetry operations which relate the positions of the atoms in the crystal extend throughout the crystal structure, in other words through space, so they are referred to as the space group. The principal objective of this chapter is to consider the nature of a space group.

Contents of the sections

9.1. Recapitulation: Point group and space lattice

In chapter 2 the concepts of elements of symmetry were introduced by considering the external forms of some selected crystals. The symmetry operations consist of **reflection, rotation** and **rotation combined with inversion**. The group of symmetry elements which is exhibited by a crystal is called a **point group** because they describe symmetry about a point—the origin of the crystallographic axes, which is not repeated by the symmetry operations. The thirty-two possible point groups first identified by Hessel in 1830 were listed in table 2.1 in section 2.8.

The **32 point groups** are divided into the **seven crystal systems** which are characterized by particular axes of symmetry.

A crystal is essentially a pattern within which a particular configuration of atoms is repeated at regular intervals in 3 dimensions. The 3-dimensional pattern may be described in terms of a selected unit-cell which when repeated in three dimensions builds up the whole pattern. The dimensions of the unit cell are indicated by points so that repetitions of the unit cell builds up a 3-dimensional array of points which is called the **space lattice**. Bravais demonstrated that there are 14 kinds of space lattice and of these, six are said to be primitive because they have unit cells which consist of points at the corners only, so that within the lattice there is one point for each unit cell. In the other eight space lattices there are additional points for each unit cell.

All fourteen space lattices possess the maximum symmetry of the respective crystal systems. Bravais realized that the occurrence of crystals with lower symmetry must be due to the arrangement of the group of atoms in the crystal which is represented by a point in the space lattice. In the space lattice the points have three main characteristics, (i) **every point is identical**. (ii) **the surroundings of every point are identical** with those of all other points, and (iii) **each point has an identical orientation**. In crystals of the native elements gold, silver, copper, and lead, the atoms occur with a cubic close packing arrangement which is represented by the face-centred cubic lattice as shown in section 4.6. All the symmetry elements of the holosymmetric class are repeated at each point which in this case is represented by a single atom. Consequently if two atoms of one of the native elements are placed at the origin and at one point in the face-centred cubic lattice, the second atom will be repeated at all other points by the operation of the symmetry elements, and the continuous structure will be produced. Halite also belongs to the holosymmetric class of the cubic system, but in this case sodium and chlorine ions have to be positioned in relation to a face-centred cubic lattice as well as the appropriate ion at the origin, so that the symmetry operations can determine the positions of the other ions in the continous crystal structure of halite.

9.2. Space groups

When the structure of diamond was studied it was necessary to introduce two addition symmetry operations, the **screw axis** described in section 8.2 and the **glide plane** described in section 8.3 in order to bring the atoms into identical positions. The **translations** associated with screw axes and glide

planes are of atomic dimensions and they are not apparent in the external symmetry of the crystal. **The group of symmetry operations including translation movements on screw axes and glide planes** as well as reflection, rotation, and rotation combined with inversion, which is used to describe a crystal structure is called a **space group**. The symmetry elements of the space group extend throughout the crystal structure indefinitely, in other words through space, in contrast to the concept of the point group in which the symmetry elements are related to a point.

Consequently although the crystals of fluorite and diamond exhibit the same point group, that of the cubic holosymmetric class, they have very different space groups. The structure of sphalerite was explained in section 8.7 and it was found that the atoms of Zn and S are arranged in the same pattern as the carbon atoms in diamond. However, because there are two kinds of atoms, sphalerite does not exhibit tetrad screw axes nor glide planes parallel to the cube faces. Sphalerite has a different space group and also a much lower symmetry which is exhibited by the point group and crystal forms.

The nature of the space group has been introduced here by studying the relatively simple structures of some cubic minerals which were determined by W. H. and W. L. Bragg within a few years of the discovery of the diffraction of X-rays by crystals in 1913. Actually the theory of space groups had been developed gradually following the demonstration by Bravais in 1848 that there were only 14 different kinds of space lattices in which identical points have identical surroundings and orientations.

It is more appropriate to represent a group of atoms by an irregular unsymmetrical body rather than a simple point, so that it is possible to consider a regular arrangement of bodies which are exactly alike and which have identical surroundings but which do not necessarily have the same orientations. A German physicist L. Sohncke, showed in 1879 that if screw axes were introduced there were 65 different structural arrangements, so that it was possible to account for some of the lower symmetry classes. The next problem that was considered was, how many kinds of structural arrangement are possible when glide planes are introduced? Three very different researchers approached this problem in different ways and discovered quite independently that there could be **only 230 kinds of space groups**. The mineralogist E. S. Federov completed the publication of his studies in Russian in 1890, the German crystallographer A. M. Schoenflies published his conclusions in 1891 and an English business man, William Barlow published his work in 1894.

The 230 space groups are listed in the International Tables for X-ray crystallography, vol. 1 *Symmetry Groups*, 1952, Kynoch Press. The notation of the space group consists of the Herman–Mauguin symbol preceded by a letter which indicates the lattice type. For example consider fig. 9.1 which represents a portion of a triclinic space lattice viewed along one of the lines of points which are not perpendicular to the plane of the figure and which have a repeat distance different from the repeat distances on the other two lines. In fig. 9.2 the asymmetrical triangles represent identical groups of atoms which occur at a certain distance represented by the + signs, above the plane of the base of the unit cell. This diagram represents a primitive triclinic space group which has no symmetry and it is given the symbol **P1**. The number **1** axis indicates no symmetry.

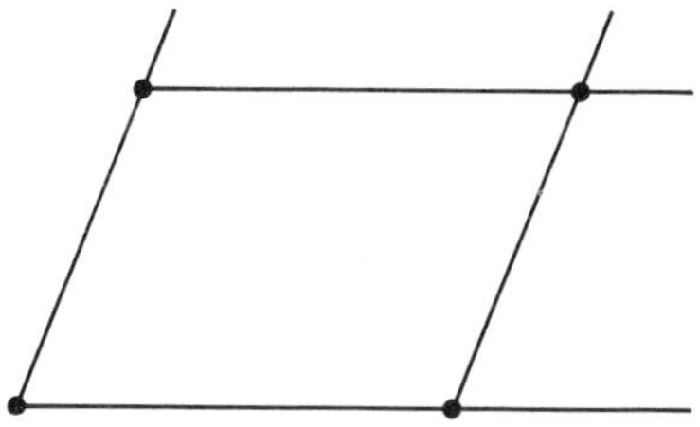

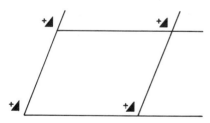

Fig. 9.1. Part of two rows of a triclinic space lattice. The third row has a different repeat distance and is not perpendicular to the diagram

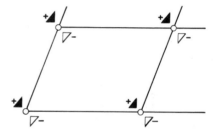

Fig. 9.3. A symmetry centre represented by the circles is placed at each lattice point and each group of atoms is inverted across the centre as indicated by the − signs

Fig. 9.2. Asymmetrical triangles represent groups of atoms which occur at a certain distance represented by the + signs, above the base of the unit cell. This represents a primitive triclinic space group with no symmetry, **Pl**

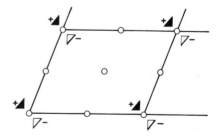

Fig. 9.4. The additional symmetry centres implied by the arrangement in fig. 9.3. This primitive triclinic space group has the symbol **P1̄**

If a centre of symmetry represented by a circle, is placed at each lattice point as shown in fig. 9.3, each group of atoms represented by a black triangle is inverted across a centre and is shown as a white triangle with a negative sign to indicate its position below the base of the unit cell. This structural arrangement also implies symmetry centres at other points which are shown in fig. 9.4. This primitive triclinic space group is given the symbol **P$\bar{1}$**, the inversion axis representing the symmetry centre. These are the only two possible triclinic space groups, **P1** and **P$\bar{1}$**. The other space groups were established by similar reasoning but the representation in two dimensions of those belonging to higher symmetry classes becomes increasingly difficult. A derivation of spaces groups is given in an *Introduction to Crystallography* by F. C. Phillips, Longmans, 1955.

The theory of space groups is important because it relates crystal symmetry on an atomic scale to the possible atomic arrangements which possess that symmetry. Consequently if a mineral is known to have cubic symmetry and **n** atoms in its unit cell, space group theory lists all possible arrangements of **n** atoms which will show the same symmetry. This greatly limits the number of postulated structures which could account for the observed symmetry and the intensities of the diffracted X-rays.

9.3 Questions for recall and self assessment

1. What is a point group?
2. How many point groups are possible?
3. How are the symmetry classes classified?
4. Can the symmetry of the structures of crystals in the lower symmetry classes be described by appropriate space lattices?
5. What are the characteristics of the points in a space lattice?
6. What are the characteristics of the additional symmetry operations which were introduced to describe the symmetry of crystal structures?
7. What is a space group?
8. Which characteristic of the space lattice is changed in order to define the nature of space groups.
9. What were the approximate dates of the first determinations of crystal structures and the identification of all the possible space groups?
10. What is the value of space group theory?

10

The Stereographic Projection of Crystals

The variation in the sizes and shapes of the faces of a particular crystal, frequently obscures the symmetrical relationships of the faces which are revealed by recording the interfacial angles as studied in section 2.3. The representation of the 3-dimensional symmetry characteristics of crystals by 2-dimensional drawings is of limited value, because when making drawings of combinations of forms it is necessary to make arbitary judgements of the relative sizes of the faces of the different forms as illustrated in sections 3.9 and 3.10.

A more useful and accurate method of representing the 3-dimensional angular symmetry of crystal forms makes use of projections similar to those used to produce maps of the surface of the earth. The principal objectives of this chapter are as follows.

1. To understand the use of the stereographic projection for the representation of the angular orientation of crystal faces.
2. To prepare stereograms to illustrate the angular orientations of the faces and the symmetry elements of a crystal.
3. To reorient a stereogram to indicate a particular symmetry element.
4. To study the use of stereograms to indicate the symmetry operations.

Contents of the sections

155

**10.1 The normals and poles of the
       crystal faces of the
       rhombdodecahedron**

The variation in the sizes and shapes of the
faces of a particular mineral frequently
obscures the symmetrical relationships of
the faces. The symmetrical relationships
shown by faces of different size making up
zones in a crystal are revealed by recording
the interfacial angles as shown in section
2.3. It is useful to represent the 3-
dimensional symmetry characteristics of a
crystal on a 2-dimensional plane such as a

sheet of paper, and the shapes of the ideally
symmetrical simple forms of cubic crystals
have already been represented by drawings
on an axial cross in chapter 3. When this
method is used to make drawings of combinations of forms, it is necessary to make
arbitary judgements of the relative sizes of
the faces of the different forms as illustrated
in sections 3.9 and 3.10.

A more universal method of representing
the 3-dimensional angular symmetry of the
forms occurring in a crystal makes use of
projections. Garnet crystals exhibiting the
simple rhombdodecahedron form are illustrated in fig. 2.34. The symmetry elements
of the rhombdodecahedron were considered
in section 2.5. and a drawing of this form
was constructed on an axial cross in section
3.6. Consider the rhombdodecahedron

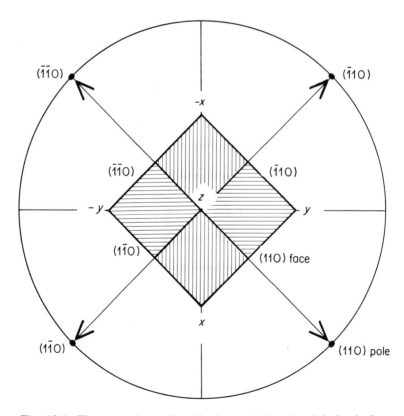

Fig. 10.1. The normals to the side faces of the rhombdodecahedron
represented by points known as poles on a circle

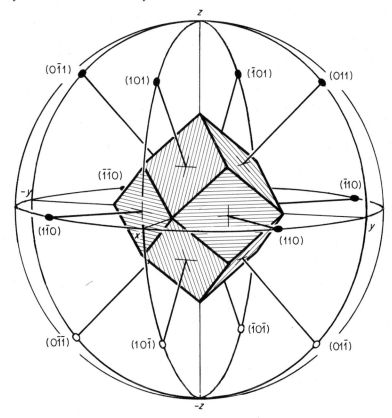

Fig. 10.2. The representation of the orientation of the faces of the
rhombdodecahedron by normals to the faces passing through the centre of
a sphere and intersecting the surface of the sphere at poles

when viewed along the tetrad axis of symmetry which has been selected as the vertical z axis and placed at the centre of a circle
as shown in fig. 10.1. The z axis is a zone
axis and the four faces (110), ($\bar{1}$10), ($\bar{1}\bar{1}$0)
and (1$\bar{1}$0) are all parallel to the z axis as is
indicated by the indices. The angular position of a face can be represented by a line
called the **normal** which is perpendicular to
the face. In fig. 10.1 the angular positions of
the four faces (110), ($\bar{1}$10), ($\bar{1}\bar{1}$0), and (1$\bar{1}$0)
are represented by the normals to the faces,
with each normal extending from the centre
to the circumference of a circle. The point
on the circle which represents the normal to
a face is called a **pole** and it is labelled with

the indice of the face. The relationship
between the interfacial angles and the angular positions of the poles to the faces is
clearly seen in fig. 10.1 and the interfacial
angle **(110)** $\widehat{}$ **(1$\bar{1}$0)** = 90°.

The angular orientation of all the faces of
the rhombdodecahedron can be represented
by poles on a sphere as illustrated in fig.
10.2. *Does the angular position of the pole
depend on the size of the face?*

The pole is the projection of the normal
which passes through the centre of the
sphere and therefore its position on the
sphere depends only on the angular orientation of the face relative to the zones axes.

10.2 The stereographic projection of the rhombdodecahedron

Maps are made by projecting onto a flat sheet of paper, the positions of the features occurring on a sphere presenting the Earth. The cartographer is concerned with representation of shapes and areas as well as their angular positions and the method of projecting is of necessity a compromise because some distortion is inevitable.

Figure 10.3 represents the globe within which the equatorial plane has been choosen as the plane for the map projection. The shapes of the land masses are shown in grey on the globe and the projection of the arrangement of the land masses in the northern hemisphere is shown in black on

the map. In this construction the position of points on the northern hemisphere of the globe, for example London and the tip of India in fig. 10.3, have been projected onto the map along lines which join the points to the South pole of the globe. This is known as a **stereographic projection.**

In crystallography the shapes and sizes of the crystal faces are not important, and the objective is to produce a diagram of the angular positions of the crystal faces in such a way that the true symmetry of the crystal is recorded.

The stereographic projection is the most useful for crystallographic work and it was brought into general use by W. H. Miller. The plane of the stereographic projection is the equatorial plane of the sphere and for

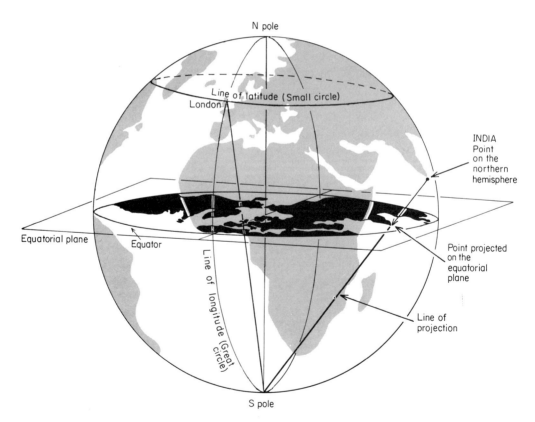

Fig. 10.3. The production of a map of the northern hemisphere of the Earth by the projection of points onto the equatorial plane along lines joining the points to the S pole of the Earth

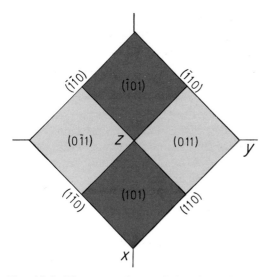

Fig. 10.4. The upper faces of the rhombdodecahedron

cubic forms such as the rhombdodecahedron shown in fig. 10.4, this will contain
the *x* and *y* axes while the *z* axis will be
vertical through the centre. Fig. 10.5 is an
oblique drawing representing the orientation of the **(011)** face of the rhombdodecahedron within a sphere. **The normal to the
face passing through the centre of the sphere
is projected to produce the pole on the
surface of the sphere.** The face normal lies in
the plane of the great circle which is shaded
in fig. 10.5. **The line joining the pole of the
(011) face on the sphere to the pole of
projection, gives the position of the pole of
the (011) face on the stereographic projection
of the rhombdodecahedron.**

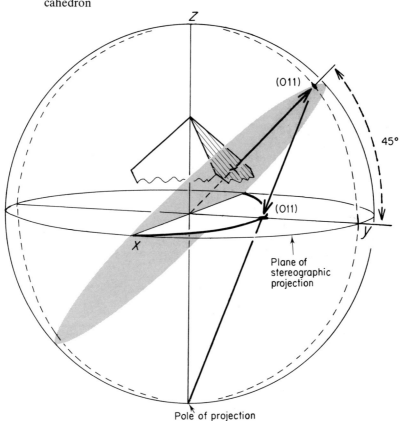

Fig. 10.5. The representation of the angular position of the (011) face of the
rhombdodecahedron, by poles on the sphere and on the plane of stereographic projection

10.3 The representation of the (011) crystal face of the rhombdodecahedron on a stereogram: Great circles

If two angular co-ordinates of each crystal face have been determined, the orientations of the crystal faces may be represented by plotting the face-poles on a stereographic projection which is then called a **stereogram**. The position of a face pole on a stereogram may be obtained by a geometrical construction, but the plotting of a pole is made very much easier by the use of a stereographic network of great and small circles known as the **stereographic or Wulff net** illustrated in fig. 10.6. Looking down on the stereographic net the point of projection on the -z axis equivalent to the pole of projection in fig. 10.5 lies below the centre

of the circle The outer circle represents the equator of the sphere and is known as the **primitive circle**.

Planes passing through the centre of the sphere intersect the surface of the sphere to form **great circles**. On a globe of the Earth the lines of longitude are great circles produced by planes which intersect on the line joining the N and S poles. On the stereographic net, fig. 10.6, the lines representing great circles are formed by planes which all intersect along the vertical diameter, and the intersections of the great circle lines on the horizontal diameter occur at 2° intervals. The horizontal diameter also represents a great circle.

In fig. 10.7 the vertical diameter of the stereographic net is selected as the x crystallographic axis. The horizontal diameter re-

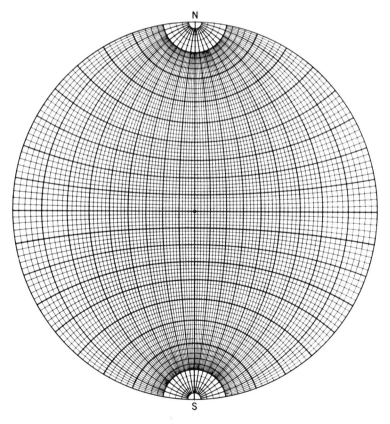

Fig. 10.6. The stereographic net

presents the y axis. The z axis is vertical and its intersection with the top of the sphere is represented by the centre point of the stereographic net.

The interfacial angle between the (011) and (0$\bar{1}$1) faces of the rhombdodecahedron is 90° so that the angle between the y axis and the (011) face is **45°** as shown in fig. 10.5. The pole of the (011) face can be located on the stereographic net, fig. 10.7, by measuring 45° along the horizontal diameter from the y axis. It is important to visualize fig. 10.7 as a hemisphere resting on the paper so that the pole can be imagined to be on the surface of the hemisphere at a point radially outwards from the pole on the stereogram. The actual position of the (011) pole on the stereogram is determined by the

projection of the pole towards the point on the lower hemisphere immediately below the centre.

It is important to remember that **every point on the stereogram represents a direction in space**.

10.4 The representation of the (101) face of the rhombdodecahedron on a stereogram: Small circles

Circles on a sphere formed by planes which do not pass through the centre of the sphere are called **small circles**. The lines of latitude on a globe of the Earth are small circles. Small circles are marked on the stereographic net fig. 10.6 and they intersect the vertical diameter at 2° intervals.

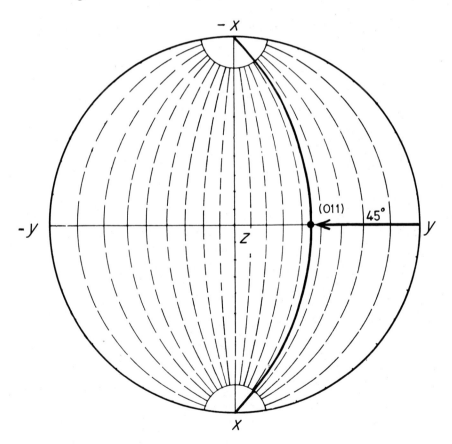

Fig. 10.7. The pole of the (011) face on the stereographic projection

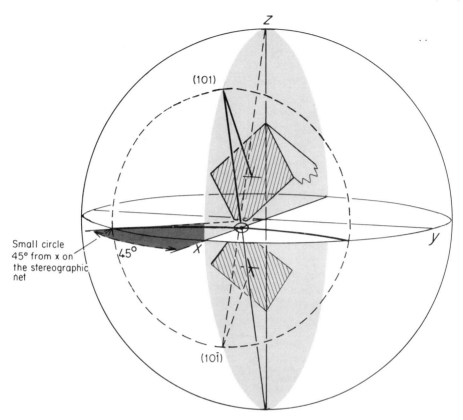

Fig. 10.8. The representation of the (101) and (10$\bar{1}$) faces of the rhombdodecahedron by poles on a sphere and on the plane of stereographic projection. The pole to the (101) face is projected along a line to the pole of the upper hemisphere and is represented by the open cricle on the stereographic projection

The **(101)** and **($\bar{1}$01)** faces of the rhomb-dodecahedron are indicated in fig. 10.4. The angle between the normal to the **(101)** face and the *x* axis is **45°**.

In fig. 10.8 the plane containing the *x* and *z* axes, and the poles of the **(101)** and **(10$\bar{1}$)** faces is shaded. The small circle at 45% from the *x* axis is shown by the dashed line and its position on the plane of stereographic projection is shown by the heavy line. The position of the **(101)** pole on the stereogram, fig. 10.9 can be located by measuring **45°** from the *x* axis along the great circle which also contains the *z* axis.

The **(10$\bar{1}$)** face of the rhombdode-

cahedron is also shown in fig. 10.8. The normal to the **(10$\bar{1}$)** face intersects the lower hemisphere, but it is the practice to plot the pole to the **(10$\bar{1}$)** face on the stereogram by projecting along the line towards the pole of the upper hemisphere as shown by dashed lines in fig. 10.8. The poles that occur on the lower hemisphere are shown by open circles. The pole to the **(10$\bar{1}$)** face coincides with the **(101)** pole in fig. 10.9.

The four faces **(101)**, **(10$\bar{1}$)**, **($\bar{1}$0$\bar{1}$)** and **($\bar{1}$01)** constitute a **zone** with the **zone axis parallel to** *y*. The angular orientations of the faces are represented by the poles on the stereogram fig. 10.9 with the lower faces

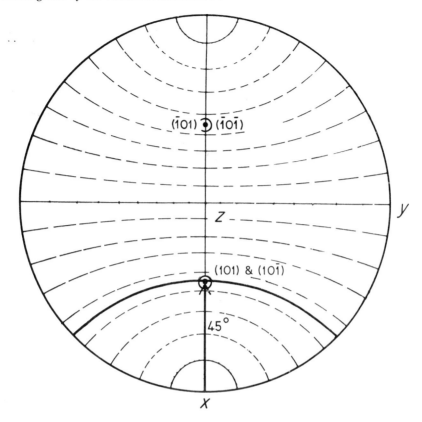

Fig. 10.9. Stereogram of the poles of the faces of the rhombdodecahedron which are parallel to the *y* axis

indicated by the open circles. The poles to the four faces fall on the great circle which is perpendicular to *y*.

The stereographic net shown in fig. 10.6 has great and small circles at 2° intervals. The poles to the faces of a crystal may be plotted on tracing paper placed over the stereographic net. *It is a valuable exercise to plot the poles to all the faces of the rhombdodecahedron form which is illustrated in fig. 10.2.*

10.5 The stereogram of the rhombdodecahedron

The complete stereogram representing the poles to the twelve faces of the rhombdode-

cahedron is shown in fig. 10.10. The four vertical faces are represented by poles on the primitive circle. The poles of the four faces on the upper side of the rhombdodecahedron are represented by the full circles and the larger labels. The poles of the four faces on the under side of the rhombdodecahedron are represented on the stereogram by the open circles with smaller labels.

The tetrad symmetry of the *z* axis is clearly displayed by the stereogram of the rhombdodecahedron, fig. 10.10. The coincidence of the poles of the faces on the top with the poles of the bottom faces indicates that the *x* and *y* axes are also tetrad axes of symmetry.

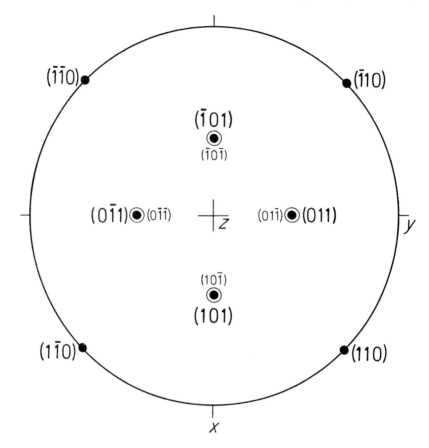

Fig. 10.10. Stereogram of the faces of the rhombdodecahedron form

10.6 The stereogram of the symmetry elements of the cube, class 4/ m 3̄2/ m, cubic holosymmetric

The rhombdodecahedron and the cube both belong to the cubic holosymmetric class. *Using the stereonet fig. 10.6 prepare a stereogram on a piece of tracing paper to illustrate the angular orientations of the symmetry elements of the cube which were studied in section 2.4.*

The stereogram is prepared by fixing the positions of the **tetrad axes** which are used as the crystallographic axes and are indicated by squares in fig. 10.11. The three planes of symmetry which are parallel to

two crystallographic axes are indicated by full lines and the primitive circle.

The **diad axes** bisect the angles between the crystallographic axes and are indicated by the ellipse symbols.

The great circles which contain a single diad axis are shown by dashed lines and these represent the six **diagonal planes of symmetry**. On a perfectly symmetrical cube the diagonal planes of symmetry join opposite edges so that three diagonal planes of symmetry intersect at each corner point of the cube. Consequently the **triad axes** of symmetry are located at the intersections of three diagonal planes of symmetry. The triad axes lie within planes of diagonal symmetry and they are between the tetrad and

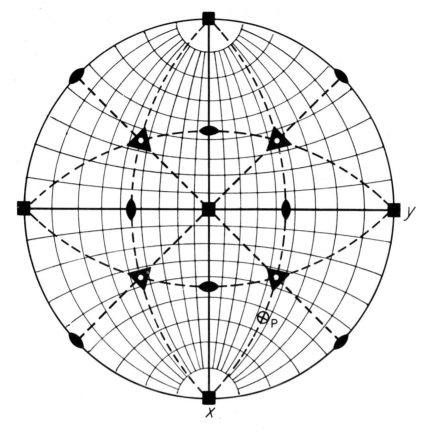

Fig. 10.11. Stereogram showing the orientation of the elements of symmetry of the cubic holosymmetric class. The point P refers to a problem posed in the text

diad axes. The dot at the centre of the triangles representing the triad axes indicates that they are **triad inversion axes**, which represent **triad axes** combined with a **centre of symmetry**.

10.7 The point-group symmetry operations

The translation-free symmetry operations of a point group can be represented on stereograms. The stereogram with the single point in fig. 10.12 represents no symmetry and the direction represented by the point is unique. If a reflection plane indicated by *m* is present the direction is repeated and two points occur on the stereogram. The series of

stereograms shown in fig. 10.13 illustrates respectively the repetition of a single direction by the operation of diad, triad, tetrad, and hexad rotation axes placed perpendicular to the stereograms.

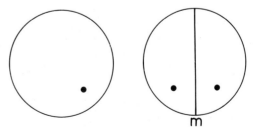

Fig. 10.12. Stereograms illustrating a single direction with no symmetry, and the repetition of a direction due to a mirror plane. *m*

The operation of a diad inversion axis is illustrated in fig. 10.14. A rotation of 180° is followed by inversion so that the point appears on the lower hemisphere and is represented by the open circle about the original point. The diad inversion axis has the same function as mirror plane normal to $\bar{2}$ and in fig. 10.14 the mirror plane is parallel to the plane of the figure.

The operation of a triad inversion axis $\bar{3}$ is shown in fig. 10.15. The first rotation of 120°

followed by inversion places the pole on the lower hemisphere. The second operation beginning with clockwise rotation returns the pole to the upper hemisphere, so that when complete there are three poles on the upper hemisphere and three poles on the lower hemisphere.

Determine the effects of the operations of tetrad and hexad inversion axes.

The operation of the tetrad inversion axis $\bar{4}$ produces a rotation of 90° followed by

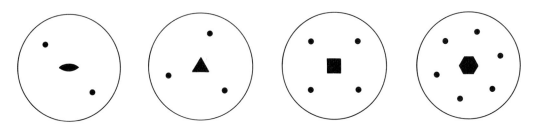

Fig. 10.13. Stereograms illustrating the repetition of a direction by the operation of diad, triad, tetrad and hexad rotation axes placed perpendicular to the stereograms

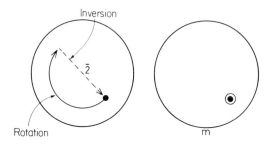

Fig. 10.14. The operation of a diad inversion axis

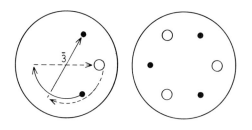

Fig. 10.15. The operation of a triad inversion axis

inversion so that two poles occur on the upper hemisphere and two on the lower hemisphere as shown in fig. 10.16. The operation of a hexad inversion axis $\bar{6}$ is illustrated in fig. 10.17. A rotation of 60° followed by inversion has the same effect as a triad axis combined with a perpendicular reflection plane $(3/\,m)$.

A stereogram of the faces of the tetrahedron may be produced by locating the pole of the **(111)** face and then repeating it by the operation of tetrad inversion axis as shown in fig. 10.16.

Problem. *The pole to one crystal face is represented by the isolated point P on the stereogram fig. 10.11. The operation of the symmetry elements will produce a number of faces with similar orientations relative to the crystallographic axes. Use the planes of reflection to complete the stereogram to illustrate the orientation of all the other faces of*

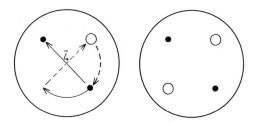

Fig. 10.16. The operation of a tetrad inversion axis

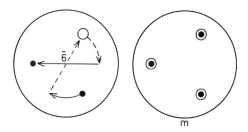

Fig. 10.17. The operation of a hexad inversion axis

the simple form and then study the operation of the symmetry axes.

The completed stereogram is shown in section 10.11.

10.8 The triad zones of the rhombdodecahedron

A stereogram of the faces of the rhombdodecahedron was prepared in fig. 10.10. The interfacial angle between the poles of two faces may be measured by rotating the tracing paper on which the stereogram has been plotted until the two poles lie on a great circle. The faces of one of the inclined zones of the rhombdodecahedron are indicated in fig. 10.18. In fig. 10.19 the stereogram has been rotated so that the poles to the $(1\bar{1}0)$, (101), (011) and $(\bar{1}10)$ faces lie on a great circle. *What are the interfacial angles between these four faces?*

It can be seen in fig. 10.19 that the interfacial angles $(1\bar{1}0)\widehat{\ }(101)$, $(101)\widehat{\ }(011)$, and $(011)\widehat{\ }(\bar{1}10)$ are all **60°**. The poles to the

faces referred to, lie on the upper hemisphere and they belong to the zone which is indicated in fig. 10.18 by the parallelism of the edges between the faces. *Which point on the stereogram will represent the direction of the zone axis?*

The zone axis will be perpendicular to the great circle within which lie the normals to the faces of the zone. The zone axis is located by measuring 90° along the EW great circle from the great circle on which the face poles lie, and it is indicated by a cross on the stereogram fig. 10.19.

Study where the $(0\bar{1}\bar{1})$ and $(\bar{1}0\bar{1})$ faces occur on the lower side of the rhombdodecahedron in fig. 10.18, and where the poles to these faces occur on the stereogram fig. 10.19. The great circle containing the poles of the zone faces on the lower side of the crystal is represented by the dashed line on the stereogram. Note the symmetry of the indices of the faces which are opposite each other in the zone. *What is the direction of the zone axis shown in fig. 10.18 and on the stereogram fig. 10.19?*

The direction of the zone axis shown on the stereogram is $[\bar{1}\bar{1}1]$. The lower end of

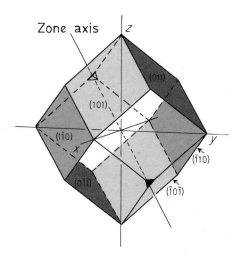

Fig. 10.18. The inclined zone of the rhombdodecahedron with the zone axis parallel to a triad axis

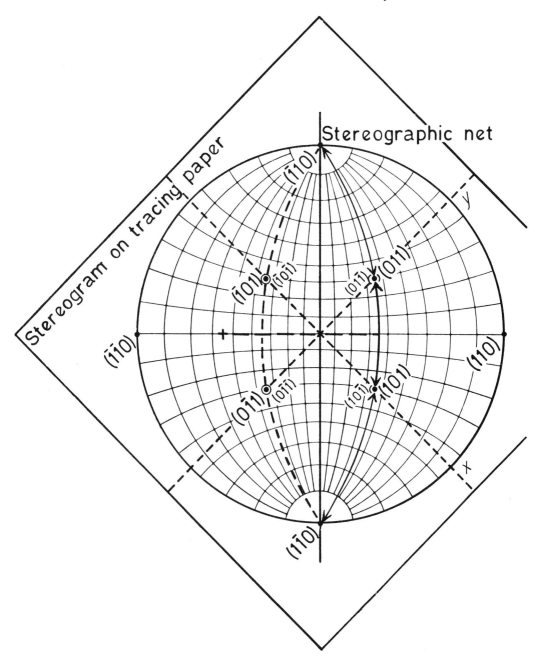

Fig. 10.19. Stereogram rotated to place the poles of the inclined zone faces on a great circle

the zone axis is in the direction **[111̄]**. *What degree of symmetry will be shown by the zone axis?* The zone consists of six faces and the zone axis is one of the four triad symmetry axes of rhombdodecahedron.

10.9 Small circles on the stereogram of the rhombdodecahedron

The angular relationship of **[111]** triad axis to the adjacent face normals is illustrated in fig. 10.20. Determine the values of the angles between the triad axis **[111]** and the poles to the **(110)**, **(101)**, and **(011)** faces shown on the stereogram fig. 10.10.

By placing the tracing paper on which the stereogram of the rhombdodecahedron has been plotted, on the stereographic net so

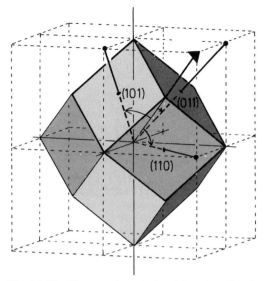

Fig. 10.20. The angular relationships of the [111] triad axis and the (110), (011), and (101) faces

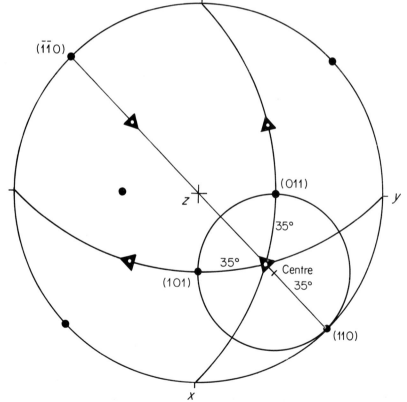

Fig. 10.21. Stereogram of the rhombdodecahedron with the small circle linking the (110), (011), and (010) poles. Note that the centre of the small circle does not coincide with the projection of the [111] triad axis

that the [111] triad axis and the pole to the (011) face fall on a great circle as in fig. 10.21, it can be seen that the angle between the [111] triad axis and the normal to (011) is approximately 35°. It will be found that in each case the angle between the triad axis and the normals to the (110), (101) and (011) faces is approximately 35°. Note that the triad axis [111] is 55° from the z axis.

On the upper hemisphere a small circle passes through the (110), (101), and (011) poles, and the [111] triad axis falls at the centre of this small circle. Fig. 10.22 represents a section of the sphere along the diameter which passes through the (110) and (1̄1̄0) poles. The triad axis [111] lies 55° from z. The small circle which passes through the (110), (101) and (011) poles is centred on the [111] axis and it has a radius of 35°. The section of the cone of projection of the small circle is shaded in fig. 10.22. The small circle linking the poles on the sphere projects on the stereogram as a small circle as shown in fig. 10.21. However it is

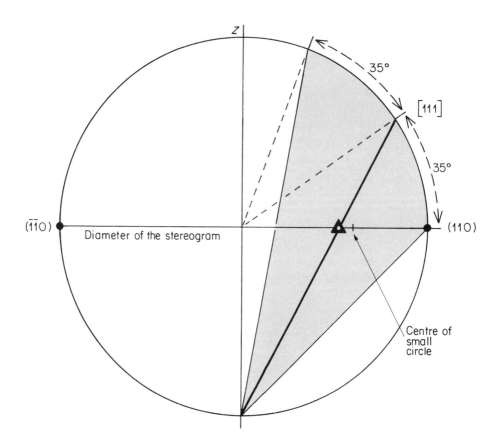

Fig. 10.22. Vertical section of the sphere along the diameter joining the (110) and (1̄1̄0) poles, to show the projection of the diameter of the small circle shown in fig. 10.21

important to note that the stereographic plot of the centre of the small circle on the sphere, in this case the [111] triad axis, does not coincide with the geometrical centre of the projected circle.

One of the valuable properties of the stereographic projection is that great and small circles on the sphere all plot as circles or arcs of circles on the stereogram. Therefore if three points on a small circle can be located on a stereogram, the small circle can be drawn accurately on the stereogram and the poles of associated crystal faces can be located. Note that the stereographic net can be used to draw constructions representing vertical sections of the sphere.

Problems

The symmetry of the pyritohedron form was considered in section 2.6 and it was drawn in section 3.7. *Construct a stereogram of the pyritohedron to show the poles to the faces and the elements of symmetry.* Find the centre of the small circle on the stereogram on which the (210), (021) and (102) poles lie. *Which direction intersects the centre of the small circle on the sphere that links the poles of these faces?*

The stereogram of the pyritohedron is considered in section 10.12.

10.10 Reorientation of the stereogram of the rhombdodecahedron to illustrate the triad symmetry of the form

Symmetry axes are easy to recognize on a stereogram provided that the axis coincides with the centre of the stereogram. The tetrad symmetry of the z axis is obvious in fig. 10.10. but the triad symmetry is not easy to recognize. Consequently it is often useful to reorient the poles so that a different direction is brought to the centre of stereogram. Consider the problem of rotating the [111] triad axis on the stereogram. The stereogram is rotated on the stereographic net so that the [111] direction lies in the EW great circle as shown in fig. 10.23. *What is the angle between the [111] direction and the tetrad axis which is labelled z?*

The angle between the [111] direction and the z axis is approximately 55°. *About which axis must the [111] direction be rotated so that it comes to the centre of the stereogram?*

The direction [111] is brought to the centre of the stereogram by a 55° rotation on the axis which is perpendicular to the EW great circle, since this plane contains the z axis and [111] direction as shown in fig. 10.23. The axis of rotation coincides with the ($\bar{1}10$) and ($1\bar{1}0$) poles. *During this rotation of the [111] direction to the centre of fig. 10.23, along which line does the [1$\bar{1}$1] direction travel on the stereogram?*

It can be seen from the stereogram fig. 10.23 that the direction [1$\bar{1}$1] lies in the NS plane and it is 55° from the z axis. Fig. 10.24 shows that during the rotation of the [111] direction to the vertical position, the [1$\bar{1}$1] direction traces out a cone which intersects the sphere as a small circle. Consequently the rotation on the axis normal to the EW plane of the stereogram will result in the movement of the [1$\bar{1}$1] direction through 55° along the small circle which is 55° from the z axis, as shown in fig. 10.24 and on the stereogram fig. 10.25.

As an exercise replot all the poles and directions on the stereogram of the rhombdodecahedron, fig. 10.10, so that the direction [111] is vertical, and take particular care with the movement of the poles which represent faces on the underside of the crystal. Use the replotted stereogram to make a drawing of the rhombdodecahedron as viewed along the [111] direction.

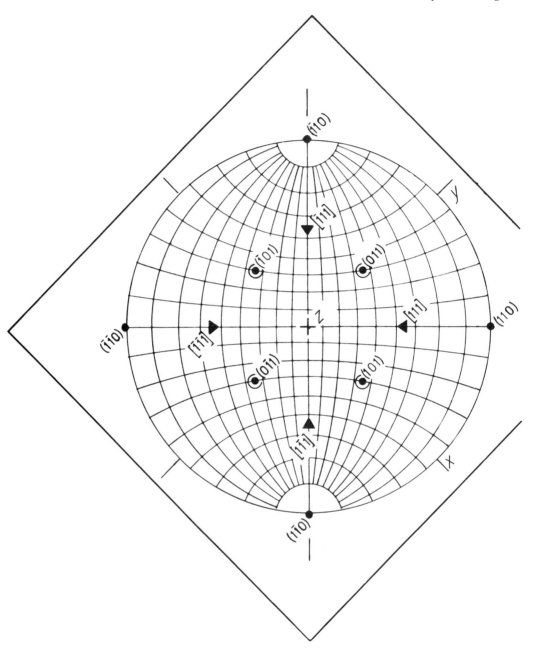

Fig. 10.23. Stereogram of the rhombdodecahedron rotated so that the [111] symmetry axis is on the
EW great circle

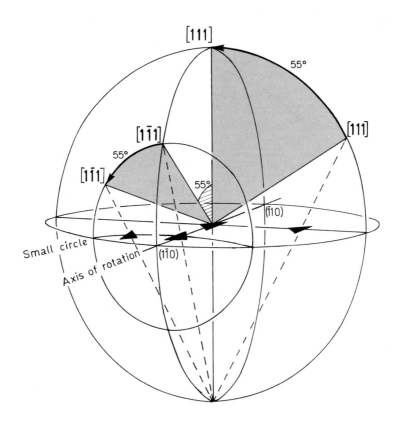

Fig. 10.24. Oblique view of the sphere to show the rotation of the
[111] and [1Ī1] symmetry axes about the axis which is parallel to the
normals to the (1Ī0) and (Ī10) faces

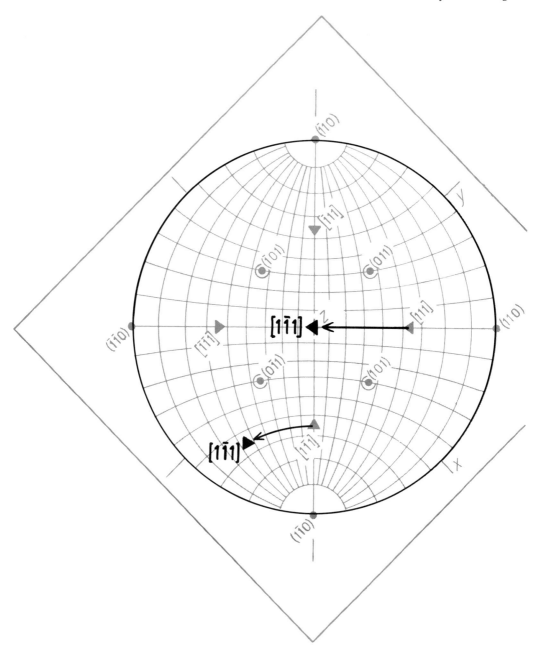

Fig. 10.25. Stereogram of the rhombdodecahedron showing the rotation of the [1$\bar{1}$1] direction along a small circle

Fig. 10.26 is a stereogram of the rhomb-dodechaedron oriented with the **[111]** direction vertical. A diagram of the rhombdode-cahedron as viewed along the **[111]** direction is superimposed on the stereogram. The triad symmetry of the **[111]** direction is quite obvious from the stereogram and the diagram of the rhombdodecahedron which has been produced from the stereogram.

The original positions of the poles occurred at the ends of the arrowed lines on the stereogram fig. 10.26. Note that whereas all the poles on the upper hemisphere have been displaced to the left on the stereogram. The poles of the **(0ĪĪ)** and **(Ī0Ī)** faces on the underside of the rhombdodecahedron have been displaced to the right and are now on the primitive circle.

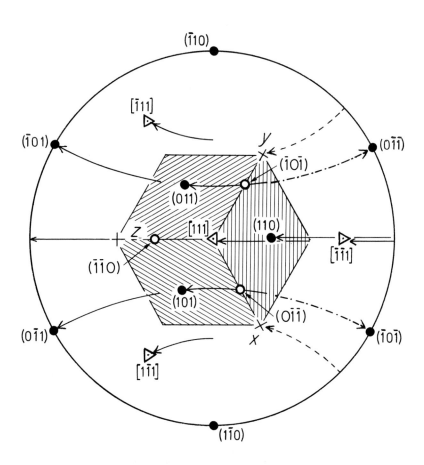

Fig. 10.26. Stereogram of the rhombdodecahedron reoriented so that the [111] direction is vertical and the triad symmetry is displayed. Note that the poles on the upper hemisphere have been displaced to the left while the poles on the lower hemisphere have been displaced to the right by the rotation

10.11 The stereogram of the icositetrahedron

The symmetry of the icositetrahedron was studied in sections 2.5 and 3.4. Fig. 10.27 is a view of the icositetrahedron along the z crystallographic axis, and fig. 10.28 is a stereogram showing the poles to the faces and the positions of the triad axes. **This is the complete stereogram that would be produced by the operation of the symmetry elements on the point P, (211), set as a problem in section 10.7 and fig. 10.11;**

The poles of the sets of 3 faces around the triad axes are joined by small circles representing approximately 19° angular distance from the triad axes.

Note that in any zone, adding the Miller indices of two faces which are not opposite to each other gives the indices of a third face

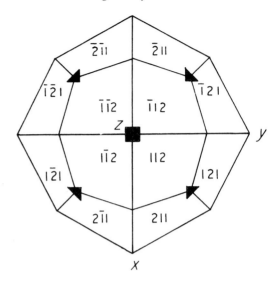

Fig. 10.27. The icositetrahedron form as seen along the direction of the z crystallographic axis

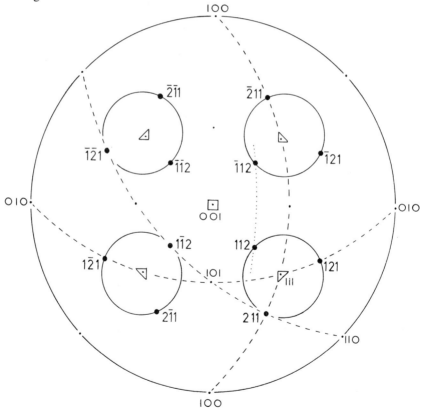

Fig. 10.28. Stereogram of the faces and triad axes of the icositetrahedron form. The brackets have been omitted for simplification

which lies between them in the zone. For example **(101)** + **(110)** = **(211)** and **(111)** + **(100)** = **(211)** also.

10.12 The stereogram of the pyritohedron

In the pyritohedron form, fig. 10.29, the angle between the *x* axis and the normal to the **(210)** face is one half the interfacial angle **(210)** $\widehat{}$ **(2Ī0)** = 53° 08'. Consequently the angle between the normal to the **(210)** face and the *x* axis is 26° 34'. The poles to the faces of the pyritohedron can be located on the stereogram by measuring 26° 34' in

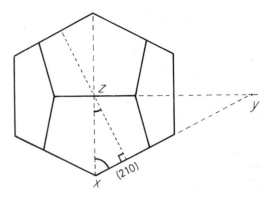

Fig. 10.29. The angle between the normal to the (210) face of the pyritohedron and the *x* axis

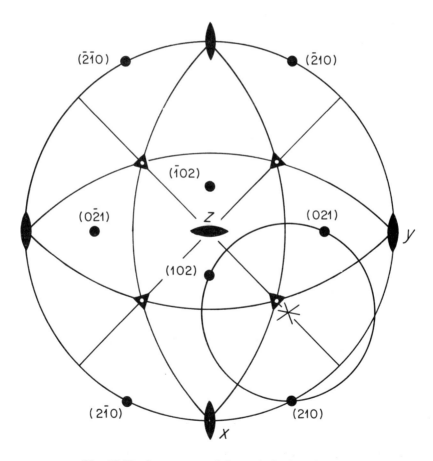

Fig. 10.30. Stereogram of the pyritohedron form

the appropriate directions from the crystal-lographic axes as shown in fig. 10.30.

The stereogram demonstrates clearly that the zone axes of the pyritohedron which are used as the crystallographic axes are also diad axes of symmetry. On the stereogram the triad axes have been located by drawing the great circles produced by the planes which bisect the angles between the diad axes, but the stereogram shows that these diagonal planes are not planes of symmetry. The centre of the small circle on which the **(210)**, **(021)**, and **(102)** poles lie, has been found by drawing perpendicular lines from the centres of the lines joining pairs of poles. On the stereogram the centre of this small circle does not coincide with the plot of the **[111]** triad axis, although this axis is at equal angle distances from the normals to the three adjacent faces, fig. 10.31. The

angle between the **[111]** axis and the three poles can be determined from the stereogram and it will be found to be approximately 40°.

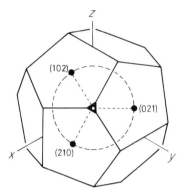

Fig. 10.31. View of the pyritohedron form along the [111] triad axis

PLATE 1. Specimens of the iron oxide pigments, red **haematite,** and yellow **limonite,** which were used by early man. Haematite also occurs as radiating fibres described as **reniform masses** or **kidney ore** ($\times 1$)

PLATE 2. The deep blue stone is **lapis lazuli,** which often contains small crystals of **pyrite** known as 'fools gold'. The small specimens are **turquoise.** The brilliant green **malachite** usually consists of bands of different shades, and the group of spheroidal masses of malachite is described as **botryoidal** from the Greek word for a bunch of grapes ($\times \frac{2}{3}$)

PLATE 3. Green **malachite** and blue **azurite** are **copper carbonates,** which are good indicators of the presence of **copper** shown here as the native element ($\times \frac{2}{3}$)

PLATE 4. **Quartz** crystals showing the characteristic purple colour of **amethyst.** The lower specimen is part of a vein containing **agate** formed by crystallization in a fracture in rock ($\times 1$)

PLATE 5. Irregular masses of natural **gold** enclosed i milky quartz ($\times 1$)

PLATE 6. Crystals of **pyrite** showing the **striated cube** and **pyritohedron** forms ($\times 1\frac{1}{2}$)

PLATE 7. **Diamond** and **sapphire gemstones** mounted on a gold ring. Two impure natural crystals of diamond exhibit the octahedron form. Sapphire is a blue transparent variety of the mineral **corundum** shown here as six-sided, columnar crystals and as rounded grains obtained from a river gravel ($\times 2$)

PLATE 8. Red, six-sided columnar crystals of **corundum** occurring in a vein in rock. **Ruby gemstones** are cut from the deep red transparent variety of corundum ($\times 1$)

PLATE 9. Water worn crystals of **red corundum** and **ruby gemstones** showing a range of colours. The purple coloured corundum gemstone is sometimes called **oriental amethyst** ($\times 2$)

PLATE 10. Gemstone varieties of **garnet** which are sometimes called **cape rubies.** The garnet crystals in the mica schist rock exhibit the **rhombdodecahedron** crystal form. The garnet crystals at the bottom exhibit the **icositetrahedron** crystal form ($\times 1$)

PLATE 11. Specimens of **sphalerite** showing the darker colour due to the substitution of some iron atoms for zinc. The more pure specimens are light brown and have a **resinous lustre.** Both specimens produce **pale brown streaks** on the porcelain plate (×1)

PLATE 12. The copper sulphide, **chalcopyrite,** has a brass yellow colour and metallic lustre but produces a black streak. **Bornite** is a copper sulphide which has a bronze colour but it tarnishes to purple colours and is known as **peacock ore** (×1)

PLATE 13. **Olivine** shows distinctive yellowish green colours. The large specimen is a **lava rock** which contains crystals of **olivine** and black crystals of **augite.** The smaller specimen shows olivine crystals in a volcanic bomb (×1)

PLATE 14. **Beryl** crystals showing the six-sided columnar forms and the characteristic green colours. The large specimen shows beryl crystals embedded in crystals of feldspar from a pegmatite vein. Beryl is found also in dark mica schists ($\times 1\frac{1}{2}$)

PLATE 15. Cube shaped crystals of **fluorite,** exhibiting cleavage planes which cause the corners of the cubes to break away easily, so that the fragments bounded by cleavage planes have the octahedron form ($\times 1$)

PLATE 16. Four specimens of **fluorite** showing a range of colours. The yellow fluorite occurs with cube crystals of **galena.** The green **autunite** in the centre is a hydrated phosphate mineral containing uranium. The white crystals are **calcite.** The other specimen is composed of the zinc bearing minerals from Franklin, New Jersey and they are **franklinite** (black), **zincite** (red) and **willemite** (white) ($\times \frac{2}{3}$)

PLATE 17. The same specimens showing **fluorescence.** Only three of the fluorite specimens show fluorescence but in two of these the fluorescence reveals growth zones which are parallel to the crystal faces ($\times \frac{2}{3}$)

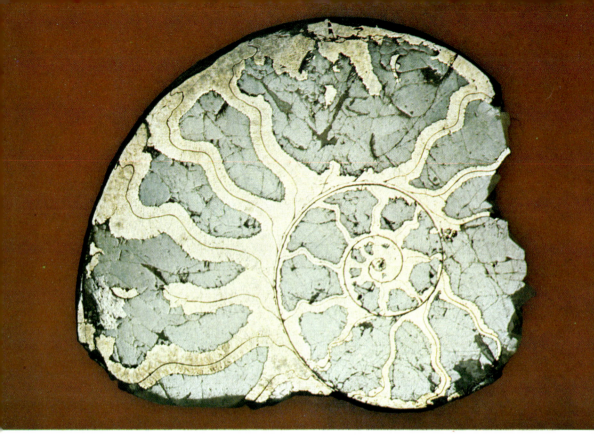

PLATE 18. An **ammonite shell** in which the original calcite crystals have been partly replaced by the metallic crystals of the iron sulphide, **pyrite** ($\times 2$)

PLATE 20. The model of the structure of sphalerite arranged with a triad axis vertical. The white rods joining the spheres in the model represent bonds between adjacent atoms in the sphalerite structure. Do the S-centred, and Zn-centred coordination groups have the same orientation?

PLATE 19. A model representing the expanded structure of sphalerite, ZnS. The blue spheres represent the positions of the Zn atoms with a face-centred unit cell shown in dark blue. The yellow spheres represent the positions of the S atoms with a unit cell shown in red. Can the tetrahedron form of the sphalerite crystals shown in fig. 2.39 be explained with the aid of this model of the crystal structure?

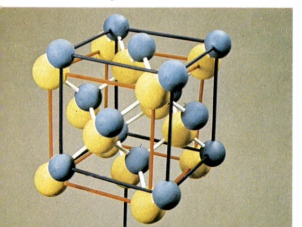

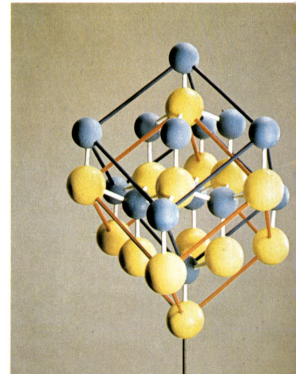

11 The Crystals of Zircon and Cassiterite from the Tetragonal System

In this chapter crystals of Zircon ($ZrSiO_4$) and Cassiterite (SnO_2) are illustrated by plans and the principal objectives are listed below.

1. To determine from specimens or the plans of the zircon and cassiterite crystals the approximate interfacial angles and the symmetry of each crystal, in order to prepare stereograms to illustrate the crystallographic observations.
2. To determine the intercept ratios and to understand the concept of axial ratio in terms of the repeat distances of the primitive tetragonal space lattice.
3. To introduce the concept of the parametral plan and the procedure for the description of a crystal.
4. To index the faces of zircon and cassiterite.

Contents of the sections

11.1 The symmetry of the crystal of Zircon, $Zr(SiO_4)$

A crystal of the zirconium silicate, **zircon** is shown in fig. 11.1 and by the oblique drawing fig. 11.2. The crystal has a **prismatic**

Fig. 11.1. A crystal of zircon, $Zr(SiO_4)$. (×2)

179

habit and plans of the crystal when viewed perpendicular and parallel to its length are shown in figs. 11.3 and 11.4 with the faces lettered.

Study the diagrams and answer the following questions

1. *What are the sequences of faces which make up the main zones exhibited by the crystal?*
2. *What are the approximate interfacial angles between the faces which constitute the main zones illustrated in figs. 11.3 and 11.4?*
3. *Does the crystal exhibit any axes of symmetry, if so how many are present?*
4. *Does the crystal possess any planes of symmetry? and if so how many are present?*
5. *Does the crystal have a centre of symmetry?*
6. *How many forms are present?*
7. *Which axes would be the most appropriate crystallographic axes for the description of the zircon crystal?*

Prepare a stereogram to show the orientation of the faces and the elements of symmetry of the zircon crystal. The answers to these questions will be found in section 11.2.

11.2 The stereogram of the zircon crystal

The zircon crystal is composed of two forms. One form consists of four faces **a, b, c, d** in a zone with the zone axis parallel to the longest edges of the crystal. The interfacial angles between these four side faces are all 90° and clearly this zone axis is also a tetrad symmetry axis.

There are two zones producing parallel edges which are perpendicular to the tetrad zone axis. The two zones are composed of the faces **d, e, f, b, g, h** and **a, j, k, c, m, n** respectively. The zone axes are perpendicular to each other and each is a diad symmetry axis.

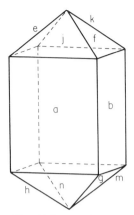

Fig. 11.2. An oblique drawing of the zircon crystal

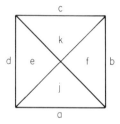

Fig. 11.3. The zircon crystal viewed along its length

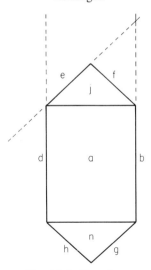

Fig. 11.4. The zircon crystal viewed perpendicular to its length

The zircon crystal exhibits three zones with mutually perpendicular zones axes. When preparing a stereogram the most appropriate orientation would be to place the tetrad axis vertical and call it the z crystallographic axis. The two equivalent horizontal zone axes are designated the x and y crystallographic axes as shown in the stereogram fig. 11.5.

The poles to the side faces **a, b, c, d** of the zircon crystal coincide with the zone axes x and y on the stereogram fig. 11.5. The indices of face **a** are (**100**) and the form

consisting of 4 faces is given the symbol {**100**} and it is called a **tetragonal prism form**.

The interfacial angle between a prism face and an adjacent inclined end face for example $\widehat{d\ e}$ and $\widehat{b\ f}$ in fig. 11.4 is **47° 50′** and a value close to this could be obtained from fig. 11.4. There are eight interfacial angles with this value indicating that the form consists of **8** faces which intersect one horizontal crystallographic axis and the vertical tetrad axis. This form is called the **tetragonal pyramid**. Since the interfacial angle is equal

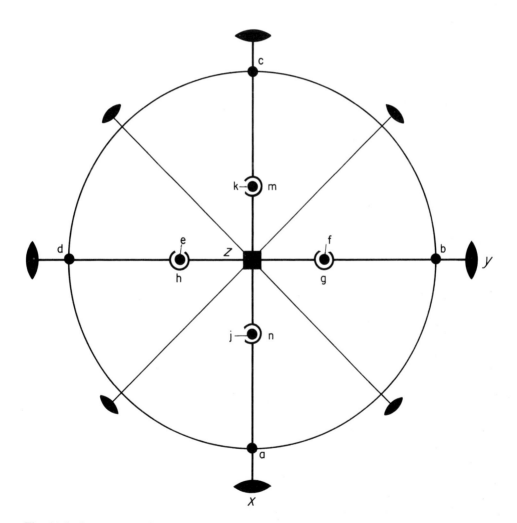

Fig. 11.5. Stereogram of the faces and symmetry elements exhibited by the zircon crystal

to the angle between the normals to a prism face and an adjacent pyramid face, the poles to the pyramid faces can be located on the stereogram by measuring **47° 50′** from the poles of the prism faces. The faces on the lower end of the zircon crystal are indicated on the stereogram by open circles.

It can be seen from the stereogram (fig. 11.5) that there are two diad axes which bisect the angles between the two diad zone axes which were selected to be the x and y crystallographic axes. The diad axes which are also zone axes are shown by the larger symbols on the stereogram.

Clearly there are 4 planes of reflection symmetry each parallel to the vertical tetrad axis and one of the diad axes represented by lines on the completed stereogram, fig. 11.5. There is a fifth plane of reflection symmetry which is perpendicular to the tetrad axis and parallel to the plane of the stereogram. The crystal has a symmetry centre so that the complete elements of symmetry are **1A₄, 4A₂, 5M,C** and the Hermann Mauguin symbol is $4/m, 2/m, 2/m$.

11.3 The Intercept ratios of the zircon crystal

The symmetry of the zircon crystal and in particular the presence of a single tetrad axis indicates that its structure can be described in terms of a tetragonal space lattice in which the repeat distance of points in the direction of the tetrad axis is different from the repeat distances along the two mutually perpendicular diad axes.

The shape of the structural unit of zircon determines the value of the interfacial angle $\widehat{a\ j}, = 47° 50′$ between the prism and pyramid faces. *What is the ratio of the intercept of face* **j** *on the z axis compared with the intercept on the x axis?*

It can be seen in fig. 11.6 that the intercept ratio $z/x = $ tangent $(90° - 47° 50′) = $ **0.9057**.

X-ray studies of zircon indicate that the

repeat distance a in the directions of the x and y crystallographic axes is **6.6 Å**, but the **repeat distance** c in the direction of the z crystallographic axis is **5.98 Å**. This gives the actual dimensions of the tetragonal unit cell. The ratio $c/a = $ **0.9060** is very close to the intercept ratio $z/x = z/y = $ **0.9057** calculated from the pyramid faces of the zircon crystal. Consequently for pyramid face **j** the intercept ratio z/x is the same as the ratio of the repeat distances along these axes in the tetragonal space lattice shown in fig. 11.7.

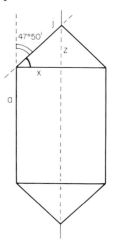

Fig. 11.6. The intercept ratio z/x of the zircon crystal

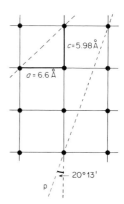

Fig. 11.7. A tetragonal space lattice indicating the repeat distances in the zircon crystal structure

Zircon crystals sometimes exhibit two different pyramid forms as shown in figs. 11.8, 11.9, and 11.10. The interfacial angles $\widehat{a\ j} = 47°\ 50'$, and $\widehat{a\ p} = 20°\ 13'$. Fig. 11.11 is a stereogram of the two pyramid forms of zircon. *What are the intercept ratios z/x of the pyramid face labelled* **p**? *and what is the ratio of the intercepts of pyramid faces* **j** *and* **p** *on the z axis?* The numerical calculations are given below in tabulated form.

to the ratio of three times the repeat distance of the structure in the z direction to unit repeat distance on the x axis as shown at the bottom of fig. 11.7.

11.4 The axial ratio of the crystal of cassiterite (SnO₂)

The crystals of cassiterite belong to the tetragonal system and the common habit of

Face	Interfacial angle	Intercept ratio z/x	Ratio of intercepts on z
Pyramid **j**	$\widehat{a\ j} = 37°\ 50'$	$\tan(90 - 47°\ 50') = 0.9056$	1
Pyramid **p**	$\widehat{a\ p} = 20°\ 13'$	$\tan(90 - 20°\ 13') = 2.7155$	3

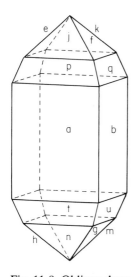

Fig. 11.8. Oblique drawing of a zircon crystal exhibiting two pyramid forms

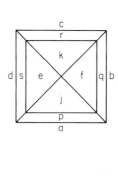

Fig. 11.9. View of the zircon crystal along the tetrad axis

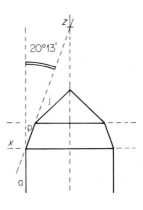

Fig. 11.10. The intercept ratios of the pyramid forms of zircon

It can be seen that the pyramid faces **j** and **p** have intercepts on the z axis in the ratio of **1:3**. This is an example of the **Law of Rational Ratios of Intercepts** and it is a direct consequence of the nature of the repeat distances in the crystal structure. The angular orientation of the pyramid face **p** is given by the intercept ratio z/x which is equivalent

the crystal is shown as an oblique view in fig. 11.12 and a plan fig. 11.13. There are two prism forms parallel to the z axis, each consisting of four faces. In one prism form, the faces **a,c,e,g** intersect one horizontal crystallographic axis and are parallel to the other horizontal crystallographic axis. This form is called **second-order tetragonal prism**

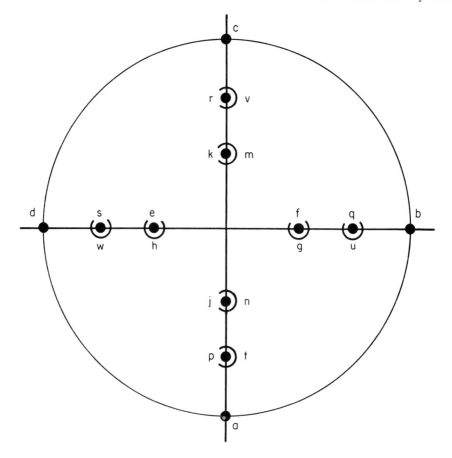

Fig. 11.11. Stereogram of the two pyramid forms of zircon

and it is represented by the indices {**100**}. The other prism form consists of four faces **b,d,f,h** which intersect both horizontal crystallographic axes. This form is called a **first-order tetragonal prism** and it is represented by the indices {**110**}. The interfacial angle between adjacent prism faces such as $\widehat{a\ b}$ = 45°.

At the ends of the cassiterite crystal there are two **pyramid forms** each consisting of 8 faces. The pyramid form which is parallel to one on the crystallographic axes is called the **second-order tetragonal pyramid**, and the interfacial angles $\widehat{a\ k}$, $\widehat{c\ n}$, $\widehat{e\ q}$ and $\widehat{g\ s}$ are all **54° 04'** as shown in figs. 11.12. and 11.14. *What is the intercept ratio z/x of the pyramid face* **k***?*

The intercept ratio z/x of face k = tan (90° − 56° 04′) = 0.6728.

The pyramid form which intersects all three crystallographic axes is called the **first order tetragonal pyramid**, and the interfacial angles $\widehat{b\ m}$, $\widehat{d\ p}$, $\widehat{f\ r}$ and $\widehat{h\ t}$ are all **46° 25'**. *What are the intercept ratios* **OA:OB :OC** *of the first order pyramid face* **m** *which is shown in fig. 11.14?*
It can be seen from fig. 11.14 that

$$\tan (90° - 46°25') = \frac{OC}{OD} = \frac{OC}{OA \sin 45°}$$

Therefore the intercept ratio OC/OA = tan 43°35′. sin 45° = 0.6728. Since OA = OB the intercept ratios for pyramid face *m* are ***x:y:z* = 1:1:0.6728.**

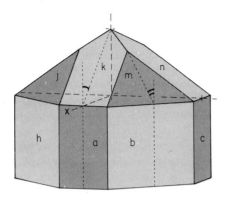

Fig. 11.12. An oblique drawing of the crystal of cassiterite

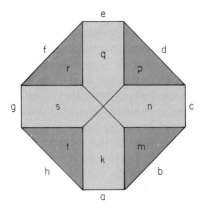

Fig. 11.13. Plan of the faces of the cassiterite crystal

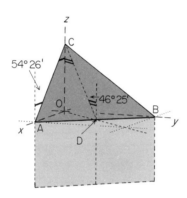

Fig. 11.14. The angular relationship of the pyramid face **m**

X-ray determinations of the repeat distances in cassiterite indicate that the tetragonal unit cell has the dimensions $a = 4.73$Å, $c = 3.18$Å. The ratio $c/a = 0.6723$ indicates that the intercept ratios of pyramid face **m** are equivalent to the ratios of the repeat distances in the tetragonal unit cell of the cassiterite structure.

11.5 The parametral plane

Before the development of X-ray techniques the procedure for the description of the external forms of crystals involved the following steps,

1. **Identification zones and zone axes, including the measurement of interfacial angles.**
2. **Selection of reference axes** called the crystallographic axes, which were placed parallel to symmetry axes or zone axes if possible.
3. **The selection of a plane parallel to one of the crystal faces which intersects all three crystallographic axes to act as a reference plane.** For example the pyramid

face **m** in figs. 11.12 and 11.14 would be an appropriate reference plane, and it has been determined that the ratios of the intercepts of face **m** with the crystallographic axes are **1:1:0.6728**. **If the intercept ratios of the faces of the cassiterite crystal are divided by the respective intercept ratios of face m, simple sets of intercept ratios and Miller indices are produced which clearly describe the orientation of the faces in relation to the reference plane** as illustrated in the table below.

sions. However, the procedure of dividing the intercept ratios of any face on the crystal by the intercept ratios of the selected parametral plane always produces simple rational indices, which clearly indicate the orientation of the face relative to the crystallographic axes and the parametral plane.

Return to the end of section 11.3 and determine the Miller indices of the pyramid faces j and p on the zircon crystal in fig. 11.10.

The zircon crystal does not exhibit a parametral plane but if face **j** is regarded as a

Face on the cassiterite crystal	Actual intercept ratio	Intercept ratios ÷ intercept ratios of reference plane *m*.	Miller indices of the face
m	1 : 1 : 0.6728	1 : 1 : 1	(111)
a	1 : ∞ : ∞	1 : ∞ : ∞	(100)
b	1 : 1 : ∞	1 : 1 : ∞	(110)
k	1 : ∞ : 0.6728	1 : ∞ : 1	(101)
n	∞ : 1 : 0.6278	∞ : 1 : 1	(011)

The reference plan (**111**) is called the **parametral plane** because it determines the parameters or ratios of the units of intercept on the three crystallographic axes. The ratio of intercepts of the parametral plane are known as **the axial ratios** referring to the ratios of the unit intercepts on the crystallographic axes.

The form which includes the **parametral plane (111)** is called **the unit form** and it consists of eight faces. The unit form is one of the possible pyramid forms.

In the cassiterite crystal, fig. 11.12 the pyramid face **m** is the only suitable face which can be used as a parametral plane. The X-ray observations indicate that the intercept ratios of face *m* are the same as the ratio of the repeat distances *c/a* of the tetragonal unit cell of the structure of the cassiterite crystal.

It is not always possible to select a parametral plane which will give axial ratios equal to the ratio of the unit cell dimen-

reference plane, the indices of the pyramid face **p** can be obtained by dividing its intercept ratios by those of the face **j**. The division results in the simple intercept ratios $1:\infty:3$ for the pyramid face **p**. The reciprocals are $1:0:1/3$ and when the fractions are cleared the Miller indices of the pyramid face **p** are (**301**).

Prepare a stereogram of the faces of the cassiterite crystal show in fig. 11.13 and label the poles with the indices. This stereogram will be used in a problem in the following section.

11.6 Indexing the tetragonal dipyramid form of the cassiterite crystal

The faces shown on the cassiterite crystal in fig. 11.13 have been plotted on the stereogram fig. 11.15. Cassiterite crystals sometimes exhibit a pyramid form which consists of 16 faces, eight at each end of the crystals as shown in fig. 11.16. This form is known as

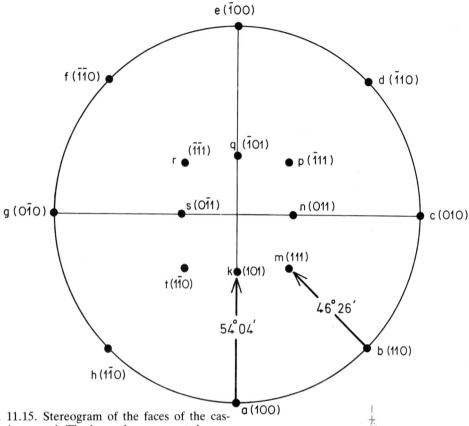

Fig. 11.15. Stereogram of the faces of the cassiterite crystal. The lower faces are not shown

the **tetragonal dipyramid** and the faces occur in pairs in the sectors bordered by the crystallographic axes. In fig. 11.16 the faces adjacent to the dipyramid face **P** have been indexed. The interfacial angles measured from face **P** in fig. 11.16 are tabulated below.

$$\widehat{P\,(100)} = 39°$$
$$\widehat{P\,(010)} = 59°$$
$$\widehat{P\,(101)} = 42°$$
$$\widehat{P\,(110)} = 24°$$

Using small circles, locate the pole of the dipyramid face **P** *on the stereogram of the cassiterite crystal which was prepared in the preceding section, and then plot the poles of the eight dipyramid faces which occur at the top of the cassiterite crystal.*

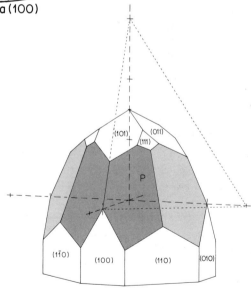

Fig. 11.16. A cassiterite crystal exhibiting the tetragonal dipyramid form. The diapyramid faces are shaded and they occur in pairs in each sector bounded by the three crystallographic axes

The pole of the dipyramid face **P** can be plotted on the stereogram by making the following measurements.

1. Since the pole to the **P** face lies at 39° from the (**100**) pole, place the (**100**) pole on the vertical diameter of the stereogram and draw the small circle which is 39° from the (**100**) pole as shown in fig. 11.17.
2. Since the interfacial angle **P** $\widehat{}$ (**010**) = 59° place the (**010**) pole on the vertical diameter of the stereogram and draw the small circle which is 59° from the (**010**) pole. The intersection of the two small

circles gives the pole of the dipyramid face **P**, but this can be confirmed by using the interfacial angles **P** $\widehat{}$ (**101**) and **P** $\widehat{}$ (**110**).

3. It can be seen from fig. 11.16. that the three faces (**101**), **P** and (**110**) occur in a zone because the edges between these faces are parallel. Consequently if the stereogram is rotated so that the poles of the (**101**), **P** and (**110**) faces are placed on a great circle, the interfacial angles **P** $\widehat{}$ (**101**) = 42°, and **P** $\widehat{}$ (**110**) = 24° can be used to confirm the location of the pole to the dipyramid face **P**.

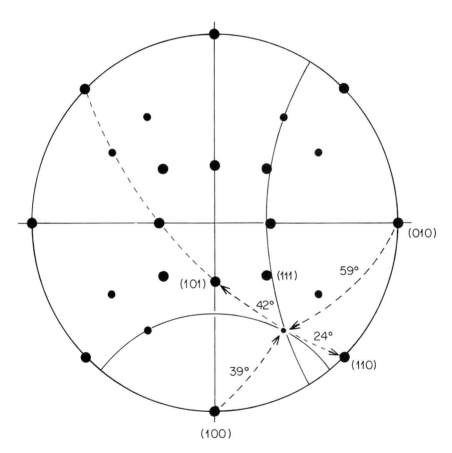

Fig. 11.17. On the stereogram of the cassiterite crystal the pole to the dipyramid face P is located by drawing small circles from the (100) and (010) poles, and the great circle through the poles (101) and (110)

Plot the other faces of the dipyramid form which occur at the top of the cassiterite crystal.

How can the indices of the dipyramid face P be determined?

In fig. 11.18 XYZ represents the intersections of face **P** with the crystallographic axes *xyz*. OP is the normal to the plane XYZ. The intercepts of plane XYZ are

$$OX = \frac{OP}{\cos X\hat{O}P}$$

$$OY = \frac{OP}{\cos Y\hat{O}P}$$

$$OZ = \frac{OP}{\cos Z\hat{O}P}$$

The intercept ratios are

$$\frac{OX:OY:OZ}{OX:OX:OX} = 1 : \frac{\cos XOP}{\cos YOP} : \frac{\cos XOP}{\cos ZOP}$$

How can the angles XÔP, YÔP and ZÔP between the normal to the face P and the crystallographic axes be obtained from the stereogram fig. 11.17.

By placing the pole to **P** and a crystallo-graphic axis on a great circle as shown in fig. 11.19 the required angles are easily obtained and they are

$$\overline{X\hat{O}P = 39°}$$
$$Y\hat{O}P = 59°$$
$$\overline{Z\hat{O}P = 68°}$$

Consequently the intercept ratios of the face **P** are

$$1 : \overline{\frac{\cos 39°}{\cos 59°}} : \frac{\cos 39°}{\cos 68°}$$

$$1 : \frac{0·7771}{0·5150} : \frac{0·7771}{0·3746}$$

$$\mathbf{1 : 1·508 : 2·07}$$

What are the intercept ratios and Miller indices of dipyramid face P with reference to the parametral plane m in fig. 11.12, which has intercept ratios of 1 : 1 : 0.6729?

The indices of the dipyramid face **P** are obtained by following steps

1. division $\dfrac{1}{1} : \dfrac{1·508}{1} : \dfrac{2·07}{0·6729} = 1 : 1·5 : 3$

2. taking reciprocals $1:2/3:1/3$

3. clearing fractions (321)

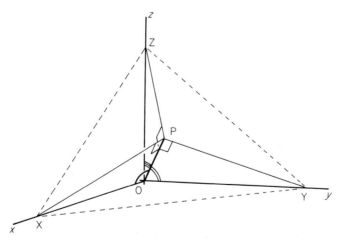

Fig. 11.18. The angles between the normal to face P and the crystallographic axes

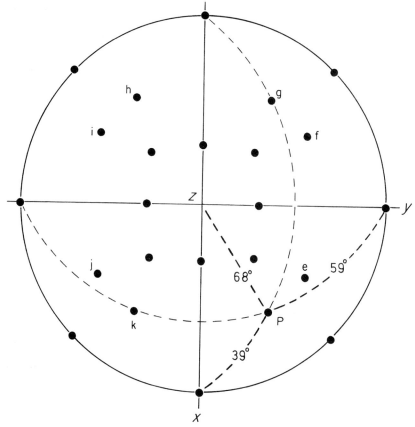

Fig. 11.19. The angles between the normal to the dipyramid face P and the
crystallographic axes obtained from the stereogram

This exercise demonstrates again the **Law of
Rational Ratios of Indices** on corresponding
crystallographic axes.

*Plot on a stereogram, the other faces of the
dipyramid form, and index the faces which
occur on the top of the crystal.*

The indices of the sequence of dipyramid
faces which occur at the top of the cassiterite
crystal read in an anticlockwise direction are
(321), (231), $(\bar{2}31)$, $(\bar{3}21)$, $(\bar{3}\bar{2}1)$, $(\bar{2}\bar{3}1)$,
$(2\bar{3}1)$, $(3\bar{2}1)$.

11.7 Questions for Recall and Self
Assessment

1. What is involved in the first stage of the
   procedure for the description of the external forms of a crystal?

2. What is the purpose of the crystallographic axes?
3. What is the parametral plane?
4. What are the Miller indices of the parametral plane?
5. What is the unit form?
6. What are the axial ratios ?
7. How can the axial ratios be determined
   with the aid of a stereogram.
8. How are the indices of the crystal faces
   obtained?
9. What is the name of the form on the
   zircon crystal which has the indices {110}
   and how many faces make up the form?
10 What is the general name of the form on
   the zircon crystal which has the indices
   {111}, and how many faces make up the
   form?

12 The Crystals of Barytes and Olivine from the Orthorhombic System

In this chapter plans of the crystals of barytes and olivine are used to achieve the objectives listed below.

1. To determine from specimens or the plans of the barytes and olivine crystals, the approximate interfacial angles in order to prepare stereograms of each crystal to display the crystallographic symmetry.
2. To study the characteristics of the orthorhombic system.
3. To appreciate the full meaning of the law of rational ratios of intercepts by studying the dome faces of the barytes crystal.
4. To appreciate the concepts of parametral plane and unit form by determining the axial ratios of olivine from the stereogram and indexing the forms displayed by the crystal.

Contents of the sections

12.1 The crystals of barytes, $BaSO_4$

Two crystals of barytes with similar orientations are viewed from an oblique direction in fig. 12.1 and perpendicular to their largest faces in fig. 12.2. A plan based on the largest crystal is shown in fig. 12.3 and sections of this crystal parallel to the shorter and longer dimensions are shown in figs. 12.4 and 12.5 respectively. The faces of the barytes crystal illustrated in the diagrams are identified by letters.

Study crystals of barytes similar to those illustrated on the plans shown here and answer the following questions.

1. *How many zones are exhibited by the crystal of barytes and what are the sequences of faces which make up the zones?*
2. *Do the barytes crystals exhibit sets of cleavage planes.*
3. *What are the interfacial angles between the faces which constitute the zones?*
4. *How many forms are exhibited by the crystal illustrated in fig. 12.3?*
5. *How many axes of symmetry are present?*

191

6. *Do the crystals show reflection planes and a centre of symmetry?*
7. *Which axes would be the most appropriate crystallographic axes?*

Prepare a stereogram to show the orientation of the faces and the elements of symmetry of the barytes crystal. The answers to these questions will be found in section 12.2.

12.2 The symmetry of the barytes crystal

The prominent crystal edges reveal the presence of two zones each consisting of the largest faces and one set of the inclined faces. One zone is composed of the faces **a,b,c,d,e,f**, and the interfacial angles of this zone are shown in fig. 12.6. The interfacial angles **a b**, **a f**, **d c**, and **d e** are all **52° 42′** and a value near this could have been obtained from fig. 12.4.

The second zone consists of the faces **a,g,h,d,m,n** and the interfacial angles are shown in fig. 12.7. The interfacial angles **a g**, **a n**, **d m**, and **d h** are all **38° 51′**.

The interfacial angles between the large faces and the inclined faces in the two zones are very different. The two zone axes are parallel to the largest faces and at right angles to each other. It can be seen from figs. 12.6 and 12.7 that the interfacial angles are symmetrically arranged so that the zone axes are diad axes of symmetry.

There are also prominent sets of cleavage planes which are parallel to the crystal faces labelled **p,q,r,s** in fig. 12.8. The cleavages and faces **p,q,r,s** are perpendicular to the largest crystal faces and they form a third zone. The interfacial angle **p q = 78° 20′** is the angle between the two sets of prominent cleavage planes seen in the crystals in figs. 12.1 and 12.2. There is also a well developed cleavage parallel to the largest crystal faces of barytes. The symmetrical arrangement of the interfacial angles in fig. 12.8 indicates that the zone axis which is perpendicular to the large face **a** is also a diad axis of symmetry.

Consequently it can be seen that the barytes crystal exhibits three mutually per-

Fig. 12.1. Crystals of barytes seen from an oblique direction (×1)

Fig. 12.2. Crystals of barytes viewed perpendicular to their largest faces (×1)

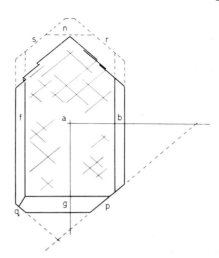

Fig. 12.3. Plan of a barytes crystal on a plane parallel to the largest faces

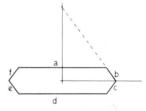

Fig. 12.4. Section of a barytes crystal perpendicular to the largest faces and the more steeply inclined faces

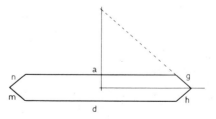

Fig. 12.5. Section of a barytes crystal perpendicular to the largest faces and the more gently inclined faces

pendicular zone axes which are also diad axes of symmetry, and this places it in the orthorhombic system. These diad axes may be used as the crystallographic axes and they are labelled in figs. 12.6, 12.7, and 12.8. There are six possible ways of labelling the crystallographic axes but the conventional orientation is illustrated in the diagrams.

The complete symmetry of the barytes crystal is best displayed by a stereogram. When preparing a stereogram it is most

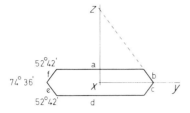

Fig. 12.6. Plan of a barytes crystal parallel to the y and z crystallographic axes

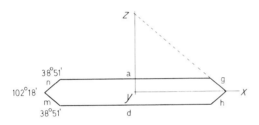

Fig. 12.7. Plan of a barytes crystal parallel to the x and z axes

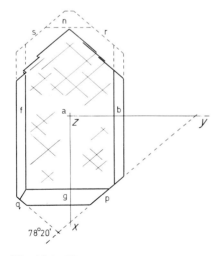

Fig. 12.8. Plan of a barytes crystal parallel to the x and y axes

convenient to place in the vertical position the zone axis which is perpendicular to the large crystal faces and this is labelled z. A stereogram showing the orientation of the barytes crystal faces and cleavages is shown in fig. 12.9. The poles to the inclined faces are plotted by measuring their interfacial angles $\widehat{a\ b}$ and $\widehat{a\ g}$, etc., from the z axis at the centre of the stereogram. The poles to the vertical side faces such as **p** and **q** are plotted by measuring the interfacial angle $\widehat{p\ q}/2$ along the primitive circle from the x axis.

The stereogram shows clearly that the zone axes are also diad symmetry axes and these are used as the crystallographic axes. There are planes of reflection symmetry perpendicular to each diad axis and the crystal also possesses a symmetry centre. The complete elements of symmetry may be summarized by the symbols **3A₂, 3M,C** and the Hermann–Mauguin symbols are **2/m 2/m 2/m**. Barytes belongs to the holosymmetric class of the orthorhombic system.

The common barytes crystals illustrated in fig. 12.1 do not exhibit a face which intersects all three crystallographic axes. Consequently a parametral plane cannot be selected but the axial ratios can be calculated from the other faces.

*What are the intercept ratios of the face **b** if the intercept on the y axis is unity?*

The angle between the face **b** and the y axis is **52° 42′** as shown in fig. 12.6 and the tangent gives the ratio $z/y = \mathbf{1.3127}$. The intercept on the z axis is **1.3127** times the unit intercept on the y axis and it can be regarded as the unit intercept on z.

*What are the ratios of the intercepts of face **g** on the x and z axes in terms of the unit intercept on the y axis?*

The angle between the face **g** and x axis is **38° 51′** as shown in fig. 12.7; therefore the ratio $z/x = \tan \mathbf{38°51′} = 0.8054$. But the intercept on the z axis is **1.3127** times the unit intercept on the y axis, therefore the ratio of the intercept on the x axis is 1.3127/

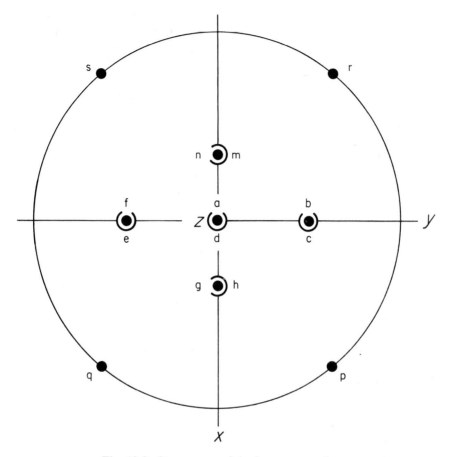

Fig. 12.9. Stereogram of the barytes crystal

$0.8054 = \mathbf{1.629}$. Consequently the ratios of the unit intercepts on the crystallographic axes $x:y:z$ may be defined as $\mathbf{1.629:1:1.3127}$. **These are the axial ratios for barytes**. It is the convention to regard the intercept on the y axis as unity.

The definition of the axial ratios allows the faces to be indexed with simple rational numbers. For example, the angle between face **p** and the y axis is $\mathbf{78°\ 20'/2}$ as shown in fig. 12.8, and therefore the intercept ratio $x/y = \tan\ \mathbf{39°10'} = \mathbf{0.8147}$. The indices of the three faces **b, g, and p** may be obtained by dividing the observed intercept ratios by the axial ratios as shown in tabulated form below.

Face	Intercept Ratios	Intercept Ratios ÷ Axial ratios	Miller indices
b	∞ : 1 : 1.3127	∞ : 1 : 1	(011)
g	1.629 : ∞ : 1.3127	1 : ∞ : 1	(101)
p	0.8146 : 1 : ∞	$\frac{1}{2}$: 1 : ∞	(210)

Clearly if the three zone axes had been labelled in one of the other five ways, different axial ratios and different indices would have been calculated, but these would be equally valid for the description of the orientation of the faces of the barytes crystal. The orientation used above is the one which is generally accepted because it is related to the unit cell of the barytes structure determined by X-ray studies. The dimensions of the orthorhombic unit cell of barytes are $a = 8.878$ Å, $b = 5.450$ Å, $c = 7.152$ Å and these values have the ratios, $a:b:c = 1.6289:1:1.3122$.

12.3 The forms characteristic of the orthorhombic system

Crystals which are characterized by the presence of three mutually perpendicular diad axes belong to the orthorhombic system. The crystallographic axes are placed parallel to the diad axes and they are sometimes given the following names, x = brachy axis, y = macro axis, and z = vertical axis. Forms which intersect all three crystallographic axis, e.g. {111} consist of eight faces and are known as the **pyramid** forms.

Because the intercepts on the horizontal axes are unequal, the forms cutting one horizontal axis and the vertical axis consist of 4 faces. These forms are called **domes** and they are named after the horizontal axis to which they are parallel.

{101} . . . . . Macro-dome ‖ macro-axis y
(4 faces).
{011} . . . . . Brachy-dome ‖ brachy-axis x
(4 faces).

The forms which intersect the two horizontal axes and are parallel to the vertical axis are called **prism** forms and they consist of four faces. The {210} form of barytes is a prism form and the most prominent cleavage planes in the crystal are parallel to this form.

Forms which intersect only one axis con-

sist of only two faces because the unit intercepts are different on all 3 axes of the orthorhombic system. These forms are called **pinacoids** and they are named after one of the axes to which they are parallel or the base.

{100} . . . . . Macro-pinacoid ‖ z and y
(2 faces).
{010} . . . . . Brachy-pinacoid ‖ z and x
(2 faces).
{001} . . . . . Basal pinacoid ‖ x and y
(2 faces).

The prominent faces **a** and **d** of the barytes crystals are regarded as the basal pinacoid form {**001**} and a good cleavage occurs parallel to this basal plane.

12.4 The dome forms of barytes

Some crystals of barytes have a series of dome faces parallel to the y axis as shown in fig. 12.10. An oblique drawing of the dome faces is shown in fig. 12.11 and the dome face **g** already studied is indicated by shading, and its intercept ratios on x and z. Fig.

Fig. 12.10. Dome faces on a barytes crystal ($\times$2)

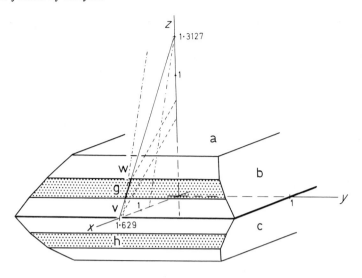

Fig. 12.11. Oblique diagram to show the orientation of dome faces which intersect the *x* and *z* axes

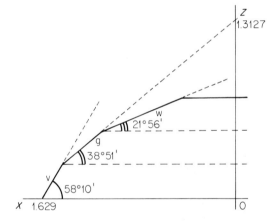

Fig. 12.12. Section of a barytes crystal to show the interfacial angles of the dome faces with the basal pinacoid (001)

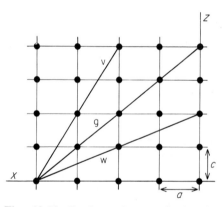

Fig. 12.13. Section of an orthorhombic space lattice to show the angular relationships of the dome faces of barytes

12.12 is a section showing the interfacial angles between the dome faces v,g,w with the basal pinacoid (001). What are the intercept ratios z/x of the three dome faces?

The tangents of the interfacial angles with the (001) face give the ratios z/x as shown in tabulated form below.

This demonstrates the **Law of Rational**

Face	Interfacial angle with base (001)	Tangent = z/x	Intercept ratios on z
w	21° 56′	0.4027	1
g	38° 51′	0.8054	2
v	58° 10′	1.609	4

Ratios of Intercepts. The intercepts of the dome faces on the *z* axis are related to the repeat distance *c* of the orthorhombic unit cell in the barytes structure as illustrated by fig. 12.13. *What are the Miller indices of the dome faces* **w** *and* **v** *if* **g** *is regarded as* (**101**)?

Since the intercept ratios of faces **w** and **v** are $\frac{1}{2}$ and **2** of the intercept made by face **g**, the Miller indices are as follows:

5. *Can the axes of the zones which contain the faces labelled* $\mathbf{p_{1-8}}$ *be located on the stereogram?*

6. *Which elements of symmetry are shown by the olivine crystal?*

The observations that should have been made in the first part of the study of the olivine crystal can be checked by reading section 12.6.

Face	Intercept ratios	Reciprocals	Miller indices
w	$1:\infty:\frac{1}{2}$	$1:0:2$	(**102**)
g	$1:\infty:1$	$1:0:1$	(**101**)
v	$1:\infty:2$	$1:0:\frac{1}{2}$	(**201**)

12.5 The crystals of olivine, $(Mg,Fe)_2SiO_4$

Olivine is an important mineral in certain igneous rocks and its characteristic green colour is shown in plate 13. Crystals of olivine are not common but typical habits of the crystals are shown in figs. 12.14 and 12.15 and the faces are labelled with letters or words such as top, front, etc. Plans which are parallel to the top, front, and side faces of the olivine crystal are shown in figs. 12.16, 12.17, and 12.18.

Study the diagrams and if possible models of olivine crystals with the objective of making observations so that the questions listed below can be answered. Read all the questions first and then carry out the first part of the study of the olivine crystal by answering the first six questions.

1. *What are the sequences of faces in the main zones exhibited by the olivine crystal?*
2. *What are the approximate interfacial angles between the faces which constitute the three zones?*
3. *Prepare a stereogram to show the orientation of the faces with the exception of those labelled* $\mathbf{p_{1-8}}$.
4. *Can the poles to the faces labelled* $\mathbf{p_{1-8}}$ *be located by drawing on the stereograms the zones within which these faces occur?*

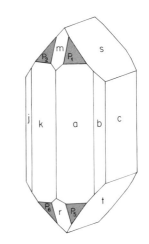

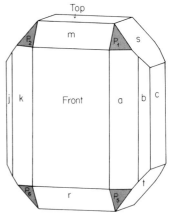

Fig. 12.14 and 12.15. Drawings of characteristic habits exhibited by olivine crystals

The next steps in the study of the olivine crystal are indicated in the questions listed below.

7. *Which face on the olivine crystal can be selected to define the parametral plane, and what are the axial ratios of olivine which can be determined from this crystal face? The procedure has been explained in section 11.6 and the answer to this question is given in section 12.7.*

8. *What are the indices and names of all the forms exhibited by the olivine crystal represented by the stereogram, fig. 12.19?* The answers to this series of problems will be found in section 12.8.

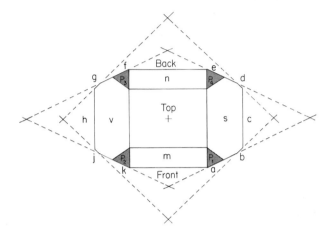

Fig. 12.16. Plan of the olivine crystal parallel to the top face

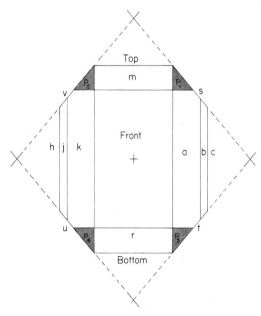

Fig. 12.17. Plan of the olivine crystal parallel to the front face

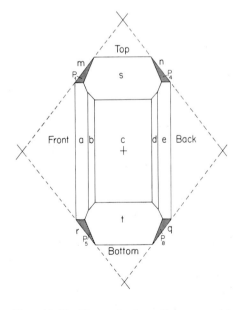

Fig. 12.18. Plan of the olivine crystal parallel to the side face

12.6 The symmetry of the olivine crystal

The most prominent edges of the olivine crystal shown in fig. 12.15 are vertical and they are produced by a zone which is composed of the sequence of faces best seen in fig. 12.16.

Zone 1: front, a,b,c,d,e, back, f,g,h,j,k. This can be called zone 1 and since the zone axis is perpendicular to fig. 12.16. estimates of the interfacial angles can be obtained and are listed below.

front $\widehat{~}$ a	25°		back $\widehat{~}$ f	25°
a $\widehat{~}$ b	18°		f $\widehat{~}$ g	18°
b $\widehat{~}$ c	47°		g $\widehat{~}$ h	47°
c $\widehat{~}$ d	47°		h $\widehat{~}$ j	47°
d $\widehat{~}$ e	18°		j $\widehat{~}$ k	18°
e $\widehat{~}$ back	25°		k $\widehat{~}$ front	25°

On the stereogram, fig. 12. 19 the poles to the faces of **zone 1** have been plotted on the primitive circle so that the zone axis is· perpendicular to the stereogram.

The second prominent zone exhibited by the crystal shown in fig. 12.15 produces horizontal edges due to the intersection of the following sequence of faces which are best seen in fig. 12.18.

Zone 2: front, m, top, n, back, q, bottom, r. The zone can be called zone 2 and estimates of the interfacial angles can be obtained from fig. 12.18, which is perpendicular to the zone axis. The interfacial angles of zone 2 are listed below.

front $\widehat{~}$ m	38° 30′
m $\widehat{~}$ top	51° 30′
top $\widehat{~}$ n	51° 30′
n $\widehat{~}$ back	38° 30′
back $\widehat{~}$ q	38° 30′
q $\widehat{~}$ bottom	51° 30′
bottom $\widehat{~}$ r	51° 30′
r $\widehat{~}$ front	38° 30′

On the stereogram fig. 12.19 the poles of the faces in zone 2 are plotted on the NS diameter and the zone axis is parallel to the EW diameter.

The third prominent zone of faces produces edges which are perpendicular to the front face. The sequence of faces composing zone 3 are best seen in fig. 12.17 because the zone axis is perpendicular to the diagram. **Zone 3: top, s,c,t, bottom, u,h,v.** The interfacial angles can be measured from fig. 12.17 and they are listed below.

Top $\widehat{~}$ s	49° 30′
s $\widehat{~}$ c	40° 30′
c $\widehat{~}$ t	40° 30′
t $\widehat{~}$ bottom	49° 30′
bottom $\widehat{~}$ u	49° 30′
u $\widehat{~}$ h	40° 30′
h $\widehat{~}$ v	40° 30′
v $\widehat{~}$ top	49° 30′

On the stereogram fig. 12.19 the poles to the faces of zone 3 are plotted on the EW diameter and the zone axis is parallel to the NS diameter.

The three zone axes which are at right angles to each other are selected as crystallographic axes and have been labelled *x, y,* and *z* on the stereogram fig. 12.19.

The stereogram can be completed by plotting the orientations of the eight faces which compose the form labelled **p**. It can be seen from fig. 12.15. that the edges between the faces p_1, **m**, and p_2 are parallel, and they lie in the plane which contains the *x* and *z* axes. Consequently the side faces *c* and *h* also occur with p_1 and p_2 in zone 4. It can be seen from figs. 12.18 and 12.19 that zone 4 consists of the faces

Zone 4: p_1, m, p_2, h, p_7, q, p_8, c.

Zone 4 is represented by the great circle on the stereogram fig. 12.19. The full line represents the zone on the upper hemisphere and the broken line represents the zone on the lower hemisphere.

Clearly there is another zone, number 5, which is composed of the faces,

Zone 5: p_3, n, p_4, c, p_5, r, p_6, h.

On the upper hemisphere of the stereogram zone 5 is represented by the broken

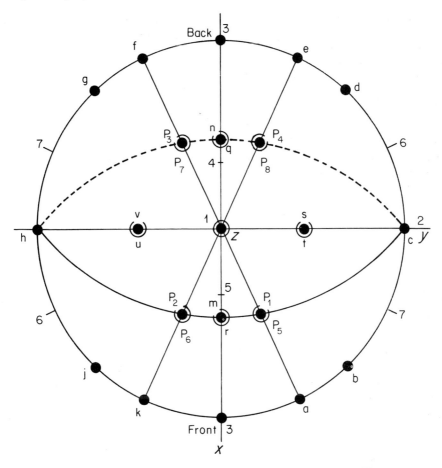

Fig. 12.19. Stereogram of the poles to the faces of the olivine crystal. The zones are shown by great circles and the zone axes are numbered

line and on the lower hemisphere it is represented by the full line. The axes of zones 4 and 5 can be located by measuring 90° along the NS diameter from the point where the zone great circles intersect the diameter. Zone axes 4 and 5 are symmetrically arranged with respect to the z axis (Zone 1).

It can be seen from fig. 12.15 that the edges between the faces p_1, a, p_5 are parallel and horizontal so that they are also parallel to the top and bottom faces. Consequently two additional zones can be identified and great circles drawn on the stereogram.

Zone 6: p_1, top, p_3, f, p_7, bottom, p_5, a.

Zone 7: p_2, top, p_4, e, p_8, bottom, p_6, k.

Since each of the faces labelled **p** occurs on two zones, the poles to the faces are located at the intersections of the appropriate zones as shown in fig. 12.19.

All the face poles have been plotted on the stereogram. It can be seen from the stereogram and the three plans of the crystal which are each perpendicular to a crystallographic axis, that these zone axes are diad axes of symmetry. It is also obvious that there are planes of reflection symmetry perpendicular to each diad axis, and the stereogram indicates that the crystal has a symmetry centre. The olivine crystal belongs to the holosymmetric class of the orthorhombic system and is described by the symbols **2/m, 2/m, 2/m.**

The study of the olivine crystal can be developed by undertaking the tasks involved in questions 7 and 8, which were presented in section 12.5.

12.7 The parametral plane of the olivine crystal

The next step in the study of a crystal is the selection if possible, of a face which intersects all three crystallographic axes, and such a face occurs on the olivine crystal studied here. Face p_1 when extended intersects all three crystallographic axes and there are 8 similar faces which are numbered 1 to 8 and shaded in the figures. The eight faces compose a pyramid form, and if face p_1 is selected as the **parametral plane** it has indices **(111)** and the form **{111}** may be called the **unit form.** Calculate the axial ratios from face p_1 on the stereogram, fig. 12.19.

It will be recalled from fig. 12.20 and the explanation in section 11.6 that the intercepts for the parametral plane are given by

$$OX = \frac{Op}{\cos X\hat{O}p}$$

$$OY = \frac{Op}{\cos Y\hat{O}p}$$

$$OZ = \frac{OP}{\cos Z\hat{O}p}$$

The axial ratios are expressed as

$$\frac{OX}{OY} : \frac{OY}{OY} : \frac{OZ}{OY}$$

Therefore $\dfrac{\cos Y\hat{O}p}{\cos X\hat{O}p} : 1 : \dfrac{\cos Y\hat{O}p}{\cos Z\hat{O}p}$

The angles between the pole to the parametral plane and the crystallographic axes

may be determined from the stereogram by placing pole p_1 on great circles containing each of the crystallographic axes in turn as shown in fig. 12.21.

	XÔp		YÔp		ZÔp
Angle	43°		70°		54° 20′
Cosine	0.7314		0.3420		0.5840
Axial ratios	0.4675	:	1	:	0.5856

The axial ratios define the ratios of the unit intercepts on the three crystallographic axes.

X-ray studies indicate that the olivine structure has an orthorhombic unit cell with the dimensions $a = $ **4.76 Å,** $b = $ **10.20 Å** and $c = $ **5.98 Å.** The ratios of the dimensions of the unit cell are **0.4666 : 1 : 0.5862.** The orientation of the olivine crystal illustrated in this section was selected so that the axial ratios calculated from the parametral plane would be approximately equal to the ratios of the dimensions of the orthorhombic unit cell. If one of the other five possible orientations of the zone axes, 1, 2, and 3 had been selected, the axial ratios would have different values, but the indices defined from them would be simple rational numbers.

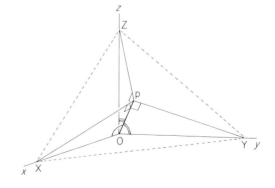

Fig. 12.20. The angles between the normal to the parametral plane and the crystallographic axes

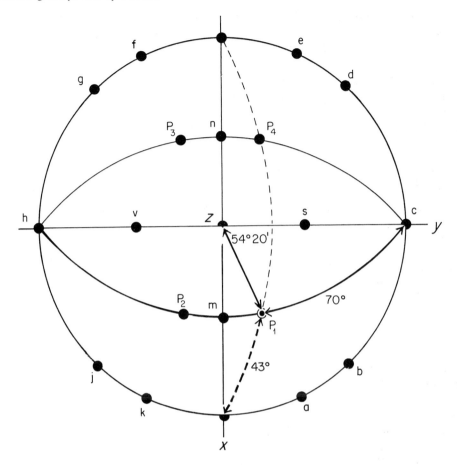

Fig. 12.21. Stereogram showing the angles between the normal to the parametral plane and the crystallographic axes. The poles to the upper faces only are shown

12.8 Indexing the forms of the olivine crystal

Simple rational indices for the faces are derived by dividing the intercept ratios of each face by the axial ratios as defined by the parametral plane. The determinations of the indices of the forms are shown in tabulated form below and are explained in the succeeding paragraphs.

The front and back faces are parallel to the y and z axes so these indices will be (100) and ($\bar{1}$00) respectively. The form has the indices {100} and is called the front or macro-pinacoid. The y axis is called the macro axis.

The side faces c and h are also pinacoids because they are parallel to the z and x axes. They are called the side or brachy-pinacoid form {010}, after the x axis.

The top and bottom faces compose the basal pinacoid with indices {001}.

The faces a,e,f,k are all parallel to the z axis and it can been seen from fig. 12.16 that the intercept ratio x/y = tan 25° = 0.4663. Consequently these faces make up a prism form which has the indices {110}.

The faces b,d,g,j also make up a prism form, but the intercept ratio x/y = tan 43° = 0.9325, fig. 12.16. When divided by the axial ratio this gives an intercept of

Form	Face	Intercept ratios	Intercept ratio ÷ Axial ratio	Miller indices of form	Form name
p_1 to p_8	p_1	$0.4675 : 1 : 0.5856$	$1 : 1 : 1$	$\{111\}$	Pyramid (Unit form)
Front, back	Front	$0.4675 : \infty : \infty$	$1 : \infty : \infty$	$\{100\}$	Pinacoid
c,h	c	$\infty : 1 : \infty$	$\infty : 1 : \infty$	$\{010\}$	Pinacoid
Top, bottom	Top	$\infty : \infty : 0.5856$	$\infty : \infty : 1$	$\{001\}$	Basal pinacoid
a,e,f,k	a	$0.4663 : 1 : \infty$	$1 : 1 : \infty$	$\{110\}$	Prism
b,d,g,j	b	$0.9325 : 1 : \infty$	$2 : 1 : \infty$	$\{120\}$	Prism
m,n,q,r	m	$0.4657 : \infty : 0.5856$	$1 : \infty : 1$	$\{101\}$	Dome
s,t,u,v	s	$\infty : 1 : 1.1708$	$\infty : 1 : 2$	$\{021\}$	Dome

approximately 2 on the x axis. The intercepts ratios are $2:1:\infty$ and consequently the indices of this prism form are $\{120\}$.

The faces **m,n,q,r** are parallel to the y axis and it can be seen from fig. 12.18 that for face **m**, $\tan 38° 30' = 0.7954 = x/z$. Since tha axial ratios indicate that the intercept on z is 0.5856, the intercept of **m** on the x axis is **0.4657**. Consequently the intercept ratios of face **m** can be regarded as $1:\infty:1$ and the indices are (**101**). This form is called a **dome** and because it is parallel to the macro axis y it is sometimes called a **macro-dome.**

The faces **s,t,u,v** also make up a **dome** form but they are parallel to the x axis so they are sometimes called a **brachydome**. It can be seen in fig. 12.17 that the ratio of the intercepts of face **s**, $z/y = \tan 49°$ $30' = $ **1.1708**. Since the axial ratio $z = $ **0.5856**, it is concluded that the intercepts of face **s** are ∞, 1, 2 and the indices of this dome form are $\{021\}$.

This study of the olivine crystal commenced with a drawing representing a geometrically symmetrical crystal, and it was easy to recognize the elements of symmetry and select the diad axes of symmetry as the crystallographic axes. **With real crystals it is not always easy to recognize the symmetry and the procedure adopted is as follows.**

1. Recognize zones by the presence of sets of parallel edges.
2. Measure interfacial angles of the faces in the zones.
3. Plot a stereogram by placing one zone axis perpendicular to the stereogram so that the poles to the faces plot on the primitive circle.
4. Rotate about zone axis 1 to determine the presence of planes of symmetry parallel to zone axis 1.
5. Plot the poles to faces in any other zones.
6. If possible select zone axes to act as crystallographic axes. It may be necessary to rotate the poles on the stereogram to obtain a different orientation.
7. Select a parametral plane and determine the axial ratios.
8. Index the forms exhibited by the crystal.

13 The Crystals of Orthoclase, Hornblende, and Augite from the Monoclinic System

Orthoclase, hornblende, and augite are members of three very important groups of rock forming minerals, which are known as the feldspars, amphiboles, and pyroxenes respectively. The principal objectives of this chapter are as follows.

1. To become familiar with the characteristic crystal forms of the three important rock forming minerals, orthoclase, hornblende, and augite.
2. To determine from specimens or the plans of the crystals of orthoclase, hornblende, and augite, estimates of the interfacial angles in order to prepare stereograms.
3. To become familiar with the main characteristics of the monoclinic system by studying the symmetry of these three crystals.
4. To use the stereograms to calculate the axial ratios of orthoclase, hornblende, and augite and to index the forms exhibited by the crystals.

Contents of the sections

13.1 The crystals of orthoclase feldspar

Orthoclase feldspar has two prominent planes of cleavage which are perpendicular to each other and the cleavages are illustrated in fig. 1.18. Simple crystals of orthoclase seen from four different views are shown in fig. 13.1.

A crystal showing additional faces which are lettered is represented by the oblique drawing in fig. 13.2. This is the orientation which is usually adopted so that the z axis will be placed parallel to the vertical edges. Plans of this crystal of orthoclase from which estimates of the interfacial angles may be obtained are shown in figs. 13.3, 13.4, and 13.5. *Study these diagrams and if possible specimens of orthoclase and determine the following.*

1. *The sequence of faces in any zones exhibited by the crystal of orthoclase illustrated in the diagrams.*
2. *The approximate interfacial angles of the faces in each zone.*
3. *The elements of symmetry exhibited by orthoclase.*

Construct a stereogram to illustrate the orientation of the faces and the symmetry of the crystal. The prominent vertical zone axis shown in fig. 13.2 should be placed vertical and the horizontal zone axis should be placed parallel to the EW diameter of the stereogram. It will be sufficient to study in addition the zone which contains the faces lettered **g** and **b**.

The observations made above may be checked by reading section 13.2, and then the study of the orthoclase will be developed further.

Fig. 13.1. Crystals of **orthoclase feldspar** seen from four different directions.
(a) oblique view,
(b) view parallel to the vertical edges,
(c) view perpendicular to the side faces,
(d) view parallel to the inclined edges (×2)

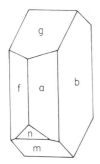

Fig. 13.2. An obli-
que drawing of an
orthoclase crystal
showing the orienta-
tion in which it is
usually studied

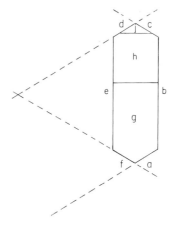

Fig. 13.3. Plan of the orthoclase crystal on a plane perpendicular to the vertical
edges

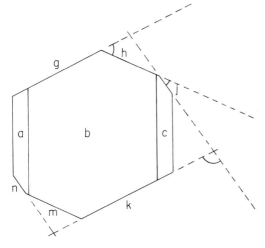

Fig. 13.4. Plan of the orthoclase crystal parallel
to the largest crystal faces **b** and **e**

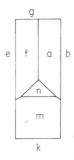

Fig. 13.5. Plan of the
orthoclase crystal on
a plane perpendicu-
lar to the forward in-
clined edges

$\widehat{a\ b}$	59° 22'
$\widehat{b\ c}$	59° 22'
$\widehat{c\ d}$	61° 16'
$\widehat{d\ e}$	59° 22'
$\widehat{e\ f}$	59° 22'
$\widehat{f\ a}$	61° 16'

13.2 The symmetry of the orthoclase crystal

The most prominent edges on the crystal
illustrated in fig. 13.2 are vertical and the
faces which produce these edges can be
regarded as zone **1**. Fig. 13.3 is a plan of the
orthoclase crystal perpendicular to the **zone
1** axis, the faces composing the zone are
a,b,c,d,e,f. The interfacial angles of zone **1**
are listed below with greater accuracy than
can be determined from fig. 13.3.

It is appropriate to construct the stereogram with the **zone 1** axis vertical so that the
poles to the zone faces fall on the primitive
circle as shown in fig. 13.6. The plan, fig.

13.3 indicated that there is a plane of symmetry which is parallel to the largest faces **b** and **e**, and it is convenient to place this plane on the NS diameter of the stereogram. It can seen from figs. 13.2 and 13.4 that the axis of **zone 1** is not an axis of symmetry. The **zone 1** axis is selected to be the *z* crystallographic axis.

The top and bottom edges of the crystal shown in fig. 13.2 indicate the presence of a second zone with a horizontal zone axis. Fig. 13.4 is a plan perpendicular to the **zone 2** axis and it can be seen that **zone 2** is composed of the faces **g,h,j,k,m,n**. The axis of **zone 2** is perpendicular to the reflection plane and the **zone 1** axis. Estimates of the

interfacial angles of zone 2 can be obtained from fig. 13.4 and the values are listed below.

$\widehat{g\,h}$	50°
$\widehat{h\,j}$	30°
$\widehat{j\,k}$	100°
$\widehat{k\,m}$	50°
$\widehat{m\,n}$	30°
$\widehat{n\,g}$	100°

The **zone 2** axis is horizontal and perpendicular to face **b** so it lies parallel to the EW diameter of the stereogram as suggested in the directions given in section 13.1.

The poles to faces **g,h,j** fall on the upper

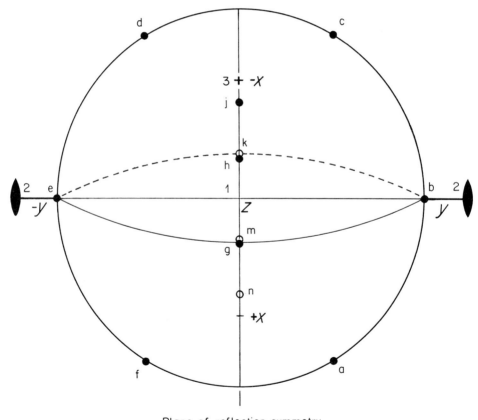

Plane of reflection symmetry

Fig. 13.6. Stereogram of the orthoclase crystal. The zone axes are indicated by numbers. The +*x* axis plots on the lower hemisphere

hemisphere as shown in fig. 13.7, which is a section parallel to the plane of reflection symmetry. The poles **g,h,j** are plotted on the NS diameter of the stereogram by projection to the pole of the lower hemisphere.

The faces **k,m,n** which occur on the lower side of the orthoclase crystal have poles on the lower hemisphere, and these are plotted on the stereogram by projection to the pole of the upper hemisphere as shown by the broken lines in fig. 13.7.

On the stereogram fig. 13.6 the poles to the faces **g,h,j** are plotted as full circles, and the poles to the lower faces **k,m,n** are plotted as open circles.

The **zone 2 axis** is a **diad axis** of symmetry and it is perpendicular to the plane of reflection symmetry. The diad axis of the monoclinic system is always selected as the *y* crystallographic axis.

It can be seen in fig. 13.2 and 13.4 that there are prominent edges which are inclined towards the front of the crystal due to the intersection of faces **b,g,e,k** which can be regarded as **zone 3**. The interfacial angles of zone 3 are all 90° and a great circle has

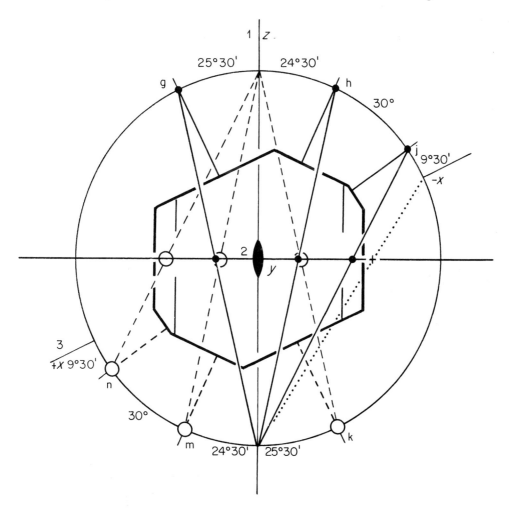

Fig. 13.7. Section of the sphere of projection parallel to the plane of reflection symmetry. The zone axes are indicated by numbers

been drawn on the stereogram, fig. 13.6, to link the poles **b,g,e,k.**

The **zone 3 axis** has been plotted by measuring 90° along the reflection plane and this can be used as the x crystallographic axis. The angle β between the vertical z crystallographic axis and the front end of the inclined x axis is 116°. The name **monoclinic** refers to the single inclined crystallographic axis. The elements of symmetry of ortho-clase feldspar consist of **one diad axis** (zone 2 axis) **one plane of reflection symmetry** which is perpendicular to the diad axis, and a **symmetry centre**. The occurrence of a single diad axis indicates that orthoclase belongs to the monoclinic system and within this system it is in the class of highest symmetry indicated by the Hermann–Mauguin symbol **2/m.**

13.3 The selection of crystallographic axes in the monoclinic system

In crystals of the monoclinic system the diad symmetry axis is selected as the y crystallographic axis. The most prominent zone axis in the plane of reflection symmetry, for example zone axis 1 in orthoclase is selected as the vertical z crystallographic axis. The x crystallographic axis is arranged in the plane of reflection symmetry but it is not perpendicular to the z axis. It is the convention to arrange the crystal so that the obtuse angle β occurs between the positive ends of the x and z axes.

Note the positions of the crystallographic axes on the stereogram, fig. 13.6. The negative end of the x axis plots on the upper hemisphere, and the positive end plots on the lower hemisphere as shown in the section fig. 13.7.

13.4 Indexing the faces on the orthoclase crystal

The orthoclase crystal shown in fig. 13.2 does not have a face which intersects all

three crystallographic axes so that it can be used as a parametral plane. However the axial ratios can be calculated if face **a** is regarded as the **(110)** face and face **h** is indexed $(\bar{1}01)$.

Estimates of the angles between the normals to face **a** and the crystallographic axes can be obtained from the stereogram fig. 13.6 by placing pole **a** and each axis in turn on a great circle, as shown in fig. 13.8.

	$+X\hat{O}a$	$Y\hat{O}a$	$Z\hat{O}a$
Angles	39° 20′	59° 26′	90°
Cosines	0.7742	0.5085	0
Intercept ratios	$\dfrac{\cos Y\hat{O}a}{\cos X\hat{O}a}$	$\dfrac{\cos Y\hat{O}a}{\cos Y\hat{O}a}$	$\dfrac{\cos Y\hat{O}a}{\cos Z\hat{O}a}$
	$\dfrac{0.5085}{0.7742}$	$\dfrac{0.5085}{0.5085}$	$\dfrac{0.5085}{0}$
	0.6568	1	∞

The angles between the pole to face **h** and the crystallographic axes are shown in fig. 13.8.

	$-X\hat{O}h$	$Y\hat{O}h$	$Z\hat{O}h$
Angles	40°	90°	24° 30′
Cosines	0.7660	0	0.9100
Intercept ratios	$\dfrac{\cos Y\hat{O}h}{\cos X\hat{O}h}$	1	$\dfrac{\cos Y\hat{O}h}{\cos Z\hat{O}h}$

$$\therefore \text{ Intercept ratio } z/x = \frac{\cos X\hat{O}h}{\cos Z\hat{O}h} = \frac{0.7660}{0.9100}$$

$$= 0.8417$$

Since $x = 0.6568$ $z = 0.5528$

Axial ratios $x:y::z = \textbf{0.6568:1:0.5528}$

X-ray studies indicate that the dimensions of the monoclinic unit cell of orthoclase are $a = 8.56$ Å, $b = 13.0$ Å, $c = 7.19$ Å. The ratios of the repeat distances calculated from these dimensions are $a:b: c = \textbf{0.6584 : 1 : 0.5530}$. The selection of the faces on the orthoclase was made so that the axial ratios would correspond with the dimensions of the unit cell. *What are the indices of the face* **j**?

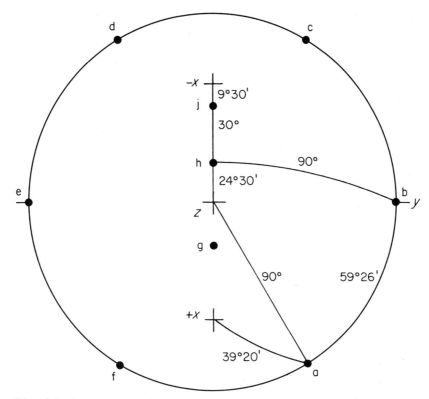

Fig. 13.8. Stereogram to show the angles between the crystallographic axes and the faces **a** and **h**

	-XÔj	YÔj	ZÔj
Angles	9° 30'	90°	54° 30'
Cosines	0.9863	0	0.5807

Intercept ratio z/x = Cos-XÔj/cos ZÔj = 1.6819.

Since x = 0.6568 the intercept ratios of **j** are:
$x : y : z = -0.6568 : \infty : 1.1047$.

After division by the axial ratios the intercept ratios of face **j** are $-1 : \infty : 2$ and the indices will be ($\bar{2}01$).

The crystallographic axes in the monoclinic system are sometimes given the names **clino axis** for the inclined x axis, and **ortho axis** for the y axis, which is also the diad symmetry axis. The faces exhibited by the orthoclase crystal in fig. 13.2 belong to the forms tabulated below.

Because of the inclination of the x axis, the ortho-domes occur as two pairs of similar faces—one pair occupying the acute angle between x and z, the other pair occupying the obtuse angle between x and z.

Form	Face	Intercept ratios Axial ratio	Indices of form	Name
b,e	b	$1 : \infty : \infty$	{100}	Clino pinacoid
g,k	g	$\infty : \infty : 1$	{001}	Basal pinacoid
a,c,d,f	a	$1 : 1 : \infty$	{110}	Prism
h,m	h	$-1 : \infty : 1$	{$\bar{1}$01}	Ortho-dome
j,n	j	$-1 : \infty : 2$	{$\bar{2}$01}	Ortho-dome

The pairs are called **hemi-orthodomes** and in the orthoclase crystal studied here, there are two hemi-orthodome forms ($\bar{1}01$) and ($\bar{2}01$), which occur in the acute angle between the z and $-x$ axes.

There are no pyramid faces represented in the orthoclase crystal of fig. 13.6, but clearly in monoclinic crystals the 8 pyramid faces will occur as two sets of 4 faces. Each form is called a **hemi-pyramid**, and one will consist of the four faces in the acute angles between x and z, while the other form consists of the four faces in the obtuse angles between x and z.

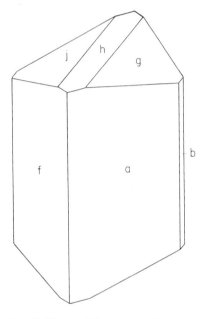

Fig. 13.10. A oblique view of a crystal of hornblende placed in the position in which it is usually studied

Fig. 13.9. A crystal of **hornblende** seen from two directions. Note the well developed cleavage planes which are parallel to the faces which produce the diamond shaped cross section of the crystal ($\times 2$)

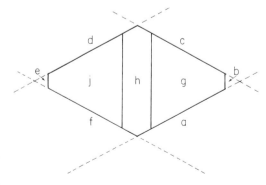

Fig. 13.11. Plan of the crystal of hornblende on a plane perpendicular to the vertical edges

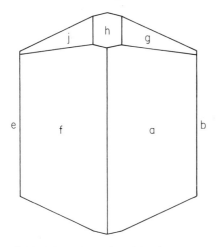

Fig. 13.12. Plan of the hornblende crystal on a plane parallel to the vertical edges and perpendicular to the faces **b** and **e**

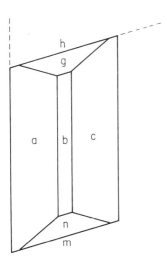

Fig. 13.13. Plan of the horn-blende crystal on a plane parallel to the faces **b** and **e**

edges are placed vertical as in the oblique drawing, fig. 13.10.

This hornblende crystal can be studied in a way similar to that used for the study of the orthoclase crystal, and consequently this provides an opportunity to recall and apply the procedures.

Plans of the hornblende crystal are presented in figs. 13.11., to 13.14 so that estimates of the interfacial angles may be obtained. The axial ratios of the crystal may be determined from the faces **a** *indexed* **(110)** *and* **g** *indexed* **(011)**.

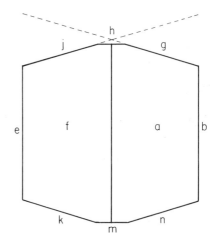

Fig. 13.14. Plane of the hornblende crystal on a plane perpendicular to the inclined edges

13.6 The symmetry of the hornblende crystal

The first tasks in the study of the hornblende crystal involve the recognition of zones and the measurement of the interfacial angles in the zones.

The most prominent edges displayed by the hornblende crystal are produced by the faces **a,b,c,d,e,f** which are seen in fig. 13.11. These faces make up the vertical **zone 1**, and

13.5 The crystal of hornblende

A crystal of hornblende exhibiting the characteristic diamond-shaped cross-section and the two sets of cleavage planes is shown in fig. 13.9. Hornblende crystals usually have a prismatic, bladed habit and the prominent

its axis can be regarded as the *z* crystallographic axis. The interfacial angles of **zone 1**
are

$\widehat{a\ b}$	62°12′
$\widehat{b\ c}$	62° 12′
$\widehat{c\ d}$	55° 36′
$\widehat{d\ e}$	62° 12′
$\widehat{e\ f}$	62° 12′
$\widehat{f\ a}$	55° 36′

It is clear from the interfacial angles and
figs. 13.11 and 13.12. that there is a plane of
symmetry parallel to faces **b** and **e**. When
constructing the stereogram, zone 1 is
arranged with the poles of face **b** and **e** at the
ends of the EW diameter so that the reflection plane is parallel to the NS diameter as
shown in fig. 13.15.

Fig. 13.13 is a plan of the hornblende

crystal on a plane parallel to face **b** and the
reflection plane of symmetry. It is clear
from this diagram that there is a diad axis of
symmetry perpendicular to the reflection
plane as in orthoclase, but in the hornblende
crystal shown in fig. 13.10 there are no
edges parallel to the diad axis of symmetry.
It is the convention to select the diad axis of
monoclinic crystals as the **y** crystallographic
axis.

The faces **h** and **m** are perpendicular to
the side faces **b** and **e**. Consequently the
pole to face **h** lies on the NS diameter of the
stereogram. The normal to face **h** is 15°30′
from the *z* axis and the pole to the face can
be located by measuring from the *z* axis. The
pole to the lower face **m** can be plotted on
the stereogram in a similar way but it falls
on the lower hemisphere and is indicated by
an open circle.

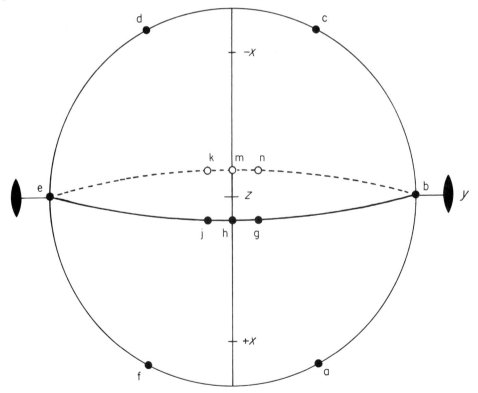

Fig. 13.15. Stereogram of the faces displayed by the hornblende crystal. +*x* plots on the lower
hemisphere

The prominent edges which are parallel to the symmetry plane but inclined to the z axis are produced by the intersection of faces **b,g,h,j,e,k,m,n** as shown in fig. 13.14. The interfacial angles of the inclined zone can be measured on fig. 13.14 and are listed below.

$\widehat{b\,g}$	74°08′
$\widehat{g\,h}$	15°32′
$\widehat{h\,j}$	15°32′
$\widehat{j\,e}$	74°08′
$\widehat{e\,k}$	74°08′
$\widehat{k\,m}$	15°32′
$\widehat{m\,n}$	15°32′
$\widehat{n\,b}$	74°08′

The next step is to draw on the stereogram the great circle on which the poles to the **b,g,h,j,e,k,m,n** faces fall, and this can be done because the poles of the faces **b,h,e,m** have already been plotted. The poles of the other faces can be located by measuring the interfacial angles $\widehat{g\,h}$, $\widehat{h\,j}$ along the zone great circle.

It is appropriate to select the inclined zone axis as the x crystallographic axis and the approximate value of the angle β between the two zones axes can be obtained from fig. 13.13. The angle β in hornblende is 105°31′. On the stereogram the position of the negative end of the x axis is obtained by measuring 74°30′ upwards from the z axis. Since the positive end of the x axis is 105° 31′ from the z axis it falls on the lower hemisphere, and its position is shown in fig. 13.13.

13.7 The axial ratios of the hornblende crystal

The hornblende crystal illustrated here does not have a face which intersects all three

crystallographic axes, and it has been necessary to regard face **a** as (**110**) and face **g** as (**011**).

The approximate angles between the normals to faces **a** and **g**, and the crystallographic axes can be obtained from the stereogram fig. 13.15, and the calculation of the axial ratios is shown in tabulated form below.

	$X\hat{O}a$	$Y\hat{O}a$	$Z\hat{O}a$
Angles	31°	62°	90°
Cosines	0.8572	0.4695	0
Intercept ratios	$\dfrac{\cos Y\hat{O}a}{\cos X\hat{O}a}$	$\dfrac{\cos Y\hat{O}a}{\cos Y\hat{O}a}$	$\dfrac{\cos Y\hat{O}a}{\cos Z\hat{O}a}$
	0.5477 :	1 :	∞

	$X\hat{O}g$	$Y\hat{O}g$	$Z\hat{O}g$
Angles	90°	74°	22°
Cosines	0 :	0.2756 :	0.9272
Intercept ratios	∞ :	1 :	0.2972

So the calculated axial ratios for hornblende are $x:y:z = $ **0.5477 : 1 : 0.2972**.

X-ray studies show that the dimensions of the monoclinic unit cell of hornblende are $a = $ **9.79 Å**, $b = $ **17.9 Å**, $c = $ **5.28 Å**. These values give the ratios **0.5469 : 1 : 0.2949** which are close to the intercept ratios determined from the crystal faces.

The forms exhibited by the hornblende crystal studied here are listed below.

Form	Indices	Name
b,e	{010}	Clino pinacoid
h,m	{001}	Basal pinacoid
a,c,d,f,	{110}	Prism
g,j,k,n,	{011}	Clino dome

13.8 The crystal of augite

The study of the crystal of augite provides a particularly good opportunity to apply the crystallographic procedures set out at the end of section 12.8. Follow this procedure in order to construct a stereogram of the augite crystal, determine the axial ratios of augite from an appropriate parametral plane, and index the forms exhibited by the crystal.

Crystals of augite seen from different directions are illustrated in fig. 13.16. Fig. 13.17 is an oblique drawing of an augite crystal exhibiting the most common combination of forms.

The single diad axis would be selected as the *y* crystallographic axis and it would be appropriate to select the obvious vertical and inclined zone axes of this particular crystal as the *z* and *x* axes respectively. With this arrangement the face labelled **q** would be regarded as the basal pinacoid form. However, some augite crystals exhibit a pair

of faces such as **s** in fig. 13.18 which represent the basal pinacoid form in relation to the most appropriate unit cell determined by X-ray analysis. Consequently **the inclination of *x* crystallographic axis is usually determined by face s** and it is this conventional orientation which has been used in the illustrations of the augite crystal to be studied here.

Figs 13.19, 13.20, and 13.21 are plans of the augite crystal from which estimates of the interfacial angles may be obtained and the symmetry of the crystal determined. Note that the interfacial angle $a \widehat{\ } s$ is slightly smaller than $m \widehat{\ } e$ but it may be difficult to detect this from fig. 13.20.

The crystallographic observations made from crystals of augite or the diagrams shown here may be checked by reading sections 13.9 and 13.10.

Fig. 13.16. Crystals of **augite** seen from different directions. (a) oblique view, (b) view parallel to the vertical edges. (c) view perpendicular to the side faces (×2)

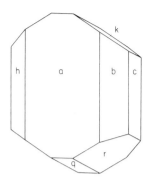

Fig. 13.17. An oblique drawing of an augite crystal showing the forms present in the fig. 13.16 photographs

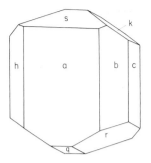

Fig. 13.18. An oblique view of an augite crystal showing an additional form composed of the two faces **s** and **t** which are regarded as the basal pinacoid

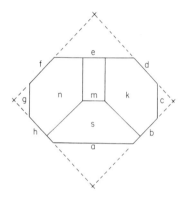

Fig. 13.19. A plan of the augite crystal on a plane perpendicular to the vertical edges

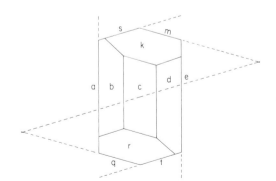

Fig. 13.20. A plan of the augite crystal on a plane parallel to the side face **c**

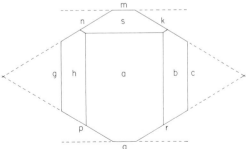

Fig. 13.21. A plan of the augite crystal on a plane perpendicular to the inclined edges such as that between faces **c** and **k**

13.9 The symmetry of the augite crystal

The prominent vertical edges seen on the augite crystal in fig. 13.18 are produced by the intersection of the faces **a,b,c,d,e,f,g,h**. This zone can be called zone 1 and it is the convention to select the zone 1 axis as the *z* crystallographic axis. Estimates of the interfacial angles of zone 1 can be obtained from fig. 13.19 and they are listed below.

$\widehat{a\,b}$	$= 46° 25'$
$\widehat{b\,c}$	$= 43° 35'$
$\widehat{c\,d}$	$= 43° 35'$
$\widehat{d\,e}$	$= 46° 25'$
$\widehat{e\,f}$	$= 46° 25'$
$\widehat{f\,g}$	$= 43° 35'$
$\widehat{g\,h}$	$= 43° 35'$
$\widehat{h\,a}$	$= 46° 25'$

On the stereogram fig. 13.22 the zone 1 axis is placed vertical so that the poles to the zone 1 faces lie of the primitive circle. The symmetry of the interfacial angles and fig. 13.19 suggest that the poles to faces **c** and **g** should be placed on the EW diameter of the stereogram.

It can be seen from fig. 13.18 that there is a set of horizontal edges parallel to the edges between faces **a** and **s**. The faces that produced these horizontal edges can be regarded as zone 2 and they are **a,s,m,e,t,q**. Estimates of the interfacial angles of zone 2 can be obtained from fig. 13.20 and they are listed below.

$\widehat{a\,s}$	$= 74° 10'$
$\widehat{s\,m}$	$= 31° 20'$
$\widehat{m\,e}$	$= 74° 30'$
$\widehat{e\,t}$	$= 74° 10'$
$\widehat{t\,q}$	$= 31° 20'$
$\widehat{q\,a}$	$= 74° 30'$

The zone 2 axis is perpendicular to faces **c** and **g** and it coincides with the EW diameter of the stereogram. It can be seen from the interfacial angles and fig. 13.20 that the zone 2 axis is a **diad symmetry axis** and so it is regarded as the *y* crystallographic axis.

The poles to the faces of zone 2 fall on the NS diameter of the stereogram and they can be plotted by measuring the interfacial angles from pole **a**.

It was stated in section 13.8 that face **s** is usually regarded as (**001**). Consequently the *x* axis must be inclined to the *z* axis and its orientation can be located on the stereogram by measuring 90° from the pole to face **s**. The pole to the basal pinacoid face **s** is plotted on the NS diameter of the stereogram at 15° 50′ from the *z* axis. Consequently the angle $\widehat{x\,z} = \beta = 105° 50'$, and the positive end of the *x* axis is plotted on the lower hemisphere. The negative end of the *x* axis occurs on the upper hemisphere and is 74° 10′ from the *z* axis. The pole to face **t** falls on the lower hemisphere. The four faces **c,s,g,t** make up zone 3 with the zone axis is parallel to the *x* crystallographic axis.

The prominent edges which are inclined to the rear of the augite crystal are produced by the sequence of faces **c,k,m,n,g,p,q,r** shown in fig. 13.21, and this can be called zone 4. Estimates of the interfacial angles of zone 4 may be obtained from fig. 13.21 and they are listed below.

$\widehat{c\,k}$	$= 60°$
$\widehat{k\,m}$	$= 30°$
$\widehat{m\,n}$	$= 30°$
$\widehat{n\,g}$	$= 60°$
$\widehat{g\,p}$	$= 60°$
$\widehat{p\,q}$	$= 30°$
$\widehat{q\,r}$	$= 30°$
$\widehat{r\,c}$	$= 60°$

The poles to faces **m** and **q** occur also in zone 2 and they have been plotted on the NS diameter of the stereogram, fig. 13.22. A great circle can be drawn through the poles **c,m,g,q**, and on the stereogram the full line shows the great circle on the upper hemisphere passing through pole **m**, while the

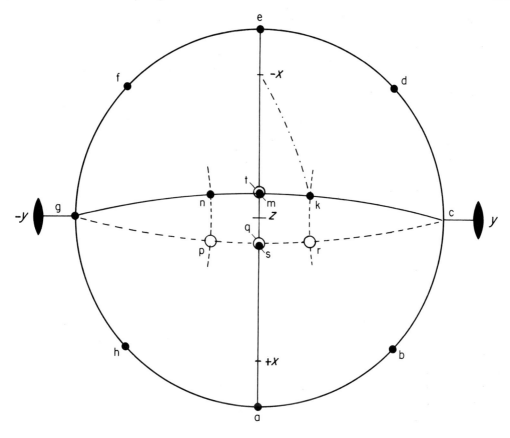

13.22. Stereogram of the augite crystal. +*x* plots on the lower hemisphere

dashed line shows the great circle on the lower hemisphere passing through pole **q**.

The poles to the four faces **k,n,p,r** may be plotted on the great circle by using the interfacial angles from **m** on the upper hemisphere, and **q** on the lower hemisphere.

13.10 The axial ratios and indices of the forms exhibited by the augite crystal

Since the augite crystal exhibits a form of four faces **k,n,p,r** which intersect all three crystallographic axes, one of these faces may be selected as a **parametral plane** to determine the axial ratios. The angles between the normal to face **k** and the crystal-

lographic axes can be obtained from the stereogram and their values are shown in the table below.

Face k	−XÔk	YÔk	ZÔk
Angles	62°	60°	32°
Cosines	0.4695	0.5	0.8480
Axial ratios	1.0649	1	0.5896

X-ray studies indicate that the most appropriate monoclinic unit cell for the augite structure has dimensions of *a* = 9.73 Å, *b* = 8.91 Å, *c* = 5.25 Å. These dimensions give axial ratios of **1.092:1:0.5892**, which are close to the axial ratios calculated from face **k** which can be indexed ($\bar{1}11$).

The indices of the other faces exhibited by the crystal can now be determined. The angles between the crystallographic axes and face **b** can be obtained from the stereogram.

	XÔb	YÔb	ZÔb
Angles	46° 30′	43° 30′	90°
Cosines	0.6884	0.7254	0
Intercept ratios	1.0537 :	1 :	∞

Consequently the indices of face **b** are (**110**).

The indices of face **m** can be determined as follows

	−XÔm	YÔm	ZÔm
Angles	58° 40′	90°	15° 30′
Cosines	0.5190	0	0.9636

Intercept ratio $z/x = 0.5386$.

Since the x intercept ratio has been calculated to be 1.0649 from face **k** the intercept ratios of face **m** are

$$1.0649 : 1 : 0.5386 \times 1.0649$$
$$= 1.0649 : 1 : 0.5735$$

Consequently the indices of face **m** are (**ī01**).

The forms shown by the augite crystal in fig. 13.18 are listed below.

Form	Indices	Name
a,e	{100}	Ortho-pinacoid
c,g	{010}	Clino-pinacoid
s,t	{001}	Basal pinacoid
b,d,f,h	{110}	Prism
m,q	{ī01}	Orthodome
k,n,p,r	{ī11}	Hemi-pyramid

Fig. 13.22 is an oblique view of the augite crystal showing the orientation of the crystallographic axes.

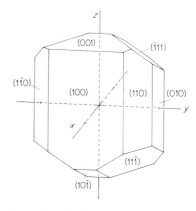

Fig. 13.23. Oblique view of the augite crystal showing the orientation of the crystallographic axes

14 *The Crystal of Albite from the Triclinic System*

Albite is one of the end members of the important rock forming group of minerals know as the plagioclase feldspars. The principal objectives of this short chapter are to become familiar with the main features of the triclinic system and in particular with the crystallographic features of the plagioclase feldspars represented by albite.

Contents of the sections

14.1 The crystal of albite

Albite is one of the end members of the plagioclase feldspars and its composition is $NaAlSi_3O_8$. The crystal of albite shown in fig. 14.1 is very similar to the crystal of orthoclase studied in section 13.1, but in albite the interfacial angle $\widehat{b\ g} = 86°24'$ as shown in fig. 14.2. Cleavages tend to develop parallel to the b (010) and g (001) faces in the feldspars, and the difference in the interfacial angles is the most useful characteristic for distinguishing between crystals of orthoclase and plagioclase feldspar in hand-specimens, since $(010)\widehat{\ }(001) = 90°$ and $86°$ respectively.

The prominent vertical zone of the albite crystal consists of the faces a,b,c,d,e,f, and the interfacial angles are shown in fig. 14.3. Construct a stereogram by plotting the poles on the primitive circle with the pole to face b

(010) at the east end of the EW diameter. The vertical zone axis is selected as the z crystallographic axis.

Zone 2 consists of the faces b,g,e,k and the intersections produce the inclined edges which can be regarded as the direction of the x crystallographic axis.

Plot the pole of face g on the stereogram by using the interfacial angles listed below. Then locate the position of the x crystallographic axis and measure $\widehat{x\ z} = \beta$.

$$\widehat{g\ b} = 86°24'$$
$$\widehat{g\ a} = 65°17'$$
$$\widehat{g\ f} = 69°10'$$

In order to locate the pole to face g, small circles are drawn with radii equal to the

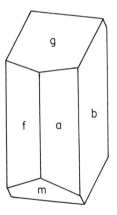

Fig. 14.1. Oblique view of an albite crystal

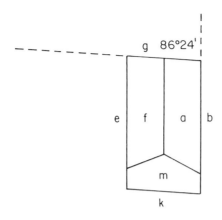

Fig. 14.2. Plan of the albite crystal viewed along the inclined zone axis

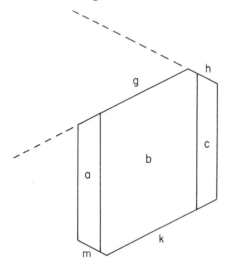

Fig. 14.3. Plan of the albite crystal viewed along the verti-cal zone axis

appropriate interfacial angles from the poles **b,a** and **f**. The three small circles will in-tersect at the position of pole **g** as shown in fig. 14.5. The pole to face **k** is plotted on the lower hemisphere directly opposite pole **g**.

Draw a great circle through the poles **b,g,e,k** and locate the zone axis by measur-ing 90° along the NS diameter of the stereo-gram from the point of intersection of the zone great circle. Since the inclined zone axis *x*, and the vertical zone axis *z*, are both

parallel to face **b**, this face is indexed (**010**), and the *x* and *z* axes will lie on the NS diameter of the stereogram. The negative end of the *x* axis intersects the upper hemi-sphere, and +*x* intersects the lower hemi-sphere as indicated on fig. 14.5. The angle $+x \hat{} z = \beta$ can be measured on the stereo-gram and it will be found to be near 116°35′.

The albite crystal has parallel edges at the top and bottom due to the intersection of the **g,h,k** and *m* faces which can be regarded as **zone 3**. The pole to face *h* can be located on the stereogram from the interfacial angles listed below.

$$\hat{h \, b} = 86°$$
$$\hat{h \, d} = 71°$$
$$\hat{h \, g} = 52°$$

A great circle can be drawn through the poles **g,h,k,m**, of zone 3, and the zone axis can be located as shown in fig. 14.6.

The zone 3 axis can be used as the *y* crystallographic axis. The negative end −*y* intersects the upper hemisphere and +*y* intersects the lower hemisphere. Estimates of the angles between the crystallographic axes may be obtained from the stereogram fig. 14.6. The orientations of the crystallo-graphic axes are shown in fig. 14.7 and the

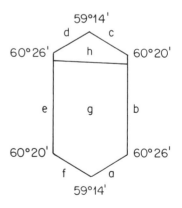

Fig. 14.4. Plan of the albite crystal on a plane perpendicular to the top and bot-tom edges

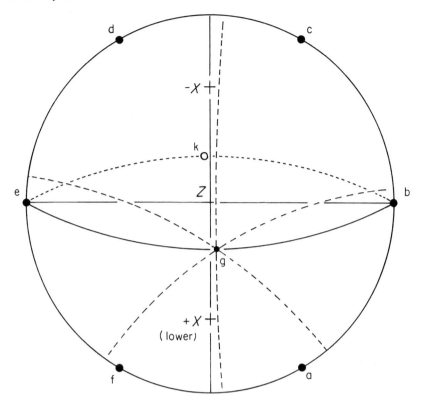

Fig. 14.5 Stereogram of the albite crystal to show the location of face **g** and the great circle through poles **b, g, e,** and **k**. The zone 2 axis is selected as the *x* crystallographic axis and the positive end intersects the lower hemisphere

angles between the axes will be found to be near the values listed below.

$$\widehat{y\,z} = \alpha = 94° 16'$$
$$\widehat{z\,x} = \beta = 116° 35'$$
$$\widehat{x\,y} = \gamma = 87° 40'$$

The name of the system refers to the fact that the three zones axes which are used as crystallographic axes are inclined in the sense that the angles between them are not 90° and are not equal.

There is no unique orientation for a triclinic crystal so for the crystal of albite studied here, the largest face **b** was regarded as the **(010)** face and the *z* and *x* axes were placed parallel to prominent edges on this face. The *y* axis was then placed parallel to a third zone axis which was not parallel to the **(010)** face.

It can be seen from the stereogram of the albite crystal, particularly the arrangement of the **g,k,** and **h,m** poles on fig. 14.6 that the crystal possesses only a centre of symmetry which is represented by an inversion identity axis $\bar{1}$. Albite belongs to the holosymmetric class of the triclinic system.

The centre of symmetry implies the occurrence of pairs of faces, and the forms exhibited by the crystal studied here are listed below.

b	(010),	**e**	(0$\bar{1}$0)	pinacoid
g	(001),	**n**	(00$\bar{1}$)	basal pinacoid
a	(110),	**d**	($\bar{1}\bar{1}$0)	hemi prism
f	(1$\bar{1}$0),	**c**	($\bar{1}$10)	hemi prism
h	($\bar{1}$01),	**m**	(10$\bar{1}$)	hemi dome

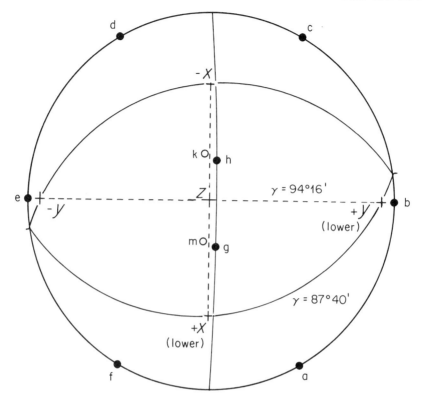

Fig. 14.6. Stereogram of the albite crystal showing the location of the *y* crystallographic axis. The *x* and *y* axes are indicated by dashed lines, and the planes containing the *x* and *y* axes is represented by the great circle

Estimates of the axial ratios of the albite crystal can be obtained by determining the angles between the crystallographic axes and the poles to faces **a (110)** and **h ($\bar{1}$01)**. The calculations are shown in tabulated form.

	XÔa	YÔa	ZÔa
Angles	40°	61°	90°
Cosines	0.7660	0.4848	0
Intercept ratios	$\dfrac{\cos \text{YÔa}}{\cos \text{XÔa}}$	$\dfrac{\cos \text{YÔa}}{\cos \text{YÔa}}$	$\dfrac{\cos \text{YÔa}}{\cos \text{ZÔa}}$
	0.6328 :	1 :	∞

	−XÔh	YÔh	ZÔh
Angles	38° 30′	90°	25°
Cosines	0.7826	0	0.9063
Intercept ratio	$z/x = 0.8635$		
	$z = 0.5464$		

Axial ratios $x : y : z = \mathbf{0.6328 : 1 : 0.5464.}$

X-ray studies indicate that the triclinic unit cell of albite has the dimensions $a = 8.144$ Å, $b = 12.787$ Å, $c = 7.160$ Å. These dimensions give the ratios **0.6368:1: 0.5599.**

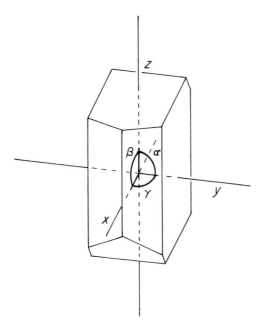

Fig. 14.7. Oblique view of the albite crystal showing the orientation of the zone axes which are selected as the crystallographic axes. The values of the angles between the axes are α = 94°16′ β = 116°35′ and γ = 87°40′

15 The Crystal of Beryl from the Hexagonal System, and the Crystals of Calcite and Quartz from the Trigonal System

The crystals which belong to the hexagonal and trigonal systems exhibit single hexad and triad symmetry axes respectively, and they are described by reference four crystallographic axes. The principal objectives of this chapter are listed below.

1. To become familiar with the main characteristics of the hexagonal system as represented by the beryl crystal.
2. To calculate the axial ratio of beryl and index the forms.
3. To study and index the common and varied forms exhibited by calcite crystals as representatives of the trigonal system.
4. To study the crystallographic features of the important rock forming mineral, quartz.

Contents of the sections

15.1 The crystal of beryl

Crystals of the mineral beryl are illustrated in fig. 15.1 and plate 14. The elements of symmetry of beryl may be determined from fig. 15.2 which is an oblique view of a beryl crystal, and fig. 15.3 which is a plan of the crystal when viewed parallel to the longest edges. The interfacial angle between a side face and the adjacent sloping face, $\hat{a\,h}$ is nearly 41°. *Plot a stereogram to show the poles to the faces, and the symmetry axes.*

It is clear that the beryl crystal is characterized by a **single hexad axis** of symmetry which is parallel to the prominent zone that

Fig. 15.1. Crystal of **beryl** (×2)

227

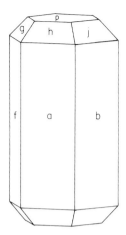

Fig. 15.2. Oblique view of the beryl crystal

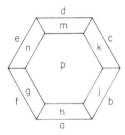

Fig. 15.3. Beryl crystal viewed in the direction parallel to the longest edges

The beryl crystal exhibits **six reflection planes** of symmetry each parallel to one diad axis and the hexad axis. The vertical planes of symmetry are represented by the lines on the stereogram. There is a **seventh plane of symmetry** which is perpendicular to the other six symmetry planes. The beryl crystal also exhibits a centre of symmetry and the Hermann Mauguin symbol is **6/m 2/m 2/m**.

Crystals of the hexagonal system are described by reference to four axes. The **hexad axis** is selected as the vertical crystallographic axis, *z*. There are three identical horizontal crystallographic axes labelled *x,y*, and *u* which are **120°** apart as shown in figs. 15.4 and 15.5, and these are placed parallel to the three diad axes which are also zone axes. The indices of the faces are given in the order *x,y,u,z*.

Since the three selected horizontal crystallographic axes are parallel to the zone axes, the indices of the front face, **a** which is parallel to *y* and intersects *x* and $-u$ are written as $(10\bar{1}0)$. The form $\{10\bar{1}0\}$ is known as a **hexagonal prism**. Note that the sum of the indices on the horizontal axes is always zero.

gives the crystal its typical prismatic habit. The hexad axis indicates that the beryl crystal belongs to the **hexagonal system**.

There are **six diad axes** which are perpendicular to the hexad axis. Three of the diad axes are also zone axes and are parallel to the edges produced by the intersection of the side and inclined faces. The other three diad axes are perpendicular to the side faces, as shown in fig. 15.4.

Fig. 15.5 is a stereogram of the faces of the beryl crystal. The hexad axis is in the vertical position and the six diad axes are horizontal. The poles to the six side faces fall on the primitive circle and they coincide with the three diad axes which are not zone axes.

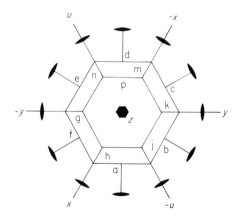

Fig. 15.4. Plan of the beryl crystal perpendicular to the hexad axis and showing the positions of the diad axes and the zone axes which are selected as the crystallographic axes

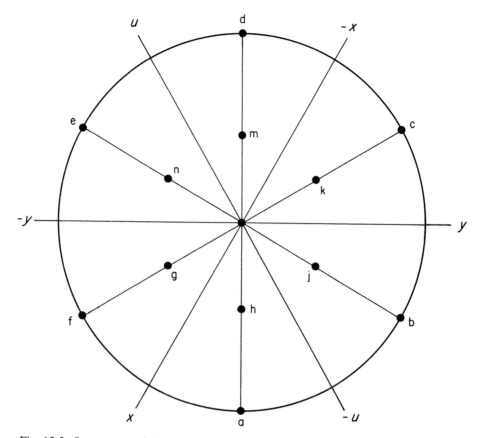

Fig. 15.5. Stereogram of the beryl crystal. The faces on the lower side of crystal are not represented because the poles coincide with the poles of the upper faces. The lines represent planes of symmetry

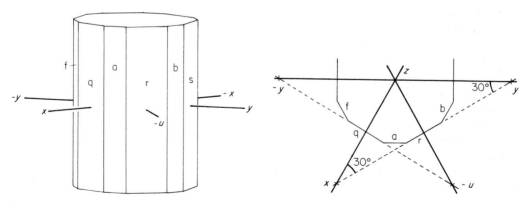

Fig. 15.6. Combination of the two hexagonal prism forms. The prism face **a** is parallel to the **y** axis, but the prism face **r** intersects all three horizontal crystallographic axes

Fig. 15.7. Plan on a plane parallel to the horizontal crystallographic axes to show the orientation of the prism face **r**

A **second hexagonal prism form** may occur as shown in fig. 15.6. *What are the indices of the prism face labelled* **r** *which intersects* +*x*, +*y*, *and* +*u*?

The intercept of face **r** on the −*u* axis = *x* **sin 30°**, as shown in fig. 15.7. Since sin 30° is ½ **the intercept on the** −*u* axis is one half of the intercepts on the *x* and *y* axes. Therefore the intercepts of this **prism** face are **1,1,−½,∞** and the indices will be (11$\bar{2}$0).

The form which consists of two faces perpendicular to the *z* axis is called the **basal pinacoid** and the indices of the faces are (**0001**) and (**000$\bar{1}$**).

15.2 The axial ratio of the beryl crystal

The faces which intersect the horizontal and vertical crystallographic axes are called **pyramid** forms.

The **parametral plane** of hexagonal crystals is chosen so that it is parallel to a **pyramid face** which is parallel to one of the horizontal axes, and has equal intercepts on the other two horizontal crystallographic axes. The ratio of the intercept on the *z* axis to the intercepts on the horizontal axes gives the axial ratio *x*:*z*.

The parametral face **h** of the beryl crystal is indexed (10$\bar{1}$1) and is shown in fig. 15.8.

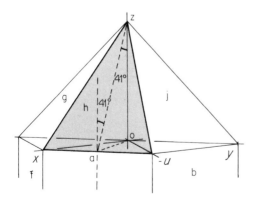

Fig. 15.8. The parametral plane for beryl is parallel to the face **h** which is indexed (**10$\bar{1}$1**)

The interfacial angle $\widehat{a\ h} = (10\bar{1}0)\widehat{}(10\bar{1}1)$ is nearly 41°. *What is the axial ratio* *x*:*z* *of the beryl crystal?*

It can be seen from fig. 15.8 that

tan 41° = a*O*/z*O* = x*O* sin 60°/z*O*

Therefore z*O* = sin 60°/tan 41° = 0.8660/0.8693 = **0.9962**

Consequently the axial ratio of the beryl crystal *x*:*z* = 1:0.9962, approximately.

The mineral **apatite** illustrated in fig. 1.10 occurs as crystals which have a very similar habit to the crystals of beryl. On the apatite crystals the interfacial angle $(10\bar{1}0)\widehat{}(10\bar{1}1) = 49°41'$. *What is the axial ratio* *x*:*z* *of apatite?*

Fig. 15.8 can be used again and it can be seen that O*z* = sin 60°/tan 49°41' = 0.8660/1.1785 = 0.7348. The axial ratio of the apatite crystal *x*:*z* = 1:0.735.

15.3 The crystal forms of calcite (CaCO₃)

Calcite is a common mineral which has several distinctive crystal habits. The simplest habit is shown in fig. 15.9 and it consists of six prominent side faces with three gently inclined faces at the ends, so that it is commonly described as **nail head spar**. Fig. 15.10 is an oblique drawing of this combination of forms. Fig. 15.11 is a plan of the crystal perpendicular to the prominent zone axis and with the broken lines showing the position of the faces of the under side of the crystal. The interfacial angle between the side face and the adjacent end face across the horizontal edge is 63°44'.

Construct a stereogram to represent the combination of forms exhibited by the crystal illustrated in fig. 15.10 and determine the symmetry of the calcite crystal.

The vertical zone axis is plotted at the centre of the stereogram fig. 15.12 and the poles to the faces of the vertical zone fall on the primitive circle. The poles to the inclined faces are plotted at 64°44' from the poles to the side faces but the alternate faces occur at the top and bottom of the crystal and are shown by full and open circles respectively.

It can seen that the single prominent vertical zone axis is a **triad symmetry axis** and this indicates that the calcite crystals belong to the **trigonal system**.

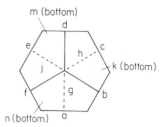

Fig. 15.11. Nail head spar viewed along the prominent zone axis. The position of the faces on the lower end of the crystal are shown by the broken lines

Fig. 15.9. Crystals of calcite exhibiting the combination of forms which is known as nail head spar (×1)

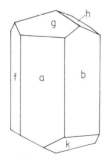

Fig. 15.10. An oblique drawing of a calcite crystal composed of the combination of forms characteristic of nail head spar

The horizontal crystal edges such as those between the faces **a,g,d,m** indicate the occurrence of three horizontal zone axes which intersect at 60°. The orientations of the horizontal zone axes are shown on the stereogram (fig. 15.12) and it can be seen that they are **diad symmetry axes**. The diad axes are used as the crystallographic axes and their arrangement on the crystal form is shown in figs. 15.13 and 15.14.

The calcite crystal exhibits **three planes of reflection symmetry**, each perpendicular to a diad axis as shown in fig. 15.12, and there is also **a symmetry centre**.

Note that the sloping faces at the ends of the calcite crystal are arranged in a manner which can be described in terms of a **triad inversion axis**. The international symmetry symbol $\bar{3}m$ indicates that the triad axis is combined with a symmetry centre and parallel mirror planes which introduces the horizontal diad axes, perpendicular to the reflection planes.

15.4 The cleavage rhombohedron of calcite

Calcite crystals fracture easily along three well developed cleavage planes to produce the form shown in figs. 1.23 and 1.24. This simple form has six rhomb-shaped sides and is called a **rhombohedron**. Fig. 15.15 is a diagram of the rhombohedron showing the

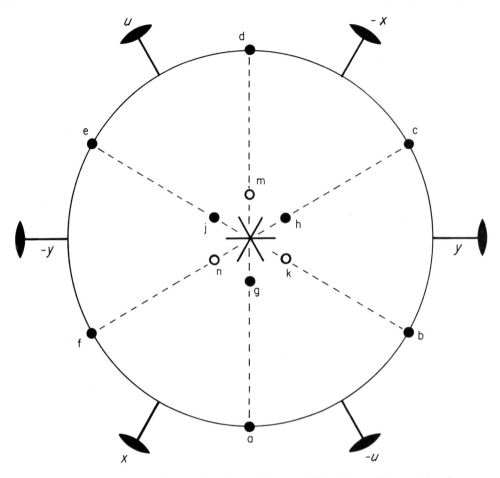

Fig. 15.12. Stereogram of the combination of forms exhibited by calcite nail head spar.
Planes of reflection symmetry are shown by the dashed lines

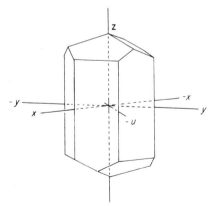

Fig. 15.13. The nail head spar habit of
calcite showing the arrangement of the
trigonal crystallographic axes

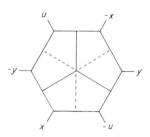

Fig. 15.14. Nail head spar
viewed along the z axis to
show the arrangement of
the horizontal crystallo-
graphic axes

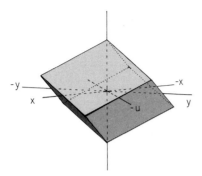

Fig. 15.15. The cleavage rhombohedron of calcite showing the arrangement of the crystallographic axes

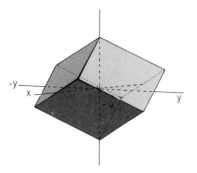

Fig. 15.16. The alternative orientation of the cleavage rhombohedron

arrangement of the trigonal crystallographic axes. The faces of the rhombohedron have a similar arrangement with respect to the crystallographic axes as the inclined faces at the ends of the crystal studied in section 15.3. The rhombohedron has the same elements of symmetry as nail head spar and in fig. 15.15 it is arranged so that the face which intersects x and $-u$ also intersects the positive end of the z axis. An alternative arrangement of the rhombohedron is shown in fig. 15.16 in which a face intersects $+x$, $-u$, and $-z$.

The faces of calcite crystals are indexed by reference to the rhombohedron cleavage form orientated as in fig. 15.15, and this is known as the unit form $\{10\bar{1}1\}$. *Plot a stereogram of the unit form using the interfacial angle* $(10\bar{1}0)\widehat{\ }(10\bar{1}1) = 45°33'$. *Index all the faces of the unit rhombohedron on the stereogram. Calculate the axial ratio $x:z$ of calcite.* The angle between the cleavage planes of calcite is an important diagnostic feature and it should be measured directly or estimated from the stereogram.

Fig. 15.17 is a stereogram of the faces of the calcite cleavage rhombohedron, and the similar arrangement of the inclined faces of the nail head spar is clearly seen by comparison with the stereogram fig. 15.12. The faces which occur on nail head spar fig.

15.13, belong to a rhombohedron form which occurs in combination with a hexagonal prism form.

The symmetry of the calcite cleavage rhombodhedron is very clearly indicated by the stereogram, fig. 15.17. Since the cleavage rhombohedron is selected to be the unit form $\{10\bar{1}1\}$, the indexing of the faces of the form involves only the determination of whether a face is parallel to an axis or intersects the negative end of a crystallographic axis.

The axial ratio $x:z$ for calcite may be determined from the relationship shown in fig. 15.8. Since the interfacial angle $(10\bar{1}0)\widehat{\ }(10\bar{1}1) = 45°22'$, the value of z may be determined from the relationship $\tan 45°22' = x \sin 60°/z$.
$z = 0.866/1.0129 = 0.8549$. Consequently the axial ratio of calcite as determined from the cleavage rhombohedron is $x:z = 1: 0.8549$.

It will be found that the interfacial angle between the rhombohedron cleavage planes of calcite is almost 75°.

15.5 The rhombohedron forms of calcite

The interfacial angle between the inclined face of nail head spar and the $(10\bar{1}0)$ face is $63°44'$. *What are the indices of the rhom-*

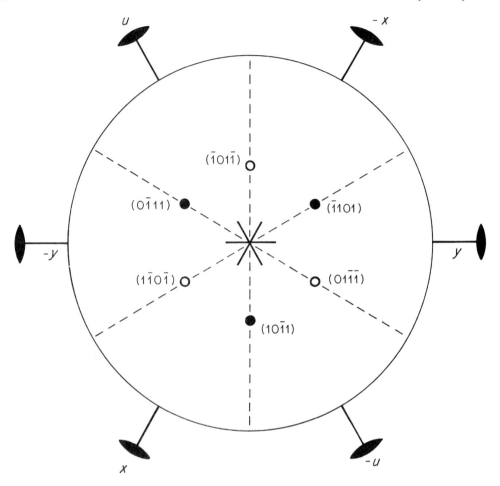

Fig. 15.17. Stereogram of the cleavage rhombohedron of calcite. The dashed lines represent
planes of reflection symmetry

bohedron form which occurs in nail head spar fig. 15.10.

It will be found from the relationship, tan $63°44' = x \sin 60°/z$ that the ratio of the intercept on $z = 0.866/2.0262 = 0.4274$. This value is one half of the axial ratio determined from the cleavage rhombohedron. Consequently the intercept ratios of the front face of the nail head spar rhombohedron are x, ∞, $-u$, $z/2$ and the indices of the form are $\{10\bar{1}2\}$. *Which form is exhibited by the calcite crystals in plates 16 and 17?*

The thin crystals of calcite shown in plates

16 and 17 are composed of the faces of the $\{10\bar{1}2\}$ rhombohedron form. Calcite crystals also occur as a form with steeply inclined rhomb faces as shown in fig. 15.18, and the interfacial angle between the rhomb face and the $(10\bar{1}0)$ prism is $14°13'$. *What are the indices of the steep rhombohedron form of calcite?*

Since tan $14°13' = x \sin 60°/z$, it will be found that the intercept ratio on $z = 3.4199$ which is 4 times the unit intercept on z as determined from the cleavage rhombohedron. Consequently the indices of the form illustrated in fig. 15.18 are $\{40\bar{4}1\}$.

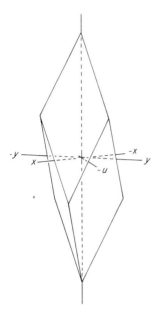

Fig. 15.18. The steep rhombohedron form of calcite

X-ray studies indicate that the most appropriate unit cell for calcite is a primitive rhombohedron cell equivalent to the $\{40\bar{4}1\}$ crystal form. The content of the primitive unit cell is $2CaCO_3$, and this is preferred to a unit cell based on the cleavage rhombohedron because its contents would be $32CaCO_3$.

15.6 The scalenohedron form of calcite

Calcite crystals frequently occur as a form composed of twelve faces each of which is a scalene triangle. The form is called the **ditrigonal scalenohedron** $\{21\bar{3}1\}$, and it is illustrated in fig. 15.19. Fig. 15.20 is a plan of the form on the **(0001)** plane and it can be seen that a pair of faces is repeated by the triad symmetry axis. Alternate edges of the scalenohedron are similar.

A stereogram of the faces of the scalenohedron is shown in fig. 15.21. It can be seen that the face indexed **(21$\bar{3}$1)** is reflected across the plane of symmetry to give the (3$\bar{1}$$\bar{2}$1) face. The poles to the faces

of the cleavage rhombohedron are also shown on the stereogram. *What are the names of the forms exhibited by the calcite crystal in fig. 1.22?*

Calcite crystals frequently show combinations of forms. The crystals in figs. 1.22 and 15.22 show the combination of the scalenohedron (s) with the unit rhombohedron (r) and the hexagonal prism (p). Fig. 15.23 is a scalenohedron form the upper half of which appears to have been rotated on the (0001) plane. Note the prominent re-entrant angles and the edges parallel to the basal pinacoid. The crystal is said to be twinned and this is the subject of chapter 16.

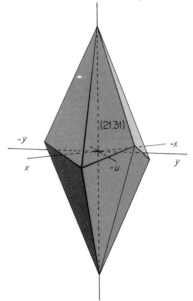

Fig. 15.19. The scalenohedron form of calcite

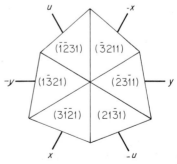

Fig. 15.20. Plan of the scalenohedron form on the (0001) plane

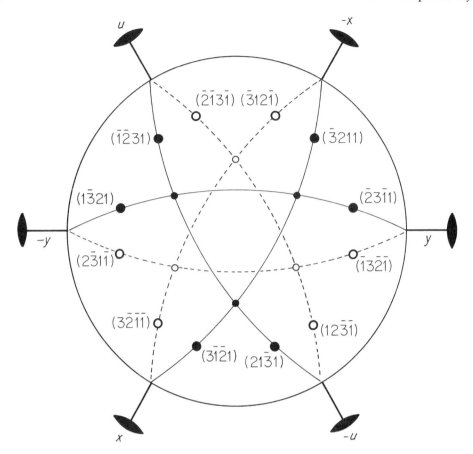

Fig. 15.21. Stereogram of the scalenohedron form $\{21\bar{3}1\}$ showing the relationship of the faces to the cleavage rhombohedron

15.7 The crystal of quartz (SiO₂)

Quartz is a very common mineral and it frequently occurs as well developed crystals in veins and cavities. The quartz crystals like those illustrated in fig. 1.1 and fig. 15.24 are often prismatic in habit and are composed of hexagonal prisms terminated by two rhombohedron forms. When the rhombohedron faces are equally developed they have the appearance of the hexagonal dipyramid form, but usually unequal growth rates results in one set of rhombohedron faces being larger than the other. The alternate faces are different in type and this may be brought out by different etching patterns when acid is placed on the faces of natural crystals.

The quartz crystal exhibits a **triad axis** of symmetry parallel to the prism faces, and **three diad axes** which are perpendicular to the triad axis. The quartz crystal also exhibits a **centre of symmetry** but there are no planes of symmetry, so that the international symbol for the class is **32**.

Trigonal pyramid forms may be present as small faces on quartz crystals. The trigonal pyramid form $\{11\bar{2}1\}$ is illustrated in fig. 15.25, and the alternative form $\{2\bar{1}\bar{1}1\}$ is illustrated in fig. 15.26. This is an example of **enantiomorphism**, the forms are mirror images of each other, but like the right and left hands they are not superposable in space. The structure of quartz is such that

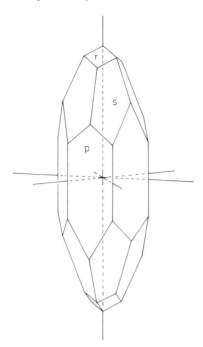

Fig. 15.22. The combination of the unit rhombohedron r {10$\bar{1}$1}, scalenohedron s {21$\bar{3}$1} and hexagonal prism p {10$\bar{1}$1}

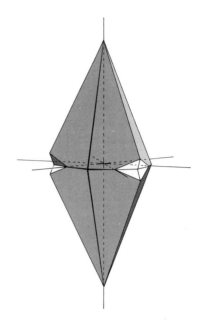

Fig. 15.23. The scalenohedron form twinned on the basal pinacoid (0001). Note the re-entrant angles

either the {11$\bar{2}$1} or {2$\bar{1}\bar{1}$1} forms may occur on a crystal.

In fig. 15.27 the {2$\bar{1}\bar{1}$1} form is present and since it truncates the upper left hand corner of the hexagonal prism face, this quartz crystal is said to be **left handed**.

The {11$\bar{2}$1} form truncates the upper right handed corner of the prism face in fig. 15.28 and this crystal is said to be **right handed**. Also in fig. 15.28 the right handed form of the **trigonal trapezohedron** {51$\bar{6}$1} is illustrated. The trigonal trapezohedron is a six sided form enclosing space as are the rhombohedron forms, but each face of the trigonal trapezohedron has sides of unequal length.

The enantiomorphic nature of quartz is determined by the structure which is described in section 17.11.

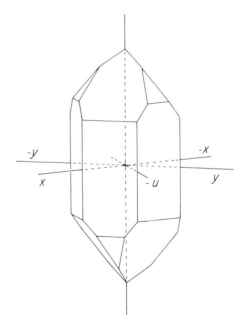

Fig. 15.24. Quartz crystal exhibiting the hexagonal prism and two rhombohedron forms

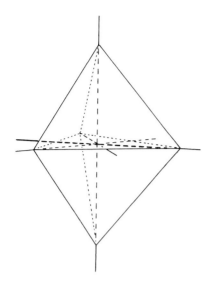

Fig. 15.25. Trigonal pyramid form
$\{11\bar{2}1\}$

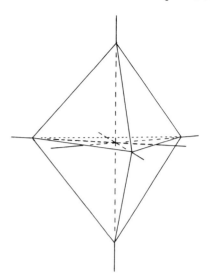

Fig. 15.26. Trigonal pyramid form
$\{2\bar{1}\bar{1}1\}$

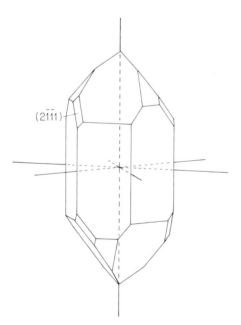

Fig. 15.27. Left handed quartz due to the occurrence of a face of the trigonal pyramid form $\{2\bar{1}\bar{1}1\}$ at the top left hand corner of the prism face

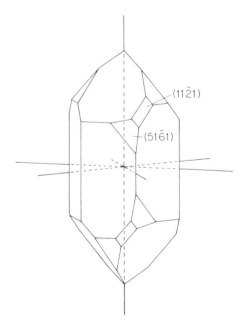

Fig. 15.28. Right handed quartz showing faces of the trigonal pyramid form $\{11\bar{2}1\}$ and the trigonal trapezohedron form $\{5\bar{1}\bar{6}1\}$

16 *Crystal Growth and Twinned Crystals*

The objectives of this chapter are listed below.

1. To introduce the concepts of dislocations in crystal structures, and consider the role of screw dislocations in the crystal growth mechanism.
2. To consider the characteristics of parallel and normal twins exhibited by the feldspar minerals.
3. To appreciate the nature of a growth twin by studying the structure of aragonite.

Contents of the sections

16.1 Crystal growth

In a magma, a solution, or volcanic gas, there is an almost completely random arrangement of the atoms which are in continuous motion. While it is probably true that over a long period of time the atomic positions in a liquid are completely random, at any instant some atoms may be arranged with regard to their neighbours exactly as they would be in a crystal structure. Thus the instantaneous structure of the liquid might be one in which some atoms form part of a crystal-like structure called a **cluster**, with some distance between it and neighbouring clusters. Due to the random translational movements of the atoms in the liquid, these clusters of atoms disperse as easily as they form. If during the cooling of a melt some clusters continue to grow in size, they are called **nuclei** from which crystal forms develop.

A cluster of atoms must grow to a critical size before its free energy is lower than that of the melt or solution so that it can continue to grow. The formation of nuclei requires a lowering of the temperature below the equilibrium temperature at which crystals can exist with melt, or a degree of supersaturation of a solution. A crystal grows outwards from the nucleus and distinctive growth zones can be seen in many crystals. A fine example of growth zones in fluorite is revealed by the fluorescence illustrated in plate 17.

Early theories of crystal growth were concerned with the rates at which atoms moved from a vapour phase to the surface of the crystal and in the reverse direction. These rates depend on the concentration of atoms in the vapour which gives rise to a vapour pressure. At the saturation vapour pressure the rates of movement of atoms from vapour to crystal and in the reverse direction, are equal. Calculations suggested that a vapour pressure at least 50% greater than the saturated vapour pressure would be required for crystal growth, but experiment showed that growth could occur if the super-saturation was as low as 1%. This is a very great discrepancy between prediction from theory and experimental observation.

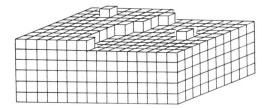

Fig. 16.1. The growth of a crystal structure by the formation of layers of additional atoms here represented by cubes

A crystal grows by the addition of layers of atoms and the process can be represented by the stacking of the small cubes as in fig. 16.1. An atom which comes in contact with a plane crystal surface is bonded only by the forces acting across the surface, and these may not be sufficient to prevent the atom from escaping again into the solution. If the atom comes to rest against the edge of a layer or in a corner, it is bonded by forces acting in several directions and is much more likely to stay bonded in the crystal structure. The problem of crystallization is concerned with the formation of a layer which allows the bonding of additional atoms on two or more sides when they migrate against the step formed by the new layer.

The explanation of crystallization at very low levels of supersaturation was provided by the English physicist F. C. Frank in 1949. Frank pointed out that crystal structures are rarely perfect and that steps may be present on the surface of the crystal because of minute defects in the structure. A missing atom in a crystal structure is described as a

vacancy or a **point defect**. If a lattice plane ends along a line as shown in fig. 16.2 the line along the end of the incomplete plane is known as **dislocation**. This **line defect**, which may be curved is called an **edge dislocation**.

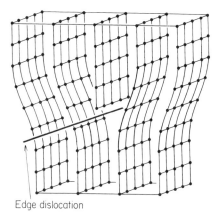

Edge dislocation

Fig. 16.2. A line defect known as an **edge dislocation** in a crystal structure due to the presence of an internal in-complete lattice layer

The defect of particular interest here is known as a **screw dislocation** and it is repre-sented by fig. 16.3. It can be seen that the lattice layer forming the top surface of the structure, twists down to a lower level around the screw dislocation. Crystal growth occurs most easily when atoms can fit into a stable position against a step of the kind that occurs on a crystal face due to the presence of a screw dislocation. Frank sug-gested that crystal growth could be ex-plained if the layers on a crystal surface were built up in a spiral manner due to the rotation of the step around a screw disloca-tion as shown by the series of diagrams in fig. 16.4. If the rate of attachment of atoms is approximately equal along the step then the step will grow into a spiral, and electron micrographs have revealed the presence of spiral growth patterns on the faces of many crystals.

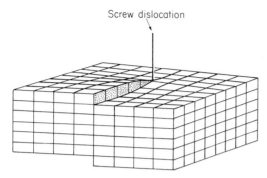

Fig. 16.3. **A screw dislocation** in which the lattice layer twists or spirals down to a lower level around the dislocation

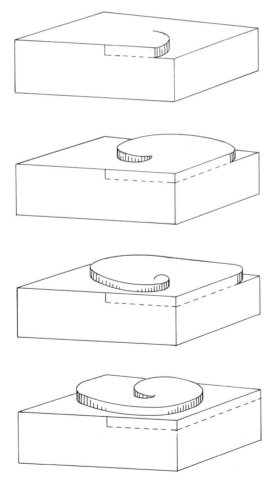

Fig. 16.4. A series of diagrams to represent crystal growth due to the spiral advance of a step extending from a screw dislocation

16.2 Twinning in the feldspars

Some crystals consist of two or more parts in which the lattice has different orientations that are related in certain geometrical ways. Composite crystals of this kind are known as **twinned crystals**. For example in the crystal of orthoclase feldspar illustrated in fig. 16.5, the *y* and *z* crystallographic axes have the same orientation but the inclined *x* axis is reversed. This arrangement is shown more clearly in fig. 16.6 in which the two parts of the crystal are shown separately.

Crystals composed of two individual parts which have a definite structural relationship to one another are said to be **simple twins**. If the two parts of a simple twin are separated by a definite surface it is described as a **contact** twin and the surface along which the two individuals are united is called the **composition plane**. In the orthoclase crystal in fig. 16.6 (**010**) is the composition plane. When the composition plane is irregular as in fig. 16.7, the crystal is said to be an **interpenetration twin**.

One part of a twin crystal may be related to the other part by rotation about an imaginary axis known as the **twin axis**. The orthoclase twins can be related by rotation through 180° about a twin axis which is parallel to the *z* crystallographic axis. When the twin axis is parallel to the composition plane and to a zone axis, the twin is described as a **parallel twin**. The indices of the faces of the part of the twin which appears to have been rotated are given in terms of the reversed axes and are underlined as shown in fig. 16.5.

The description of a twin in terms of a symmetry operation is sometimes referred to as a **twin law**. The type of twin illustrated in the orthoclase crystal, fig. 16.6, with a composition plane parallel to (**010**) and related by 180° rotation on a twin axis parallel to *z*, is called a **Carlsbad twin**, or is described as twinning according to the **Carlsbad law**. The smaller crystal of orthoclase in fig. 16.8 can be compared with fig. 13.4

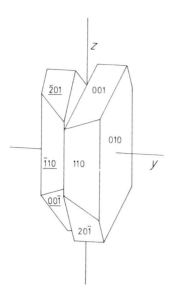

Fig. 16.5. **Carlsbad twin** in orthoclase feldspar with the composition plane parallel to **(010)**

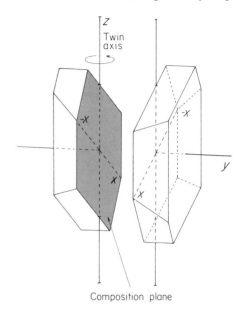

Fig. 16.6. The two parts of the Carlsbad twin can be related by 180° rotation on the twin axis which is parallel to z

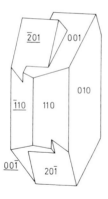

Fig. 16.7. An inter-penetration twin in orthoclase

Fig. 16.8. Crystals of **orthoclase feldspar** viewed perpendicular to the (010) face. The larger crystal is a Carlsbad twin ($\times 1$)

which was used in the study of the symmetry of orthoclase. The effect of the presence of a Carlsbad twin is shown by the larger specimen of orthoclase in fig. 16.8. Carlsbad twins occur in both orthoclase and the plagioclase feldspars such as albite.

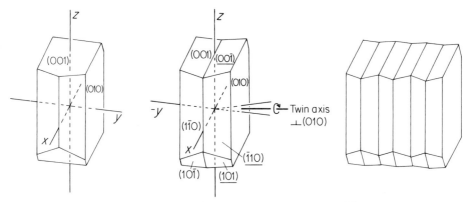

Fig. 16.9. The pla-gioclase feldspar, albite

Fig. 16.10. **A normal twin** in albite. The two parts of the albite twin can be related by 180° rotation on a twin axis which is perpendicular to the **(010)** plane

Fig. 16.11. **Repeated twinning** on the albite law in plagioclase feldspar

When the **twin axis** is normal to a crystal face which may be a composition plane, the twin is described as a **normal twin**. In the albite crystal shown in fig. 16.10 the composition plane is parallel to **(010)** and the twin axis is normal to **(010)**. The albite twin may also be described by **reflection** across the **(010)** plane which is then called a **twin plane**. A twin plane does not always coincide with the composition plane. In the plagioclase feldspars normal twinning of this kind is referred to as the **Albite twin law**, and it often occurs as repeated or multiple twins as shown in fig. 16.11. During the growth of the crystal the alternative orientation of the structure was adopted at intervals. *Why does albite twinning occur only in triclinic feldspars and not in orthoclase?*

In the monoclinic feldspar, **orthoclase**, the axis perpendicular to (010) is a diad axis of symmetry, and therefore 180° rotation results in an identical orientation. The greater the symmetry of the lattice the greater the restriction on the number of axes and planes about which twinning may occur. The absence of symmetry planes and axes in the triclinic lattice means that any plane or lattice row can be a twin plane or twin axis.

The crystallographic characteristics of twinning have been illustrated by reference to the feldspars because the relationship of the optical characteristics to the twin laws, is very useful for identification and the estimation of the composition of the feldspars in transparent thin sections.

16.3 The twinning of aragonite ($CaCO_3$)

A single perfect crystal represents the least energy state, but a crystal may grow with different orientations of the lattice on either side of a composition plane if the change in the orientation of the lattice gives rise to only a slightly higher energy state along the composition zone. The nature of a growth twin was illustrated and explained by Sir W. L. Bragg in terms of the structure of aragonite (*Proc. Roy. Soc.*, 1924, A105.16). **Aragonite** is composed of calcium carbonate and its crystals shown in fig. 1.42, belong to the **orthorhombic system**. Fig. 16.12 is a section of the crystal parallel to the (001) plane showing that the interfacial angles $(110)\widehat{\ }(1\bar{1}0) = 63°48'$ and $(110)\widehat{\ }(010) = 58°06'$.

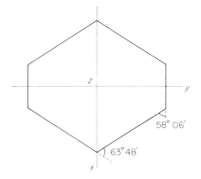

Fig. 16.12. A section of the aragonite crystal in a plane parallel to (001) indicates that the interfacial angles are close to 60°

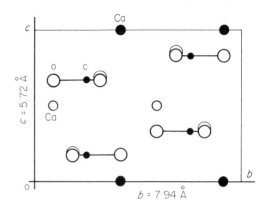

Fig. 16.13. Projection of the unit cell of aragonite on a plane parallel to the **b** and **c** directions. The carbon atoms are in coordination with three oxygen atoms and the (CO_3) groups lie in planes parallel to (**001**). The **Ca** atoms that occur at the bottom and top of the unit cell are shown in black, while the **Ca** atoms that occur at half the height of the cell ($c/2$) are shown as open circles. (after W. L. Bragg.)

The structure of aragonite consists of triangular groups (CO_3), consisting of carbon atoms in coordination with three oxygen atoms and lying in planes parallel to the (**001**) plane as shown in fig. 16.13. The orientation of the (CO_3) coordination groups are shown by triangles in fig. 16.14, which is a projection of the unit cell on the plane parallel to the **a** and **b** directions. Each **Ca** atom is bonded to six oxygen atoms. The **Ca** atoms are arranged in a hexagonal manner but the arrangement of the (CO_3) groups lowers the symmetry to orthorhombic. The plane parallel to (**100**) is a **reflection plane**, but the plane parallel to (**010**) is a **glide plane** of symmetry.

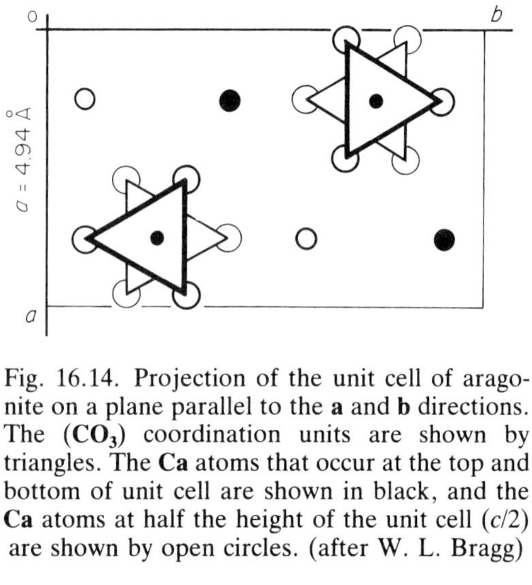

Aragonite twins on the prism faces {**110**}. The structural arrangement across a (**110**) composition plane is shown in fig. 16.15 after Sir. W. L. Bragg. In the lower part of the twin the triangles represent the (CO_3) groups in the upper half of the unit cell as shown in fig. 16.14. Across the (**110**) plane the triangles represent the (CO_3) groups in the lower part of the unit cells. The structure in the zone indicated by the two dashed lines in fig. 16.15 is consistent with the structure on both sides of the twin. The atoms in the border zone have the same inner coordination arrangements and differences arise only in the positions of the atoms

Fig. 16.14. Projection of the unit cell of aragonite on a plane parallel to the **a** and **b** directions. The (CO_3) coordination units are shown by triangles. The **Ca** atoms that occur at the top and bottom of unit cell are shown in black, and the **Ca** atoms at half the height of the unit cell ($c/2$) are shown by open circles. (after W. L. Bragg)

beyond the first coordination zone. Consequently the twinned aragonite structure represents an energy state which is only a little higher than the energy state of the untwinned crystal. The aragonite twin illustrates that the structure is continuous through the twin zone and it does not necessarily contain dislocations.

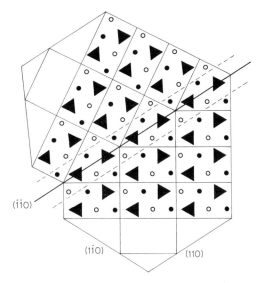

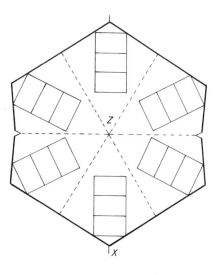

Fig. 16.15. The twinning of aragonite on the (110) plane. The (CO₃) coordination groups are represented by the triangles. The structure on the zone between the dashed lines is consistent with the structures in both parts of the twin

Fig. 16.16. Section of an aragonite crystal perpendicular to the z axis showing repeated twinning on the {110} form which produces pseudohexagonal symmetry. The orientations of the unit cells are shown by the rectangles

Fig. 16.17. **Aragonite** crystals twinned on {110} and showing pseudohexagonal symmetry (×2)

Fig. 16.18. Orthorhombic crystal of **witherite** twinned on {110} so that the intergrowth of three individuals produces pseudohexagonal symmetry

pseudohexagonal symmetry shown in fig. 16.18.

Marcasite FeS_2 is sometimes known as white iron pyrites, because it has a paler metallic yellow colour than the dimorph form, pyrite. Marcasite is orthorhombic but twinning on {101} produces attractive groups known as **spear-head twins** shown in fig. 16.19.

Staurolite occurs in certain schists with garnet or kyanite and it commonly shows **cruciform twins** as in fig. 16.20. The orthorhombic staurolite crystals on the left side of fig. 16.20 have prism faces on which the interfacial angle $(110)\widehat{\ }(1\bar{1}0) = 50°26'$. Twinning on {031} produces two individuals which cross at nearly 90° to give the cruciform twins shown in fig. 16.20.

Twinning may also develop during cooling if the crystal transforms into a structure more stable at lower temperatures. Deformation twins arise as a result of translation gliding during plastic deformation of the crystal.

Simple twins are rare and cyclic twins on the form {110} often occur in aragonite crystals. This **repeated twinning** produces aragonite crystals which are superficially very like hexagonal crystals as shown in fig. 16.16. Repeated twinning which appears to produce a higher degree of symmetry than that possessed by the structure is sometimes described as **mimetic twinning**. The presence of twinning is indicated by the presence of re-entrant angles. Aragonite crystals exhibiting twinning are shown in fig. 16.17.

Witherite $BaCO_3$ has an orthorhombic structure and the interfacial angle $(010)\widehat{\ }(110) = 59°11'$. Crystals of witherite are always twinned on {110} and the intergrowth of three individuals produces the

16.4 Questions for recall and self assessment

1. What is an edge dislocation?
2. What is a screw dislocation?
3. Why are screw dislocations important for crystal growth?
4. What is a twinned crystal?
5. What is the composition plane of a twinned crystal?
6. What is the characteristic of a parallel twin?
7. What are the characteristics of the Carlsbad twin?
8. What are the characteristics of a normal twin?
9. What is the meaning of twin plane?
10. Is the composition plane of a twinned crystal characterized by the presence of dislocations?

Fig. 16.19. Group of pale metallic **marcasite** crystals exhibiting spear-head twins which have grown by replacement within a white, fine grained limestone called chalk

Fig. 16.20. Orthorhombic crystals of **staurolite** with $(110)\,\char`^\,(1\bar{1}0) = 50°26'$. The lower crystal is viewed along the z axis. Twinning on $\{031\}$ produces the characteristic **cruciform twins**

17 *The Structures of the Common Silicate Minerals*

The eight most common elements occurring in the rocks of the outer part of the earth are **O, Si, Al, Fe, Mg, Ca, Na, and K**. Of these, oxygen can be regarded as the only negative ion and it is associated with the second most common element, silicon, to form **(SiO$_4$)** structural units. Consequently the most common rock forming minerals are silicates which contain **(SiO$_4$)** tetrahedra bonded to other **(SiO$_4$)** units or to metal cations. The silicate minerals are classified according to the manner in which the **(SiO$_4$)** tetrahedra occur and the names of selected common silicate minerals, their structure and the name of the mineral group to which they belong are summarized below.

The principal objectives of this chapter are listed opposite

1. To consider the abundances and coordination numbers of the eight most common elements in the earth's crust.

2. To understand the chemical nature of isomorphous series by considering the composition of the olivines.

3. To study the structures of selected common rock forming silicate minerals in order to understand the characteristics of the main kinds of silicate structure.

4. To appreciate the relationships between the crystal and cleavage forms of the common rock forming silicate minerals and their structures.

5. To introduce the crystallization reaction series.

Mineral	Structure	Mineral group
Olivine	Independent (SiO$_4$) tetrahedra	Olivines
Diopside	Single chains (Si$_2$O$_6$)	Pyroxenes
Tremolite	Double chains (Si$_4$O$_{11}$)	Amphiboles
Kaolin and muscovite	Sheets (Si$_4$O$_{10}$)	Clays and Micas
Quartz	Framework (SiO$_2$)	Silica minerals

Contents of the sections

17.1 The abundance of elements in the earth's crust

The earth's crust with which we are concerned forms less than 1% of the earth's
mass. Clarke and Washington estimated
that the upper 10 miles of the earth's crust
consists of rocks in the proportion; igneous
rocks, 95%; mudstone, 4%; sandstone,
0.75%; and limestone, 0.25%. Therefore
they considered that the average composition of the crust could be arrived at by
averaging 5150 analyses of fresh igneous
rocks. The composition of rocks is usually
shown in the form of oxides and from this
the abundances of the individual elements in
the crust can be calculated and are shown in
the table below.

Oxide	Weight %	Element	Weight %
		O	46.60
SiO_2	60.18	Si	27.72
Al_2O_3	15.61	Al	8.13
Fe_2O_3	3.14	Fe	5.00
FeO	3.88		
MgO	3.56	Mg	2.09
CaO	5.17	Ca	3.63
Na_2O	3.91	Na	2.83
K_2O	3.19	K	2.59

The most outstanding feature of these
calculations is that **eight elements make up
98.6% of the total weight**. The proportions
of the eight common elements are illustrated by the histogram fig. 17.1 and the
prominence of **oxygen forming 46.6% by
weight** is emphasized.

The atoms of the different elements have
different atomic weights so that different
numbers of atoms are required to make up a
particular weight. The approximate numbers of atoms of the common elements
required to equal the weight of the atom of
the heaviest common element, iron, are
shown diagrammatically in the fig. 17.2. The
histogram shown in fig. 17.2 represents the
percentages of atoms required to make up
the weight percentages shown in fig. 17.1. It
can be seen that **nearly 63% of the atoms
consist of oxygen with silicon making up
another 21%**.

The atoms of the common elements have
different sizes and the estimated radii of the
ions in angstrom units are shown in the table
below. The percentage volume occupied by
the proportions of common elements are
shown in the table and in the histogram fig.
17.3. **The atoms of oxygen are the lightest
and the largest of the atoms of the common
elements so that oxygen occupies nearly 94%
of the volume of the calculated average composition of the earth's crust.**

	Radius (Å)	Volume per cent
O	1.40	93.77
Si	0.42	0.86
Al	0.51	0.47
Fe	0.74	0.43
Mg	0.66	0.29
Ca	0.99	1.03
Na	0.97	1.32
K	1.33	1.83

These estimates may be subjected to certain criticisms but there is no doubt about
the prominence of oxygen of the earth's
crust. **Consequently it is reasonable to im-**

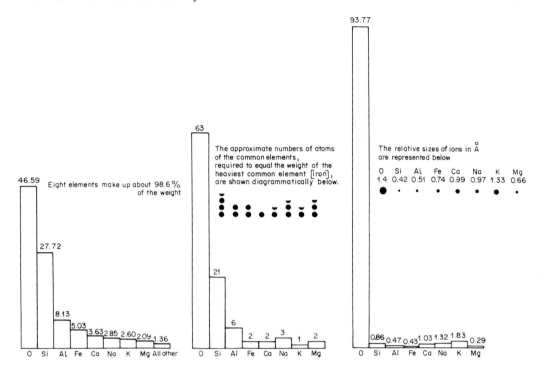

Fig. 17.1. Histogram of the estimated weight percentages of the
eight common elements in the
earth's crust, after Clarke, F. W.,
and Washington, H. S. (1924,
U.S. Geol. Surv. Profess. Paper,
127., p. 177)

Fig. 17.2. Histogram of the
percentages of the atoms of
the common elements re-
quired to make up the weight
percentages shown in fig. 17.1

Fig. 17.3. Histogram of the
percentage volume occupied
by the ions of the common
elements shown in fig. 17.1

agine the earth's crust as a mass of oxygen
atoms with the smaller, heavier, and less
abundant atoms of Si, Al, Fe, Mg, Ca, Na,
and K occurring in the spaces between the
oxygen atoms. The atoms of all the other
elements occur as trace amounts only, except where they have been concentrated by
particular geological processes.

17.2 The ionic characteristics of the
common elements

The radius ratio of the oxygen and silicon
atoms suggest that the small Si atom could
be enclosed by four oxygen atoms in a
tetrahedral unit.

The most common rock forming minerals
are **silicates** in which (SiO_4) tetrahedral units
are bonded to other (SiO_4) tetrahedral units
and to metal cations. Electronegativity was
discussed in section 5.8, and the electronegativity values of oxygen (3.5) and silicon
(1.8) indicate that **the bonding between O
and Si is approximately 50% ionic in character.** However, for the purpose of understanding the main structural arrangements
and chemical compositions of the silicate
minerals it is convenient to consider the
elements to be present as ions whose size
and charge determine their arrangement.

The position of the eight common elements is shown in the periodic chart fig. 5.1.

Both Na^+ and K^+ have single electrons in the outer shells of their structure, and they can form ionic bonds by losing the outer electron and becoming positively charged cations. Mg and Ca have two electrons in the outer shells of their structure and they can form the divalent cations Mg^{+2} and Ca^{+2}. Fe also forms a divalent cation known as **ferrous iron, Fe^{+2}**, but by losing a third electron it may form a trivalent cation known as **ferric iron, Fe^{+3}**. Al has three electrons in its outer shell and it can be regarded as a trivalent cation Al^{+3}, while **Si** which has four electrons in its outer shell can be regarded as a quadrivalent cation Si^{+4}.

Oxygen has an atomic number of 8 and it has two electrons in the first shell and six in the second shell. Consequently if the oxygen atom attains an eight electron outer shell structure by obtaining electrons from adjacent cations it can be regarded as an anion with a negative charge of two, O^{-2}.

Oxygen not only occupies most of the volume but it is the only one of the eight most common elements that forms an anion. Consequently nearly all the common rock forming minerals can be regarded as ionic structures formed by the combination of oxygen anions associated with cations of the other common elements. The compositions and the structural characteristics of the main silicate minerals can be understood by regarding the atoms as ions, and balancing the ionic charges in the formulae which represent the compositions. Sulphur and the

halogens, fluorine and chlorine, are the other common anions.

In an ionic structure the chemical law of valency is satisfied by making the total positive and negative charges balance. There are no molecules and each positive ion is surrounded by a number of negative ions in a continuous three-dimensional ionic structure. **The positive charge of the cation is divided between the surrounding anions, the number of which is determined by the relative sizes of the anion and the cation and not by the valency.**

17.3 The coordination numbers of the common elements

Coordination number and Pauling's rules concerning the structure of ionic crystals were considered in the section 7.7. The coordination units characteristic of particular radius ratios are repeated below.

Radius ratio	Coordination numbers	Structural unit
0.22–0.41	4	Tetrahedron
0.41–0.73	6	Octahedron
0.73–1	8	Cube corners
1	12	Mid-points of cube edges

The ionic radii of the common cations and the radius ratios of each with oxygen (radius 1.4 Å) are shown below with the predicted and observed coordinate numbers.

Ion	Radius	Radius/ratios	Coordination Number	
			Predicted	Observed
Si^{+4}	0.42	0.30	4	4
Al^{+3}	0.51	0.36	4	4 or 6
Fe^{+3}	0.64	0.46	6	6
Mg^{+2}	0.66	0.47	6	6
Fe^{+2}	0.74	0.53	6	6
Na^{+1}	0.97	0.69	6	6–8
Ca^{+2}	0.99	0.71	6	6–8
K^{+1}	1.33	0.95	8	8–12

Many cations occur exclusively in a particular coordination. In the silicate minerals the silicon atom always lies between four oxygen atoms in the manner illustrated in fig. 17.4. The black sphere represents the relative size of the silicon atom at the centre of the four oxygen atoms which are represented by the large circles. The centres of the four oxygen atoms occur at the corners of a tetrahedron of nearly constant dimensions.

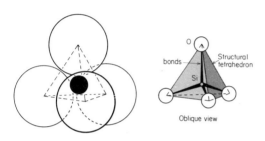

Oblique view

Fig. 17.4. **The coordination tetrahedron SiO₄.** The black sphere represents the relative size of the silicon atom enclosed by four oxygen atoms

The bonds between the silicon and oxygen ions are strong because they are partly covalent in character. Consequently in diagrams of the structures of silicate minerals, the atoms in a (SiO_4) tetrahedron will be represented by small circles and the strong Si–O bonds may be shown by the heavy black radiating lines as in fig. 17.4. The thin lines joining the centres of the oxygen atoms serve to outline the tetrahedron structural unit and they do not represent bonds. The (SiO_4) tetrahedron structural unit is sometimes represented simply as a tetrahedron.

The silicate minerals are classified according to the way in which the (SiO_4) tetrahedra occur or are linked together by the sharing of oxygen atoms. Some silicate mineral structures are characterized by the occurrence of separate (SiO_4) tetrahedra but in others the tetrahedra are linked together by shared oxygen atoms to form structures characterized by chains, sheets or frameworks of silicon and oxygen atoms. All the following diagrams of the silicate structures are based on the explanations in the original papers indicated by references. An extensive review of crystal structures is given by Sir W. L. Bragg in his book titled *The Atomic Structure of Minerals.* Cornell University Press, New York, 1937. Additional information is given in *Rock-Forming Minerals,* Vols. 1–4, by W. A. Deer, R. A. Howie, and J. Zussman, Longmans, Green and Co. Ltd., London, 1966.

17.4 The olivines: an isomorphous series

In the family of minerals known as the **olivines** the (SiO_4) tetrahedra are not linked to other tetrahedra but occur as separate tetrahedra. **Silicate minerals containing separate (SiO_4) tetrahedra are known as the orthosilicates.** Each silicon atom can be regarded as a cation which has contributed 4 electrons to adjacent oxygen anions. In order to achieve an 8 electron outer shell, each oxygen atom would have to receive two electrons from adjacent cations. Consequently the charges on the ions in the tetrahedron unit are $(Si^{+4} O^{-8})$ so that it can be regarded as a complex anion $(SiO_4)^{-4}$ which is known as the **silicate anion.**

The negative valency of the silicate anion may be balanced by the presence of divalent ions in the ratio two Mg^{+2} ions for each $(SiO_4)^{-4}$ tetrahedron in the crystal structure. This composition is represented by the formula Mg_2SiO_4 and the particular mineral which has this composition is called **forsterite.** The valencies may be balanced by the occurrence of Fe^{+2} ions to give a composition Fe_2SiO_4 and the mineral so produced is called **fayalite.**

The two minerals forsterite Mg_2SiO_4 and fayalite Fe_2SiO_4 are the end members of a whole series of minerals in which Mg^{+2} and Fe^{+2} ions occur together in different ratios but always in the combined ratio of two

(Mg^{+2} or Fe^{+2}) ions for each SiO$_4$ tetrahedron. The composition of this series of minerals can be represented by the formula **(Mg, Fe)$_2$SiO$_4$** and the minerals are known as **the olivines**. A series of minerals which have similar structures and forms, but which have compositions between those of two end members, is described as an **isomorphous series**.

When is was first realised that many minerals have variable compositions it became the practice to consider them as solid solutions or mixed crystals of the two end members in varying proportions. This concept is fundamentally incorrect because ionic structures are continuous and contain no molecules as such. However the method of representation is still used and it is more precise than the use of different names for arbitrarily defined parts of the continuous

isomorphous series. **The end members of the olivine series are represented by the symbols Fo = forsterite and Fa = fayalite.** A mineral of intermediate composition can then be represented by a molecular percentage composition such as **Fo$_{60}$Fa$_{40}$** and often only one symbol is used such as **Fa$_{40}$**.

17.5 The independent (SiO$_4$) tetrahedra structure of the olivines

In the structure of **forsterite Mg$_2$SiO$_4$**, the oxygen atoms occur in sheets which are parallel to the *yz* axial plane. Fig. 17.5 is a plan of the structure projected on the *xy* **axial plane**, and the unit cell with the dimensions *a* = 4.756 Å and *b* = 10.195 Å is indicated. At the side of fig. 17.5 the layers of oxygen atoms are numbered to indicate the distance from the *yz* **axial plane**, with the

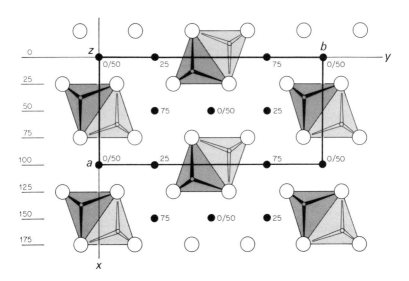

Fig. 17.5. Plan of the structure of **forsterite Mg$_2$SiO$_4$** on a plane parallel to the *x* and *y* axes. The **SiO$_4$** tetrahedra are shown and it can be seen that the oxygen atoms occur in layers parallel to the *yz* plane. The numbers at the side refer to the distances of the layers of oxygen atoms from the *yz* plane with the repeat distance *a* = 100. The black circles represent *Mg* ions and the numbers indicate the heights of the *Mg* ions above the *xy* plane which is the plane of the diagram, with the repeat distance *c* = 100. **Mg** ions occurring at **0** also occur at **100**, (after Bragg, W. L. & Brown, G. B., 1926, *Zeit, Krist.*, **63**, p. 538)

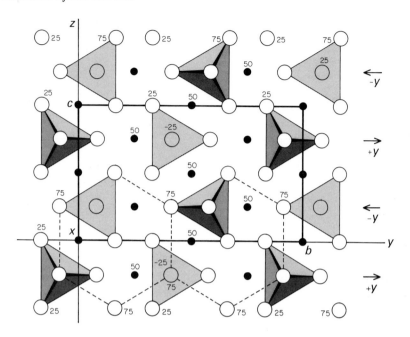

Fig. 17.6. Plan of the **forsterite** structure on the **yz plane.** The numbers by the oxygen atoms refer to distances from the **yz plane** with **100** = *a*. The approximate hexagonal close packing of the oxygen atoms in the layers is indicated in the lower part of the diagram. The **Mg** ions are shown in black, with the number **50** indicating that the ion occurs in the central plane of the unit cell, while no number indicates that **Mg** ions occur in the front and back planes of the cell (after Bragg, W. L., and Brown, G. B., 1926, *Zeit, Krist.*, **63**, p. 538)

repeat distance *a* = 100. It can be seen that the layers of oxygen atoms are labelled **25** and **75** indicating that they occur at 1/4 and 3/4 of the repeat distance *a* in the *x* direction. The **SiO₄** tetrahedra indicated with black bond lines occur at the top half of the unit cell, and the tetrahedra with the white bonds occur in the lower half of the unit cell.

Fig. 17.6 is a plan of the structure of forsterite projected on the **yz axial plane** and the unit cell has the dimensions *b* = 10.195 Å and *c* = 5.981 Å.

In fig. 17.6 the numbers by the oxygen atoms indicate the distances of the oxygen atoms from the **yz axial plane** as in fig. 17.5. The oxygen atoms are represented by small circles in the diagram but in the structure

the oxygen atoms have approximate hexagonal close packing. the hexagonal rings of oxygen are indicated in the lower part of fig. 17.6.

In the rows of **SiO₄** tetrahedra which are parallel to the *z* axis the tetrahedra point alternately either way along both the *x* and *y* directions. This arrangement of the tetrahedra gives rise to the holosymmetric orthorhombic symmetry.

The positions of the **Mg** ions are shown by black circles in fig. 17.5 and the numbers indicate the heights of the ions above the *xy* **plane** with the repeat distance *c* = **100**. In fig. 17.6 the **Mg** ions with the number **50** occur at a distance of **1/2** *a* from the **yz** axial plane, and the **Mg** ions with no numbers

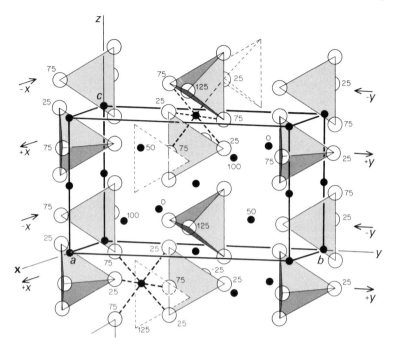

Fig. 17.7. An oblique drawing of the structure of **forsterite** to show the
position of the **Mg** ions represented by black spheres. The numbers refer
to the distances from the *yz* plane. The alternating orientations of the
(**SiO₄**) tetrahedra are shown by the arrows at the sides of the diagrams.
(after Bragg, W. L., and Brown, G. B., 1926, *Zeit. Krist.*, **63**, p. 538)

occur in the back and front planes of the
unit cell. Consequently the plans on the *xy*
and *yz* axial planes can be used in conjunc-
tion to visualize the position of the **Mg** ions
in the forsterite structure. Fig. 17.7 is an
oblique view of the forsterite structure to
show the position of the **Mg** ions more
clearly. All the numbers in fig. 17.7 refer to
distances from the *yz* **plane** with the repeat
distance *a* = 100.

The **Mg** ions have two kinds of sites. One
half of the **Mg** ions occur on the reflection
planes parallel to the (**010**) plane and lying
between the rows of (**SiO₄**) tetrahedra. The
other **Mg** ions occur between the (**SiO₄**)
tetrahedra. Each **Mg** ion has six oxygen
atoms as nearest neighbours so that they
form **octahedral coordination units** as shown
in fig. 17.7. A cleavage parallel to (010) may
be present in some crystals but irregular
cracks are more common.

The structure of olivine was determined
by W. L. Bragg and G. B. Brown in 1926
using a crystal with a composition $Fo_{90}Fa_{10}$.
The name forsterite is restricted to compositions in the range Fo_{100-90} and fayalite
is restricted to compositions in the range
Fo_{10-0}.

The term **diadochy** has been introduced to
describe the ability of the atoms of different
elements to occupy similar sites in a crystal
structure. The radius of the ferrous ion Fe^{+2}
is 0.74 Å compared with a radius of 0.66 Å
for the Mg^{+2} ion, and consequently the
dimensions of the unit cell of fayalite are
larger than those of forsterite.

	Forsterite	Fayalite
a	4.756 Å	4.817 Å
b	10.195 Å	10.477 Å
c	5.981 Å	6.105 Å

Because the **Fe–O** bonds are weaker than the **Mg–O** bonds due to the larger size of the **Fe^{+2}** ion, the melting temperature ranges from 1890 °C for forsterite to 1205 °C for fayalite. The first olivine to crystallize from a particular melt tends to be rich in **Mg** but the **Fe:Mg** ratio increases at lower temperatures.

Minerals containing independent SiO_4 tetrahedra are known as **orthosilicates**.

The garnets which form distinctive cubic crystals have structures which contain independent (SiO_4) tetrahedra. The **garnets** have a wide range of compositions and consist of a number of isomorphous series. **Zircon** which has a tetragonal unit cell, also has a structure containing independent (SiO_4) tetrahedra.

17.6 Paired tetrahedra (Si_2O_7) and ring silicate structures (Si_6O_{18})

When two (SiO_4) tetrahedra are linked by sharing one oxygen atom the pair can be represented by $(Si_2O_7)^{-6}$. Silicates containing paired tetrahedra are described as **sorosilicates**. The **epidote group** of minerals contain both independent and paired tetrahedra in their structures.

In some silicate minerals two oxygen atoms of each (SiO_4) tetrahedra may be shared with adjacent tetrahedra to form **rings** containing three $(Si_3O_9)^{-6}$, four $(Si_4O_{12})^{-8}$ or six $(Si_6O_{18})^{-12}$ tetrahedra. The ring silicates are also known as **cyclosilicates**.

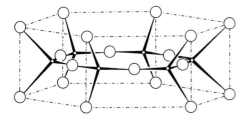

Fig. 17.8. The arrangement of the (SiO_4) tetrahedra in a **hexagonal ring** (Si_6O_{18}) which occurs in the structure of **beryl**

The structure of **beryl** contains rings of six tetrahedra arranged as shown in fig. 17.8 and these are stacked one on top of another in a manner which gives rise to the hexagonal symmetry of the crystals described in section 15.1. The composition of beryl is $Be_3Al_2 Si_6O_{18}$

17.7 The single chain structure of the clinopyroxene, diopside

In some minerals each (SiO_4) tetrahedra shares two oxygen atoms with adjacent tetrahedra to form straight chains of infinite

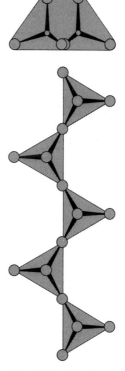

Fig. 17.9. **The single silicate chain** produced when each (SiO_4) tetrahedral unit shares two oxygen atoms. The shape of the single chain when viewed along its length is shown at the top of the diagram

extent as shown in fig. 17.9. The silicon atoms which occur at the centres of the tetrahedra are not shown. The general shape of the single chain when viewed along its length is shown at the top of fig. 17.9 and it can be seen that the bases of the tetrahedra lie in a plane. The oxygen atoms at the centre of the chain are represented by the overlapping circles. *What is the composition of the repeat unit in the single chain?*

It can be seen from fig. 17.9 that the repeat unit of the single chain is (Si_2O_6). **The pyroxenes are an important group of silicate minerals which are characterized by single chain structures.** Silicate minerals which

contain chains in their structures are sometimes called **inosilicates**.

The morphology of the crystal of **augite** has been considered in sections 13.8–13.10 and the common occurrence of the dome face $(\bar{1}01)$ which might be mistaken for the basal pinacoid is illustrated in fig. 13.20. In 1928 Warren and Bragg determined the structure of another pyroxene, the mineral **diopside**, and they established that the fundamental feature of the structure of the pyroxene group of minerals is the linkage of the tetrahedra to form single chains with a repeat unit of (Si_2O_6). The charge on the repeat unit $(Si_2O_6)^{-4}$ is balanced by the

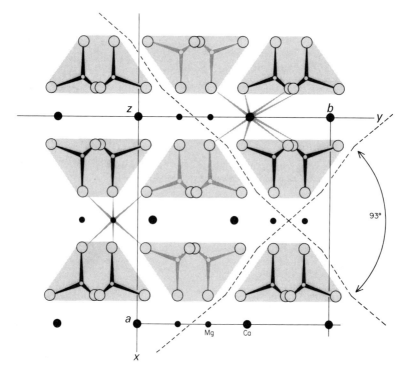

Fig. 17.10. Idealized plan of **the structure of diopside** viewed along the z direction. The single chains are parallel to the z axis and are packed together with the apices of the tetrahedra pointing in opposite directions. The larger black circles represent the Ca^{+2} ions which occur between the bases of the tetrahedra. The unit cell is based on Ca^{+2} ions. The smaller black circles represent the Mg^{+2} ions which occur between the apices of the tetrahedra. The dashed lines represent the cleavage planes which are parallel to the prism form {110}. (after Warren, B. E., and Bragg, W. L., 1928, *Zert. Krist.*, **69**, p. 168)

occurrence of two divalent cations. In diopside the cations are Ca^{+2} and Mg^{+2} so that **the composition of diopside is $Ca\ MgSi_2O_6$.**

The composition of **augite** has the same pattern but is more complex due to substitution by more elements as shown by the formula

$(Ca, Mg, Fe^{+2}, Ti, Al)_2 (Si, Al)_2O_6$.

In augite the **aluminium** occurs in **6-fold coordination** sites where it substitutes for Mg^{+2} or Fe^{+2}, and it also occurs in **4-fold coordination** with oxygen where it substitutes for Si^{+4} The aluminium ion is intermediate in size between the Si^{+4} and Mg^{+2} ions, and consequently it may occur with 4- or 6-fold coordination.

In the pyroxenes the single chains are arranged parallel to the *z* axis as shown in the idealized diagram, fig. 17.10 in which the structure of **diopside** is viewed along the *z* direction. The chains are packed together with the apices of the tetrahedra alternately pointing in opposite directions.

The chains are bonded together by Mg^{+2} and Ca^{+2} cations. The Ca^{+2} cations occur between the bases of the chains, and the unit cell outlined in fig. 17.10 is based on Ca^{+2} cations. The chains on the left side of this unit cell are shown in grey while the chains stacked on the right side of the cell are shown in black so that they can be distinguished easily in subsequent diagrams.

The **Mg** atoms lie mainly between the apices of the (SiO_4) tetrahedra and each Mg^{+2} ion is surrounded by six oxygen atoms which are bonded to only one **Si** atom, as shown on the left side of fig. 17.10. The **Ca** atoms are surrounded by eight oxygen atoms some of which are linked to two **Si** atoms in the chains. The oxygen atoms which are bonded to only one **Si** atom are linked to one **Ca** and one **Mg** atom.

The bonds between the sides of adjacent chains are very weak and the bonds to the Mg^{+2} ions between the apices of adjacent chains are also weak. Consequently two sets of cleavage planes develop parallel to the sides of the chains and passing between the apices of the adjacent chains as shown by the broken lines in fig. 17.10. The two sets of cleavage planes intersect at approximately **93°**. The prism faces of the pyroxene crystals develop parallel to the same layers of atoms as the cleavages, so that these are often described as **prismatic cleavages**. Sections of pyroxene crystals which are cut perpendicular to the *z* axis are characterized by the two sets of cleavage which intersect at **93°**, as shown in fig. 17.11.

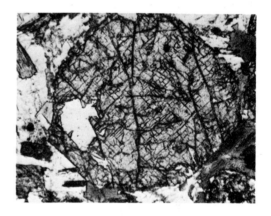

Fig. 17.11. A **pyroxene crystal** seen in thin section cut perpendicular to the *z* axis. The two sets of cleavage planes are seen as dark lines which intersect at approximately **93°** and are parallel to the prism faces (×6)

The single chain structure of diopside viewed along the *y* direction is shown in fig. 17.12. The chains are coloured in the same way as in fig. 17.10, so that those in the near half of the unit cell are black, while the chains behind them are shown in grey. It can be seen from fig. 17.12 that the chains are displaced in the *z* direction so that monoclinic symmetry results with β = 105° 50'. The pyroxene minerals which have monoclinic symmetry are described as the **clinopyroxenes**.

The unit cell of diopside has the dimensions $a = 9.73$ Å, $b = 8.91$ Å, $c = 5.25$ Å

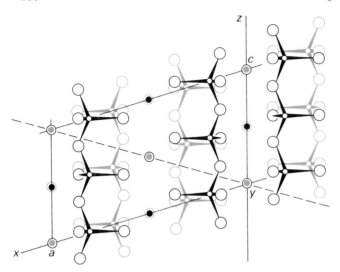

Fig. 17.12. Idealized plan of the **structure of diopside** on a plane parallel to the x and z axes. The atoms in the near side of the unit cell are shown in black, and the atoms on the far side of the unit cell are shown in grey as in fig. 17.10. The small full circles represent Mg^{+2} ions and the larger open circles represent Ca^{+2} ions. The dashed line represents the $(\bar{1}01($ **plane.** (after Warren, B. F. and Bragg, W. L., 1928, *Zeit. Krist.*, **69**, p. 168). The front planes of the unit cells shown in figs 17.12 and 17.13 are aligned

and it contains four formula units. The c dimension is the repeat distance of the single chain Si_2O_6. It can be seen from fig. 17.12 that there are well developed layers of atoms parallel to the $(\bar{1}01)$ plane which is represented by the broken line. This explains the occurrence of the dome form $\{\bar{1}01\}$ on many pyroxene crystals.

17.8 The single chain structure of the orthopyroxene, enstatite.

An isomorphous series of pyroxenes in which Mg^{+2} and Fe^{+2} ions are mutually replaceable between $Mg_{100}Fe_0$ and $Mg_{10}Fe_{90}$ exhibits orthorhombic symmetry. The end members are **enstatite** $Mg_2(Si_2O_6)$ and

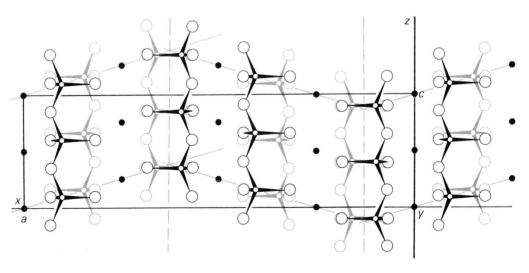

Fig. 17.13. Idealized plan of the structure of the orthopyroxene, **enstatite**, on a plane parallel to the x and z axes. The black circles represent the Mg^{+2} ions. (after Warren, B. F. and Modell, D. I., 1929, *Zeit. Krist.*, **72**, p. 42)

orthoferrosilite $Fe_2(Si_2O_6)$. Orthopyroxenes which have a composition within the range Mg_{70} to Mg_{50} are called **hypersthene** and the structure of this orthopyroxene was determined by B. E. Warren and D. I. Modell (*Zeit. Krist.*, **72**, p. 42, 1929).

A simplified diagram of the structure of **enstatite** is shown in fig. 17.13 for comparison with fig. 17.12. The Mg^{+2} ions in enstatite are approximately in the positions of the Mg^{+2} and Ca^{+2} ions in the diopside structure shown in fig. 17.12. However, because the Mg^{+2} ion is smaller than the Ca^2 ion, the stacking of the chains is different in enstatite. A comparison of fig. 17.12 (diopside) and fig. 17.13 (enstatite) shows that the stacking of the first two layers of chains is the same in both structures. In diopside the third layer of chains is stacked in the upward direction so that similar points occur along the clino-axis x, but in enstatite the third chain is displaced downwards. Consequently in enstatite the chains are stacked with a zig-zag arrangement shown by the grey dashed lines in fig. 17.13 and this gives rise to the **orthorhombic symmetry**.

The b and c dimensions of the unit cell of **diopside** and **enstatite** are very similar, but the a dimension in enstatite is almost double the a dimension in diopside as shown below.

	Diopside		Enstatite
a	9.71 Å	$\beta = 105° 50'$	18.20 Å
b	8.89 Å		8.87 Å
c	5.24 Å		5.20 Å

17.9 The double chain structure of the amphibole, tremolite

When two parallel single chains are linked together by the sharing of the basal oxygen atoms of alternate tetrahedra, a double chain structure is produced as shown in fig. 17.14. The shape of the double chain when viewed along its length is shown at the top of the diagram. *What is the composition of the silicate repeat unit in a double chain structure?*

It can be seen from fig. 17.14 that the composition of the repeat unit is (Si_4O_{11}). **The amphiboles are an important group of silicate minerals which are characterized by double chain structures.**

The structural arrangement of the amphibole, **tremolite**, is shown in the simplified diagram, fig. 17.15 (after Warren, B. E., 1929, *Zeit. Krist.*, **70**, p. 42). The double chains are aligned parallel to the z axis and they form layers parallel to the (100) plane.

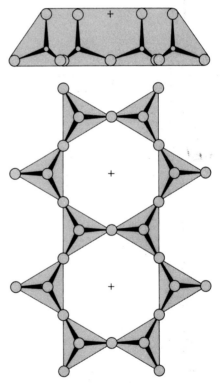

Fig. 17.14. **The double chain silicate structure** which is produced when two chains are linked by sharing alternate oxygen atoms on the basal planes of the tetrahedra. The shape of the double chain when viewed along its length is shown at the top of the diagram. The crosses at the centres of the rings indicate the positions which are sometimes occupied by the hydroxyl ion (**OH**)

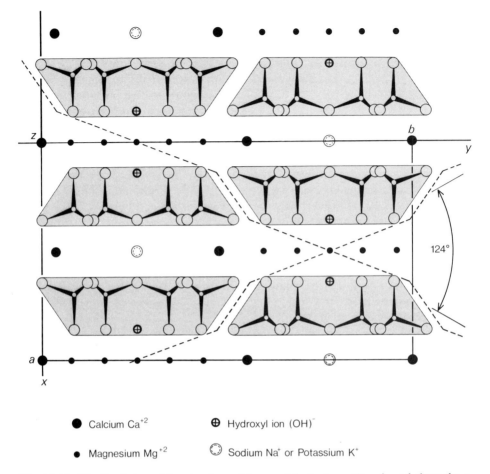

- ● Calcium Ca^{+2}
- ● Magnesium Mg^{+2}
- ⊕ Hydroxyl ion (OH)$^-$
- ◌ Sodium Na$^+$ or Potassium K$^+$

Fig. 17.15. Idealized plan of the structure of the amphibole, **tremolite,** viewed along the z direction. The double chains are parallel to z and the ends are shaded. The dashed lines indicate the planes of cleavage which are parallel to the prism form $\{110\}$. (after Warren, B. F., 1929, *Zeit. Krist.,* **70,** p. 42)

The double chains are bonded by the cations Ca^{+2} and Mg^{+2} which occupy similar positions to those which they occupy in the clinopyroxene, diopside. The Ca^{+2} ions occur mainly between the bases of the double chains while the Mg^{+2} ions occur between the apices of the tetrahedra.

The repeat distance of the double chain (Si_4O_{11}) is approximately 5.3 Å and this determines the c dimension of the unit cell. The a and c dimensions and the angle β of the unit cells of the clino-pyroxenes and amphiboles are nearly equal but the b dimension of amphiboles is nearly double the b dimension in pyroxene as shown below.

	Diopside		Tremolite	
a	9.71 Å	$\beta = 105° 50'$	9.84 Å	$\beta = 106°$
b	8.89 Å		18.05 Å	
c	5.24 Å		5.27 Å	

Prismatic cleavage planes develop in a similar manner in the pyroxenes and amphiboles, but because the double chains are much wider the cleavage planes in the amphiboles intersect at approximately **124°** and **56°** as shown by the dashed lines in fig. 17.15. Sections of amphibole crystals cut perpendicular to the z axis show two sets of cleavage lines which intersect at approximately 124° as shown in fig. 17.16.

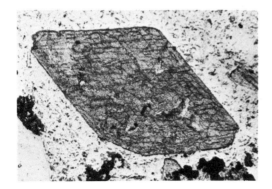

Fig. 17.16. A section of the amphibole, **hornblende,** cut perpendicular to the z direction and showing the cleavage planes intersecting at approximately 124° which are parallel to the prism faces. ($\times$6)

The double chain structure of the amphiboles allows the location of the hydroxyl ion (**OH**) in sites between the oxygen atoms at the apices of the tetrahedra along the centres of the double chains as shown by the crosses in fig. 17.14. The composition of **tremolite** is $Ca_2Mg_5Si_8O_{22}(OH)_2$ and two formula units occur in each cell.

The larger cations Na^{+1} or K^{+1} may substitute for the Ca^{+2} ions, and may also occur at the additional sites in between the two **Ca** sites shown in fig. 17.15.

If the number of the large Ca^{+2} cations is low a double chain structure with orthorhombic symmetry develops which is similar to the structure of the orthorhombic pyroxene, enstatite. **The orthorhombic amphibole anthophyllite**, has a composition

$(Mg,Fe^{+2})_7Si_8O_{22}(OH,F)_2$.

The first 25 questions in section 17.16 refer to the preceding sections.

17.10 The sheet silicates: The clays and micas

The (SiO_4) tetrahedra may share three oxygen atoms with adjacent tetrahedra to form sheets of indefinite extent of the kind shown in fig. 17.17. The oxygen atoms at the apices of the tetrahedra are bonded to the silicon atom only and are sometimes described as free oxygen atoms. An oblique view of a hexagonal ring to show the orientation of the (SiO_4) tetrahedra is shown in fig. 17.18. *What is the composition of the repeat unit in the sheet structure of fig. 17.17?*

In the silicate sheet structure the repeat unit has the composition (Si_4O_{10}), and this is the fundamental unit in the compositions of the mica and clay group minerals which are characterized by silicate sheet structures. These minerals have a single well developed cleavage parallel to the sheet structure illustrated in fig. 1.16, and they are sometimes known as the **phyllosilicates** from the Greek word for leaf or sheet. **The (001) plane is placed parallel to the single prominent cleavage.**

The hydroxyl ion $(OH)^{-1}$ may occur in the centre of the hexagonal ring of free oxygen atoms to produce a hexagonal closest packing arrangement as shown in fig. 17.19. The charges may be balanced by the occurrence of aluminium cations which rest above three ions, two of which are oxygen ions and the third is a hydroxyl ion. Three hydroxyl ions may occur above each aluminium ion so that the aluminium ion is surrounded by six oxygen or hydroxyl ions and is said to be in six-fold coordination. The six-fold coordination unit has the octahedron form and is shown in grey in fig. 17.19.

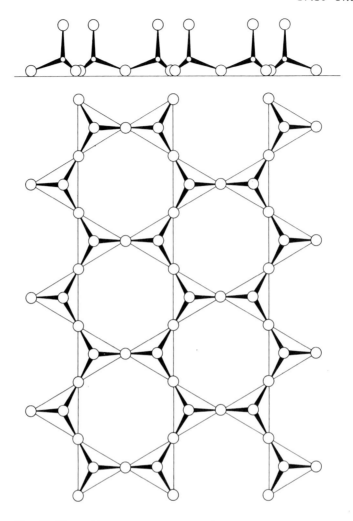

Fig. 17.17. **A silicate sheet structure** which is produced when all three oxygen atoms at the bases of the (SiO_4) tetrahedra are shared with adjacent tetrahedra. At the top of the diagram a vertical section of the sheet structure is shown with the shared oxygen atoms slightly offset

The clay mineral **kaolinite** $Al_4Si_4O_{10}(OH)_8$, is composed of the **tetrahedral sheet** (Si_4O_{10}) linked to an **octahedral layer** composed of the aluminium and additional hydroxyl ions. The hexagonal arrangements of the **Al** and **(OH)** ions in the kaolinite structure are represented by the hexagons in fig. 17.19.

The mineral **brucite** $Mg_3(OH)_6$, has a layered structure in which each layer consists of two sheets of $(OH)^{-1}$ ions in hexagonal close packing bonded by Mg^{+2} ions as represented by fig. 17.20. Each Mg^{+2} ion is in six-fold coordination and is surrounded by three $(OH)^{-1}$ ions from each layer. The location of the Mg^{+2} ions above three $(OH)^{-1}$ ions is shown by the vertical arrows in fig. 17.20. The brucite structure consists of repetitions of the layers shown in fig. 17.20, and they are held together by weak

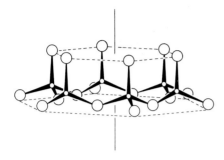

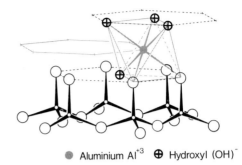

● Aluminium Al^{+3} ⊕ Hydroxyl $(OH)^-$

Fig. 17.18. An oblique view of a hex-agonal ring in the silicate sheet structure to show the orientation of the (SiO_4) tetrahedra

Fig. 17.19. **The structure of kaolinite** $Al_4Si_4O_{10}(OH)_8$, composed of a silicate sheet (Si_4O_{10}) linked to an octahedral layer composed of aluminium and additional hydroxyl ions. (after Pauling, L. 1930., *Proc. Nat. Acad. Sci.,* **16,** p. 578)

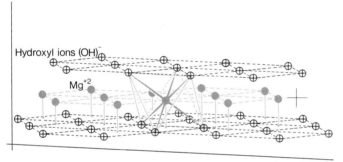

Fig. 17.20. **The structure of brucite, $Mg(OH)_2$,** composed of two sheets of $(OH)^{-1}$ ions in hexagonal close packing, bonded by Mg^{+2} ions

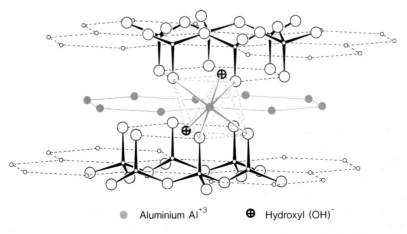

● Aluminium Al^{+3} ⊕ Hydroxyl $(OH)^-$

Fig. 17.21. **The structure of pyrophyllite** composed of an octahedral layer of aluminium ions sandwiched between two silicate tetrahedral layers forces only, so that a perfect cleavage develops parallel to the layers and the mineral is **sectile**.

The mineral **gibbsite** $\mathbf{Al_2(OH)_6}$ has a similar structure to that of brucite but because the $\mathbf{Al^{+3}}$ ion is trivalent only $\frac{2}{3}$ of the possible sites are occupied. The structure of **kaolinite** can be regard as a tetrahedral layer (Si_4O_{10}) combined with an octahedral layer similar to the structure of gibbsite.

The structure of the mineral **pyrophyllite** is illustrated in fig. 17.21. It can be seen that the octahedral layer composed of aluminium ions is sandwiched between two tetrahedral layers (Si_4O_{10}) with the apices of the tetrahedra pointing towards the centre of the composite layer. The hexagonal pattern of the upper tetrahedral layer is displaced by $\frac{1}{3}$ of the repeat distance relative to the lower tetrahedral layer. Fig. 17.22 is a plan of the structure of pyrophyllite parallel to the (001) plane which shows the positions of the hexagons formed by the free oxygens in the tetrahedral layers. The lower layer is represented by black bonds. The direction and amount of displacement is indicated by the positions of the hydroxyl ions which are at the centres of the hexagons.

The composition of **pyrophyllite** is represented by the formula $\mathbf{Al_4(Si_8O_{20})(OH)_4}$ which indicates the occurrence of two silicate tetrahedral sheets linked by a sheet of octahedrally coordinated $\mathbf{Al^{+3}}$ ions. The structure of pyrophyllite consists of repetitions of the three layers.

Aluminium may replace silicon in the tetrahedral units producing sheets represented by the composition $\mathbf{(AlSi_3O_{10})^{-5}}$. The mineral **muscovite** has a three-layered sheet structure like that of pyrophyllite but one quarter of the silicon sites are occupied by aluminium ions. The excess negative charge on the three layered sheet is balanced by the occurrence of potassium ions in between the three layered sheets as shown in fig. 17.23. The large potassium ions are located between the hexagonal rings of the shared oxygen ions in the tetrahedral sheets so that

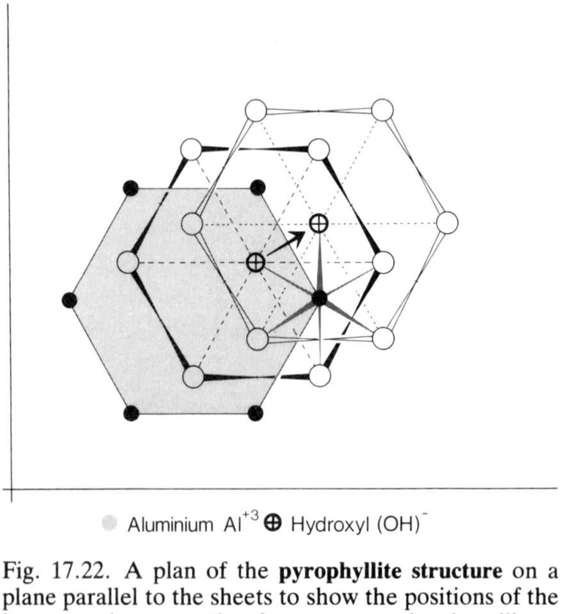

Fig. 17.22. A plan of the **pyrophyllite structure** on a plane parallel to the sheets to show the positions of the hexagons between the free oxygens in the silicate sheets. The arrow shows the apparent displacement of the upper silicate sheet. Aluminium atoms occur at the corners of the shaded hexagons

they have twelve-fold coordination. The structure of muscovite was determined by W. W. Jackson and J. West (*Zeit. Krist.*, 1930, **76**, 211; 1933, **85**, 160). It was found that the hexagonal rings are not as regular as in the simplified diagrams shown here.

There is no lateral displacement of the tetrahedral layers across the layer of potas-

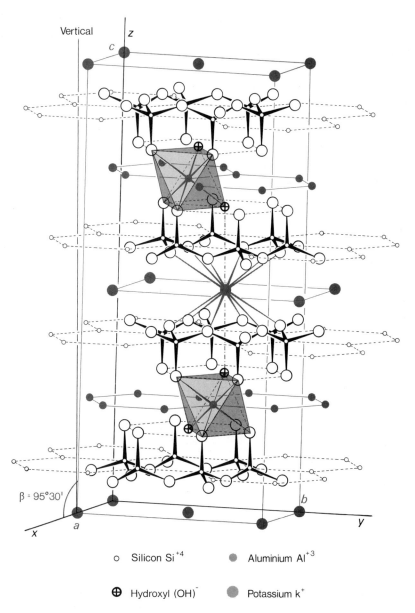

Fig. 17.23. Obligue drawing of **the structure of muscovite.** Large potassium ion **K**$^{+1}$ bond the composite three-layered sheet structure which is similar to that in pyrophyllite, (after Jackson, W. W., and West, J., *Zeit. Krist.*, **76**, 211, 1930; **85**, 160, 1933)

sium ions, and in the oblique view of the muscovite structure fig. 17.23 the central hydroxyl ions are on the same vertical line. The unit cell of muscovite extends across two of the composite sheets and the lateral displacement of the tetrahedral layers on either side of the octahedral layer is in opposite directions in the two composite sheets.

Fig. 17.24 is a plan parallel to (001) to show the stacking arrangement of the hexagonal layers in the unit cell of muscovite. The hexagon of free oxygen atoms in the lowest tetrahedral layer is represented by black bonds as in fig. 17.22. The hexagon of free oxygen atoms illustrated by white bonds represents the tetrahedral layers at the top of the lower group and the bottom of the upper group. The hexagon of free oxygen atoms in the tetrahedral layer at the top of the unit cell is shown by black bonds and the components of the two displace-

ments in planes parallel to (001) are shown by the arrows.

A plan of the structure of muscovite projected on the (100) plane of the unit cell is shown in fig. 17.25, and the component displacements of the tetrahedral layers in the *y* direction are displayed. A pronounced cleavage develops parallel to the layers of potassium ions and the crystals have a pseudohexagonal form which reflects the hexagonal nature of the sheets as illustrated in fig. 1.16.

Muscovite has monoclinic symmetry. Diad axes of symmetry parallel to the *y* axis pass through the potassium ions and are indicated in fig. 17.25. There are also **glide planes** parallel to the (010) represented by dashed lines in fig. 17.25, on which there is a translation of *c*/2. Note the positions of the K^{+1} and Mg^{+2} ions on either side of the glide plane near the *xz* axial plane in fig. 17.25.

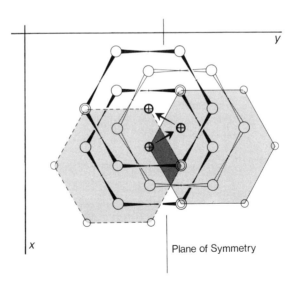

Fig. 17.24. **Plan of part of the muscovite structure on the (001)** plane to indicate the stacking arrangement of the hexagonal rings of free oxygen atoms. Aluminium atoms occur at the corners of the shaded hexagons

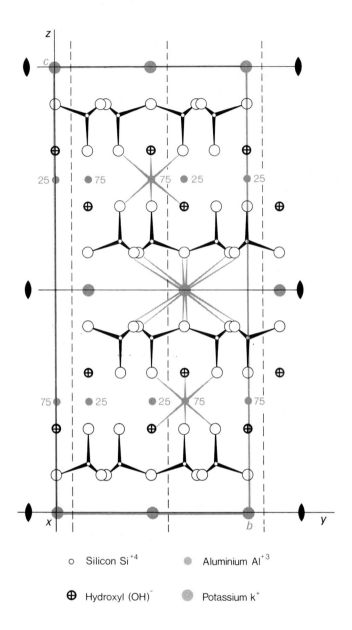

o	Silicon Si^{+4}		Aluminium Al^{+3}
⊕	Hydroxyl (OH)$^-$		Potassium k$^+$

Fig. 17.25. **Plan of the muscovite structure on a plane parallel to (100).** Diad symmetry axes parallel to *y* pass through the **K^{+1}** ions. Glide planes parallel to **(010)** on which there is a translation of *c*/2 are shown by the dashed lines. The numbers by the **Mg^{+2}** ions indicate distances from the *yz* axial plane with the repeat distance $a = 100$

17.11 The classification of the sheet silicate minerals

In the structures of **kaolinite, pyrophyllite,** and **muscovite** studied here, the octahedral layers consist of two aluminium ions for every six $(OH)^{-1}$ or oxygen ions. This is described as a **dioctahedral structural arrangement** and it is a characteristic of the mineral **gibbsite** which has a composition of $Al_2(OH)_6$.

In the mineral **brucite** $Mg_3(OH)_6$ (fig. 17.20), magnesium occurs in octahedral layers with hydroxyl ions but in the ratio of three magnesium ions to every six (OH) ions. The brucite layer structure is described as a **trioctahedral structural arrangement**.

The minerals which have sheet structures may be classified into two groups which are characterized by dioctahedral layers of the gibbsite type, or trioctahedral layers of the brucite type. The classification of the common sheet silicate minerals is shown in the table below.

Dioctahedral Layers (Gibbsite type)	Trioctahedral layers (Brucite type)
Two layer structures	
Kaolinite $Al_4Si_4O_{10}(OH)_8$	Antigorite $Mg_6Si_4O_{10}(OH)_8$
Three layer structures	
Pyrophyllite $Al_2Si_4O_{10}(OH)_2$	Talc $Mg_3Si_4O_{10}(OH)_2$
Muscovite $KAl_2(AlSi_3O_{10})(OH)_2$	Phlogopite $KMg_3(AlSi_3O_{10})(OH)_2$
	Chlorite $Mg_5Al(AlSi_3O_{10})(OH)_8$

Antigorite is one of the serpentine minerals with a sheet structure similar to that of the two layer structure of kaolinite but with Mg^{+2} forming a trioctahedral layer in place of the dioctahedral aluminium layer which is characteristic of kaolinite. In the same way

talc illustrated in fig. 1.8 has a three-layer structure similar to that of pyrophyllite but again with Mg^{+2} occurring in a trioctahedral layer.

The composition of **phlogopite** is $KMg_3(AlSi_3O_{10})(OH)_2$ and this is analogous to the composition of muscovite. In the common mineral **biotite**, some Fe^{+2} occurs as well as Mg^{+2} so that the composition is $K(Mg,Fe)_3(AlSi_3O_{10})(OH)_2$.

Chlorite contains octahedral layers in which Al^{+3} and Mg^{+2} ions occur in the ratio of one Al^{+3} ion to $5Mg^{+2}$ ions. Octahedral layers of this kind are flanked by tetrahedral layers with $(OH)^{-1}$ ions as in the talc and pyrophyllite structures. However the structure of chlorite consists of repetitions of a three layer structure with an additional octahedral layer in which the Mg^{+2} and Al^{+3} ions are bonded to $(OH)^{-1}$ ions as in the brucite structure shown in fig. 17.20.

In the minerals **muscovite, phlogopite**, and **biotite** the negative charges at the bases of the tetrahedral layers are compensated by the occurrence of cations such as K^{+1} between the three layer units and these cations give the minerals some strength. These minerals are members of an isomorphous group known as the **micas** within the phyllosilicates. In contrast the minerals **kaolinite, pyrophyllite, antigorite** and **talc** are relatively weak because the two or three layered structures are not bonded together by the presence of cations.

Kaolinite is a member of a large group of minerals known as the **clays** which are hydrous silicates mainly with aluminium. The clays have composite sheet structures and they are classified into four groups on the basis of the spacing of the (001) planes which are approximately **7 Å, 10 Å, 15 Å,** and **14.5 Å.** Kaolinite belongs to the group with a c dimension of approximately 7 Å and some important members of the other groups are illite, 10 Å, montmorillonite, 15 Å, and vermiculite, 14.4 Å when fully expanded with water.

17.12 Framework silicates: The structure of β quartz

When all four oxygen atoms of the (SiO_4) tetrahedra are shared a **three-dimensional framework structure** is produced in which the **Si:O** ratio is 1:2, so that the composition is SiO_2 and the positive and negative valencies are balanced. **Silica (SiO_2)** crystallizes in several different forms of which **quartz** is not only the most common of the polymorphs, but is also one of the most abundant rock forming minerals. There are two forms of **quartz**. The high temperature form which is stable between 573 °C and 870 °C is called β-**quartz**. β-quartz has hexagonal symmetry and the arrangement of the tetrahedra in the structure is illustrated in fig. 17.26. The structure of quartz consists of helices of Si–O–Si bonds, and one is illustrated by the black bonds in fig. 17.27.

The helices are either all left handed as those illustrated in fig. 17.27, or they are all right handed. The thumb indicates the vertical and the fingers the direction of the helix. This explains the enantiomorphic characteristics of the quartz crystals studied in section 15.7.

Fig. 17.28 is a plan of the structure projected on the (**0001**) plane. The (SiO_4) tetrahedra are oriented with diad axes parallel to the z crystallographic axis so they are represented as squares with the top edge of the tetrahedron indicated by a diagonal. The bonds between the silicon and oxygen atoms of the tetrahedra represented in fig. 17.26 are indicated, and the bonds from the silicon to the upper oxygen atoms are shown in black. A primitive unit cell is indicated by dashed lines to illustrate the hexagonal symmetry of the structure.

In fig. 17.28 the tetrahedra which occur at

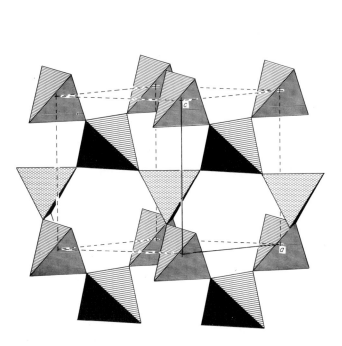

Fig. 17.26. The arrangement of the **SiO₄** tetrahedra in the **structure of β quartz**. (after Bragg, W. H., and Gibbs, R. E., 1925, *Proc. Roy. Soc.*, **A109,** 405)

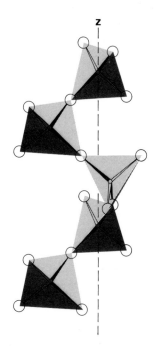

Fig. 17.27. A left-handed helix of **Si—O—Si** bonds in the structure of β **quartz** is shown by the bonds coloured black

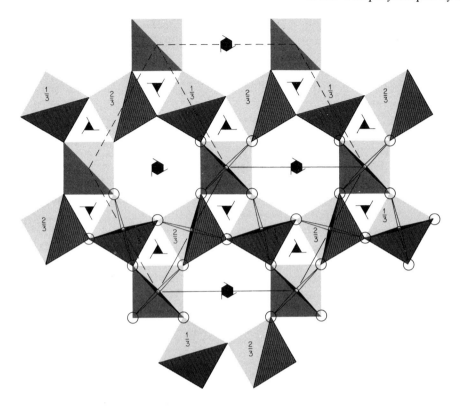

Fig. 17.28. A plan of the structure of β **quartz** projected on the **(0001)** plane. The tetrahedra which occur at the top and bottom of the unit cell are not labelled, and 1/3 and 2/3 refer to the height of the centres of the tetrahedra as fractions of *c*. The triad and hexad screw axes are indicated by symbols

the bottom and top of the unit cell are not labelled. The labels 1/3 and 2/3 refer to the height of the centres of the tetrahedra above the base of the unit cell as fractions of the *c* dimension. The positions of the vertical triad and hexad screw axes in the structure are indicated by symbols.

When quartz is subjected to directed pressure, positive and negative charges develop at the ends of the diad axes. Consequently when an alternating current is applied across a specially cut quartz plate, it oscillates and can be used to stabilize radio waves.

17.13 The polymorphs of silica

Silica occurs in nature as four distinct minerals **quartz, tridymite, crystobalite, opal**, and as rare silica glass. The crystalline forms quartz, tridymite, and crystobalite all have high and low temperature forms.

Low temperature quartz is known as α-quartz and it is stable up to a temperature of 573 °C. The structure of α-quartz is less regular than that of β-quartz, because the tetrahedra are twisted and it has trigonal symmetry only. The transformation from α to β quartz at 573 °C takes place easily because it involves relatively small modifications of the structure without the breaking of any Si–O bonds.

Tridymite is stable in the temperature range 870 °C to 1470 °C at atmospheric pressure. The structure of tridymite is less dense than that of quartz. It consists of an open network of SiO_4 tetrahedra in sheets parallel to **(0001)**. The bases of the tetrahedra lie in the **(0001)** plane but the apices point in alternate directions. Although the oxygen atoms are not closely packed, the arrangement of the oxygen atoms is that of hexagonal close packing. The stable form of tridymite has hexagonal symmetry but there are forms which can exist at lower temperatures that exhibit orthorhombic symmetry.

Crystobalite is stable in the temperature range 1470 °C to 1713 °C. Its structure also consists of sheets of tetrahedra but the oxygen atoms have a cubic close packing arrangement so that cristobalite has cubic symmetry. The open structures of cristobalite and tridymite are stable at high temperatures due to the thermal energy, and both these minerals occur in lavas. At lower temperatures they tend to invert to the more stable structures of quartz, but the inversion occurs very slowly because the rearrangement of the structure involves the breaking of bonds. These changes are called **reconstructive transformations** whereas the smaller changes which occur between the α–β forms are known as **displacive transformations.**

Coesite is a high pressure form of silica which is found in rocks subjected to the impact of meteorites.

The presence of small amounts of water greatly changes the stability fields of the silica minerals, by lowering the melting points. **Opal** is a hydrous form of silica which consists of extremely small crystals of cristobalite with water in sub-microscopic pores. Precious opal exhibits a play of colours due to diffraction of light from the structure.

17.14 The feldspars

When Al^{+3} substitutes for Si^{+4} in tetrahedral coordination with oxygen in a framework structure, additional cations are required in order to attain electrical neutrality. The cations fit into large spaces in the framework structure which is constructed of (SiO_4) and (AlO_4) tetrahedra. If one quarter of the Si^{+4} sites are occupied by Al^{+3}, the valencies may be balanced by K^{+1} or Na^{+1} ions to form **orthoclase, $KAlSi_3O_8$** or **albite $NaAlSi_3O_8$**. If one half of the Si^{+4} sites are occupied by Al^{+3}, then Ca^{+2} cations may balance the charges to give **anorthite $CaAl_2Si_2O_8$**. These minerals are the end members of a group of minerals known as the **feldspars**.

The feldspars are the most important group of rock forming minerals and they are used in the classification of igneous rocks. The range in compositions of the feldspars can be represented by the ternary diagram fig. 17.29. Members of the series between the end members **orthoclase $KAlSi_3O_8$**, and **albite $NaAlSi_3O_8$**, are known as the **alkali feldspars** and those between **albite** and **anorthite $CaAl_2Si_2O_8$**, are called the **plagioclase feldspars**. Fig. 17.29 shows that the akali

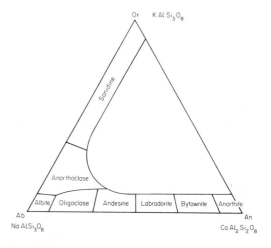

Fig. 17.29. The nomenclature of the **plagioclase series** and the **high-temperature alkali feldspars**

feldspars generally can be regarded as containing less than 10% of the anorthite component, while the plagioclase series contains less than 10% of the orthoclase component.

The potassium end member can exist in a number of different structural states. A low temperature form called **orthoclase** has monoclinic symmetry which has been studied in sections 13.1–4. The high temperature structural form of **KAlSi₃O₈** is called **sanadine** and it also has monoclinic symmetry. In sanadine the distributions of the Al^{+3} ions which substitute for Si^{+4} ions is irregular and is said to be **disordered**. In the low temperature structure of orthoclase the distribution of Al^{+3} ions is more regular and is said to be **ordered**. Most feldspars in volcanic rocks are high-temperature forms because the rapid cooling has prevented the structure modifying to a low temperature arrangement. The low-temperature feldspars either crystallized at low temperatures or developed from the high temperature structures during the slow cooling and crystallization which is characteristic of large intrusions of igneous rocks.

In the high temperature triclinic plagioclase series there is a complete range of compositions between the end members **anorthite CaAl₂Si₂O₈** and **albite NaAlSi₃O₈**. At low temperature there are four structural divisions and limited substitution of $Na^{+1}Si^{+4}$ for $Ca^{+2}Al^{+3}$.

It has been found by experiment that when an isomorphous series like the high temperature plagioclases crystallizes, the composition of the mineral is almost always different from the composition of the melt. The temperature range of crystallization of the plagioclase feldspars was determined by N. L. Bowen and is illustrated in fig. 17.30. A melt of pure anorthite would begin to crystallize at approximately 1550 °C whereas a melt of pure albite begins to crystallize at 1100 °C. In each case the crystals would have the same composition as the melt. However a melt of composition **An₅₀Ab₅₀**

begins to crystallize at 1450 °C represented by point **A** in fig. 17.30. But the crystals which start to grow in the melt at approximately 1450 °C have a composition **Ab₁₈An₈₂** represented by the point **B** on the isotherm.

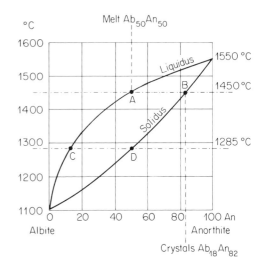

Fig. 17.30. **The compositions of the melts and plagioclase feldspar crystals** which are in equilibrium at particular temperatures. The composition of the melt is indicated by the **liquidus** curve and the composition of the plagioclase crystal in equilibrium with a liquid at a particular temperature can be obtained from the **solidus** curve. (after Bowen, N. L., 1913, *Amer. Journ. Sci.,* **35,** p. 577)

Bowen conducted a series of experiments in which a melt was held at a selected temperature until crystals had grown and equilibrium was established between the liquid and the crystals. When the compositions of the equilibrium liquid and crystals were plotted for the series of temperatures, a curve could be drawn through the compositions of the liquids. This curve is called the **liquidus** and it indicates the temperature at which a melt of particular composition begins to crystallize. A curve could be drawn through the compositions of the crystals also. This curve is called the **solidus** and it indicates the composition of the crystals which are in equilibrium with a liquid of particular composition at a particular temperature. The first crystals of the isomorphous series to grow always have a composition which is richer in the end member that crystallizes at the higher temperature.

When the temperature decreases, crystallization of the plagioclase feldspar will continue but the formation of calcium rich crystals will cause the composition of the melt to become richer in sodium. With falling temperature the changing composition of the melt is represented by the liquidus from point **A** towards point **C**. However if equilibrium is maintained, the composition of the crystals must also change and the appropriate crystal compositions at particular temperatures are indicated by the solidus from point **B** towards point **D**. Equilibrium between melt and crystals is approached by the diffusion of Ca^{+2} and Al^{+3} ions out of the crystals into the melt, and the diffusion of Na^{+1} and Si^{+4} ions in the reverse direction. If perfect equilibrium could be attained the last liquid represented by point **C** would crystallize at 1280 °C when the crystals have the composition of the original melt, $An_{50}Ab_{50}$ represented by point **D** on the solidus. Because the reaction between the crystals and the liquid is continuous, Bowen has called the plagioclase series a **continuous reaction series**.

Fig. 17.31. A microscope section of tabular shaped, **zoned plagioclase crystals** showing a gradual change in the polarization colour from light grey in the centre to black at the margin. The optic orientation of plagioclase depends on the composition, and this zonal change in the polarization colour indicates a change in composition from An_{60} in the early formed centre, to An_{10} at the margins. The remaining spaces were filled by the simultaneous crystallization of albite and quartz to form the **graphic intergrowth texture** which derives its name because of its occasional similarity to hieroglyphics. The parallel bands with different polarization colours in the plagioclase indicate **repeated albite twinning** of the kind shown in fig. 16.11. ($\times 10$)

Crystallization often proceeds at a faster rate than the reaction between the crystals and the liquid so that equilibrium is not often attained. When more Na-rich plagioclase crystallizes around a Ca-rich crystal, the composition varies from the centre outwards and the crystal is said to be **zoned** as shown in fig. 17.31. When zoned crystals are formed, the composition of the final liquid and the last plagioclase to crystallize is more rich in **Na** than the composition of the original melt, although the average composition of the **Ca**-rich centres and **Na**-rich margins of the crystals is equivalent to the composition of the original melt.

In the alkali-feldspar series there is a complete range in composition from high temperature **albite** to high temperature **sanadine**, but a change from the triclinic symmetry of albite to the monoclinic symmetry of sanadine occurs at a composition of approximately **Ab$_{63}$** as shown in fig. 17.29. At lower temperatures the degree of substitution is limited and albite layers or patches separate within the potassium rich crystal.

The crystallization of the alkali feldspar series at relatively high water vapour pressure is represented by fig. 17.32. It can be seen that when melts richer in **K** than approximately **Or$_{30}$Ab$_{70}$** crystallize, the crystals are much richer in potassium than the melt. In contrast melts with less potassium than Or$_{30}$Ab$_{70}$ produce crystals which are richer in Na than the melt. In each case crystallization proceeds towards the minimum indicated by the intersection of the two liquidus curves, and if equilibrium could be attained the composition of the final crystals would be the same as the composition of the original melt. However, at lower temperatures the structures of crystals within the composition range **Or$_{77}$Ab$_{23}$** to **Or$_8$Ab$_{92}$** are unstable, and they tend to separate into two mineral phases. When albite separates to form streaks within a potassium rich alkali feldspar, the texture is

called **perthite** or **mircroperthite**. Microperthite texture seen in a thin section of feldspar is illustrated in fig. 17.33.

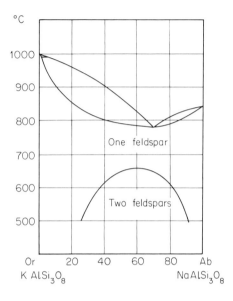

Fig. 17.32. Equilibrium diagram of the crystallization of the alkali feldspars at a water vapour pressure of 2000 bars. (after Bowen, N. L., and Tuttle, O. F., 1950, *Journ. Geol.*, **58**, p. 489)

17.15 The crystallization reaction series

The composition of **olivine (Mg,Fe)$_2$SiO$_4$** ranges from the high temperature mineral, **forsterite Mg$_2$SiO$_4$** to the lower temperature mineral, **fayalite Fe$_2$SiO$_4$**. During crystallization the early formed olivine crystals are rich in **Mg^{+2}** ions but as crystallization proceeds, **Mg^{+2}** ions in the early crystals are replaced by **Fe^{+2}** ions by reaction with the liquid. The pyroxene minerals change in a similar way from Mg-rich to Fe-rich minerals.

Many of the silicate minerals which commonly occur in igneous rocks do not undergo simple melting called **congruent melting**, in which the solid crystal phase changes to a liquid phase of the same composition at the

Fig. 17.33. Part of a large crystal of **orthoclase** which contains smaller crystals of plagioclase and biotite. The presence of a **Carlsbad twin plane** on the right side of the orthoclase crystal is indicated by the change in the polarization colour. The **microperthite texture** consists of thin, irregular layers of albite (light grey) within the orthoclase (dark grey). (×10)

melting point. Some of the silicate minerals break down at the melting point to another mineral and some liquid. For example the pyroxene, **enstatite**, breaks down at 1557 °C to crystals of **olivine** with liquid SiO_2 as represented by the equation.

$$2MgSiO_3 = Mg_2SiO_4 + SiO_2$$

Enstatite = Forsterite + Silica liquid

This process is known as **incongruent melting**.

The process is reversed during the cooling and crystallization of a magma. Olivine has a high melting point so it begins to crystallize at an early stage. After the temperature falls below the incongruent melting point of orthopyroxene, the olivine crystals are no longer in equilibrium with the liquid, and they react with the liquid until they are completely made over to the orthopyroxene mineral which is stable in the melt within this particular temperature range. The olivine and orthopyroxene are sometimes referred to as a **reaction pair**.

If the rate of cooling was very slow, the crystals of olivine would be completely changed over. Commonly the reaction between the olivine crystals and the liquid was incomplete so that gabbros often contain both olivine and pyroxene crystals. The olivine crystals are often enclosed by or are rimmed by pyroxene, indicating that the olivine crystals were of earlier formation. This texture is called a **reaction rim**. At still lower temperature the pyroxene minerals may react with the liquid to form one of the amphibole minerals such as tremolite and this in turn may alter to form a sheet silicate such as chlorite. N. L. Bowen used the term **discontinuous reaction series** to describe the change from a higher temperature isomorphous series to a lower temperature isomorphous series by reaction with the magma.

   Mineralogy is a fascinating subject in its own right, but it is also the basis of petrology—the science of the formation of rocks. Clearly the conditions of temperature, pressure and the chemical environment determine which mineral species will form. Much knowledge of the stability ranges of particular minerals and simple mineral assemblages have been obtained from the experimental crystallization of melts. However most rocks have complex chemical compositions, and are composed of isomorphous series of minerals which crystallized in response to geological processes that operated at extremely slow rates over long periods of time. Many rocks exhibit intricate sequences of crystallization of minerals, and information concerning the nature of the geological processes that have operated may be deduced from studies of the textural relationships of the mineral assemblages, and the knowledge of their stability ranges. The reader who is stimulated to develop mineralogical studies further, will certainly find a wealth of very interesting and challenging, unsolved petrological problems.

17.16 Questions for recall and self assessment

1. Which are the eight most common elements in the rocks of the earth's crust?
2. Which element occupies the greatest volume in the rocks of the earth's crust?
3. Which is the second most abundant element in the rocks of the earth's crust?
4. List the ionic characteristics of the eight most common elements.
5. What is the form of the most common coordination unit in the minerals of the earth's crust?
6. What is the nature of the bonding within the most common coordination unit?
7. What is an isomorphous series?
8. What is the characteristic of the structure of olivine, and how are the charges balanced?
9. What are the names and compositions of the end-member olivines?
10. In which way is a particular composition of an olivine represented by symbols?
11. Why do the end members of the olivine series have different melting points?
12. When olivine crystallizes in a magma, what will be the chemical characteristic of the first crystals?
13. What is the reason for the hexagonal crystal form of beryl?
14. What is the characteristic feature of the structures of the pyroxenes?
15. What are the names and compositions of the end members of the orthopyroxene series?
16. What is the composition and symmetry of diopside?
17. In diopside, where do the cations occur in relation to the silicate structure?
18. What is the characteristic cleavage of the pyroxenes and the explanation for its development?
19. Which features determine the c and b dimensions of the unit cells of the pyroxenes?
20. What are the relative values of the a dimensions of the unit cells of the monoclinic and orthorhombic pyroxenes?
21. What is the composition of the repeat unit of a silicate double chain?
22. What is the name of the most important group of rock forming silicate minerals

which is characterized by double chain structures?

23. What are the relative values of the dimensions of the unit cells of the clinopyroxenes and amphiboles?

24. What is the characteristic cleavage of the amphiboles?

25. Which ion is able to fit into the rings of free oxygen atoms in the double chains?

Questions referring to sections 17.10 to 17.16

26. What is the composition of the repeat unit of a silicate sheet?

27. What is the characteristic physical property of the phyllosilicates?

28. What are the characteristic features of the structure of kaolinite $Al_4Si_4O_{10}(OH)_8$?

29. What is the structure of brucite, $Mg_3(OH)_6$?

30. What is the characteristic of the structure of gibbsite, $Al_2(OH)_6$?

31. What are the names given to the arrangements of the ions in brucite and gibbsite?

32. What is the nature of the structure of pyrophyllite, $Al_4(Si_4O_{10})_2 (OH)_4$?

33. What is the composition of talc?

34. Muscovite is composed of layers like the structure of pyrophyllite but with $\frac{1}{4}$ of the Si^{+4} sites occupied by Al^{+3}. How are the charges balanced in muscovite?

35. What is the symmetry of muscovite?

36. If the composition of pyrophyllite is represented by $Al_2Si_4O_{10} (OH)_2$, what is the composition of muscovite?

37. What is the composition of phlogopite?

38. What is the composition of biotite?

39. What is the characteristic of the structures of the micas?

40. What are the general characteristics of the clay minerals such as kaolin?

41. What is the name which is given to the structure which is produced if all four

oxygen atoms in the SiO_4 tetrahedra are shared?

42. What are the names of the minerals which are composed of silica?

43. Name the high and low temperature forms of quartz and the characteristic symmetry of their structures.

44. What is the main characteristic of the arrangement of the SiO_4 tetrahedra in the high temperature form of quartz.

45. What is likely to happen to the structure of β quartz, if the temperature is raised above 870 °C and above 1470 °C.

46. What are the names and chemical compositions of the end members of the plagioclase feldspars series?

47. When plagioclase crystallizes from a magma, what will be the chemical characteristic of the first formed crystals and what changes might occur in these crystals as the temperature decreases?

48. What is the name which is used to describe the crystallization of an isomorphous series?

49. What are the names of the curves which indicate the compositions of the crystals and liquid which are in equilibrium at particular temperatures?

50. What are the names and symmetry of the low and high temperature forms of $KAlSi_3O_8$?

51. What symmetry is exhibited by the high temperature alkali-feldspars?

52. What are the characteristics of the first crystals to form from alkali feldspar melts?

53. Assuming that equilibrium crystals have been formed by the crystallization of a melt in the composition range $Or_{77} Ab_{23}$ to $Or_8 Ab_{22}$, what happens during the cooling of the crystals?

54. What is incongruent melting?

55. What is a reaction pair?

56. What is the meaning of discontinuous reaction series?

Appendix

A *Answers to the Self-assessment Questions*

Chapter 2: Answers to the questions posed in section 2.11

The key words or phrases are indicated in bold type.

1. The interfacial angle is the **angle between the normals** to two faces **or the external angle** between two crystal faces, measured **in a plane which is perpendicular to both faces**.
2. The interfacial angles of large crystals may be measured with a contact **goniometer**, but accurate work is done with a reflecting light goniometer.
3. Steno discovered that the **interfacial angles between corresponding faces** on the crystals of a particular mineral **are constant** regardless of the size and shape of the faces.
4. A zone is **a set of faces which intersect**, or whose extensions would intersect **to produce parallel edges, the direction** of which is defined as the zone axis.
5. A plane of reflection symmetry is defined as an imaginary plane that divides the crystal so that **the angular orientation of a face** on one side **is the mirror image** of a corresponding face on the other side of the reflection plane.

6. **During rotation** of 360° about an axis of symmetry, exactly **the same angular orientations of the faces occur** in more than one position. A 1-fold axis is used to indicate no symmetry.
7. The names of the symmetry axes indicate the number of positions in which identical angular orientations of the faces occur during 360° rotation, and they are **diad** (two), **triad** (three), **tetrad** (four), and **hexad** (six).
8. A crystal exhibits a symmetry centre if **each face is related to a similar parallel face on the opposite side** of the crystal, and the centre may be regarded as the point of intersection of zone axes.
9. The operation by which a face is placed in an identical orientation on the opposite side of a crystal is called **inversion**.
10. In crystallographic studies the term form refers to **a groups of faces** all of which have **the same angular relationship to the elements of symmetry**.
11. An axis of rotary inversion combines **rotation** about an axis with **inversion**.
12. A diad inversion axis produces the same effect as a **plane of reflection symmetry** (m).
13. **A one-fold inversion axis ($\bar{1}$) involves**

rotation through 360° combined with inversion and this produces the same effect as a symmetry centre.

14. Each symmetry class is characterized by **a particular combination of elements of symmetry** and this is the basis of the classification of the external forms of crystals.

15. Hessel demonstrated in 1830 that there are only **32 different symmetry classes**.

16. The three symmetry operations are **reflection, rotation**, and **rotation combined with inversion**.

17. **The group of symmetry elements** which is possessed by a particular crystal is described as a point group because they **describe symmetry about a single central point** which may be indicated by the intersection of zone axes.

18. The 32 symmetry classes are classified into 7 crystal systems which are characterized by **the occurrence of particular axes of symmetry**.

19. The five symmetry classes which belong to the cubic system all possess **4 triad axes**.

20. The crystal class which exhibits the highest degree of symmetry in a crystal system is called the **holosymmetric class**.

Chapter 4: Answers to the questions posed in section 4.11.

1. Haüy proposed that crystals are made up of **small units bounded by cleavage planes**, but this aspect of his hypothesis fails because the cleavage form of fluorite is the octahedron which cannot be packed together to completely fill space.

2. Both Haüy and Huygens had the idea that a crystal is composed of **a fundamental unit which is repeated to produce a three-dimensional structure**, and it is now thought that **the unit is composed of the smallest group of atoms** which when repeated in 3-dimensions, would produce the continuous crystal structure.

3. A 2-dimensional pattern can be represented by a **network** or lattice **composed of two sets of equally spaced, parallel lines which intersect at identical points**.

4. The unit of the least symmetrical 2-dimensional pattern has the shape of **a parallelogram** with a point at each corner.

5. There are **only five kinds of 2-dimensional patterns** of points, and the unit cells are the **parallelogram, rectangle, square, rhombus** or centred rectangle and the **hexagon**.

6. All 2-dimensional patterns can be represented by a network having a parallelogram shaped unit which has only a centre of symmetry, but **the other unit cells indicate greater degrees of symmetry of the patterns**. In the centred rectangle and the hexagon, additional points are used to display the symmetry.

7. Three sets of parallel, equally spaced planes can be used to divide space into cells, identical in size, shape, and orientation. The intersection of the planes produces three sets of parallel lines joining equally spaced points which have identical surroundings **The regular 3-dimensional array of points in space** is known as a space lattice.

8. A primitive space lattice has points only at the eight corners of the unit cell, but since each point is shared by eight adjacent unit cells and counts as $\frac{1}{8}$, there is only **one point for each primitive unit cell**.

9. Bravais demonstrated in 1848 that there are **only 14 kinds of space lattice**.

10. x front to back.
    y right to left.
    z top to bottom.

11. $\alpha = y \widehat{\ } z$; $\beta = z \widehat{\ } x$; $\gamma = x \widehat{\ } y$.

12. a = repeat distance in the x direction.
    b = repeat distance in the y direction.
    c = repeat distance in the z direction.
13. Primitive spaces latices
    Cubic P
    Tetragonal P
    Orthorhombic P
    Monoclinic P
    Triclinic P
14. Angles between the rows of points

Cubic	$\alpha = \beta = \gamma = 90°$
Tetragonal	$\alpha = \beta = \gamma = 90°$
Orthorhombic	$\alpha = \beta = \gamma = 90°$
Monoclinic	$\alpha = \gamma = 90°, \beta > 90°$
Triclinic	$\alpha \neq \beta \neq \gamma \neq 90°$

15. Repeat distances

Cubic	$a = b = c$
Tetragonal	$a = b \neq c$
Orthorhombic	$a \neq b \neq c$
Monoclinic	$a \neq b \neq c$
Triclinic	$a \neq b \neq c$

16.

Cubic	$x = y = z$ = tetrad
Tetragonal	$x = y$ = diad;
	z = tetrad.
Orthorhombic	$x = y = z$ = diad.
Monoclinic	y = diad.
Triclinic	No symmetry, represented by axis 1.

17. The cubic system is characterized by **4 triad axes**.
18. The multiple point, face-centred cubic lattice **emphasizes the cubic symmetry of the array of points**, and in particular it indicates the occurrence of the four triad axes which is characteristic of the cubic system.
19. Hexagonal crystals and space lattices are described by reference to **four axes** which are illustrated in fig. 4.15. **The hexad axis is** z, **and there are three axes perpendicular to** z **which are at angles of 120°** to each other, and are labelled x,y,u.
20. The trigonal system is characterized by a **single triad axis**.
21. A structure built up of the repetition of two layers of closely packed spheres is characterized by the occurrence of **hexad symmetry axes**, and the arrangement is called **hexagonal close packing**.
22. A stacking pattern which consists of the repetition of three layers of closely packed spheres produces the **tetrahedron form** which is characterized by the occurrence of **4 triad axes**. Consequently this arrangement is known as **cubic close packing**.
23. It can be seen from fig. 4.34 that when the cubic close packing tetrahedron is related to the cube, there are layers of spheres parallel to the planes of **the {111} form**.
24. In the cubic close packing structure the positions of the spheres are defined by **the face-centred cubic lattice**.
25. In the body-centred cubic space lattice **the diagonal planes such as (110) have the most closely packed atoms**, and therefore the **rhombdodecahedron form** {110} is most likely to develop.
26. The central plane of points in the face-centred cell shown in fig. 4.37 has the indices (**200**) indicating that **the spacing is half the dimension of the unit cell**.
27. The plane of points intersecting x and y but parallel to z in the face-centred unit cell has the indices (**220**).
28. It can be seen from fig. 4.38, that the plane of points which intersects all three axes in the face-centred unit cell has the indices (**111**).
29. In section 4.10 the ratio of the spacing of the planes of points parallel to the (110) plane in the primitive cell was found to be $a/\sqrt{2}$, therefore the spacing of the (220) planes of points will be $a/2\sqrt{2}$.
30. The intercept ratios determined from interfacial angles are small rational numbers **because they are directly related to the repeat distances in the crystal structure** which can be represented by the appropriate space lattice as shown in fig. 4.36.

Chapter 5: Answers to the questions posed in section 5.15

1. John Dalton postulated the **atomic theory of matter** in which the chemical combination of elements was explained as the union of atoms of the elements in simple ratios. An atom is the smallest particle of an element to possess the characteristics of that element.

2. **Protons** carry a **unit positive electrostatic charge. Neutrons** have a mass similar to the proton, but **no charge. Electrons** have a mass only 1/1850 that of the proton and carry a **unit negative electrostatic charge**.

3. Mendeleeff proposed that **the properties of the elements were periodic functions of their relative atomic weights**.

4. Mendeleeff constructed a chart in which the elements were arranged in lines in order of increasing relative atomic weight so that similar properties occurred in **vertical columns known as groups and subgroups**. Since the properties of the elements occur periodically the **horizontal rows are called periods** and the chart shown in fig. 5.1 is known as the **periodic chart.**

5. Atoms may occur with different numbers of neutrons so that they may have different relative atomic masses and are called **the isotopes of the element.** The relative atomic mass of an element is determined partly by the nature of the isotopes of its atoms, so it **is the occurrence of neutrons** that cause discrepancies in Mendeleef's chart.

6. The sequence of the elements is determined by the number of protons occurring in the atoms, and this is called **the atomic number.**

7. Bohr suggested that an electron in an orbit about the nucleus could occur only with particular values of angular momentum and consequently it would

have a particular energy level. **When an electron changes orbit to a lower energy, it emits a quantum of radiation energy which is observed as a discrete wavelength.**

8. Only **2** electrons can occur in the inner suborbital *s*, and only **6** electrons can occur in the *p* suborbital so that the quantum shell would then contain **8** electrons. The next highest energy level is that of the *s* suborbital in the next quantum shell of an atom.

9. Because the inert gases do not react or combine readily with other elements it was concluded that **these gases have stable electron structures**. With the exception of helium the inert gases have **eight electrons in their outer-most quantum shell**, and therefore it was concluded that this is the most stable electron structure for an atom.

10. Arrhenius suggested in 1887 that **in solution, NaCl dissociates into the electrostatically charged atomic particles Na$^+$ and Cl$^-$, called ions.** Positive and negative ions are called **cations** and **anions** respectively.

11. Atoms such as Na and Cl may attain an 8 electron outer shell by losing or gaining electrons and becoming ions. **The attraction of oppositely charged ions gives rise to the ionic bond**, and this explains the NaCl molecule in the vapour phase.

12. The chemical characteristics of halite are similar to those of salt solution and therefore it is concluded that **the halite crystal is characterized by ionic bonding.**

13. An atom may attain an 8 electron structure **by sharing electrons with another atom** so that they are held together by a **covalent bond** as in the Cl$_2$ molecule.

14. **Electronegativity** is the name given to **an index which measures the tendency of an atom to attract the bonding electrons.** Pauling calculated electronega-

tivity values from bond strengths indicated by heats of formation, and these **may be used to predict whether the bonding in a compound is predominantly ionic or covalent**.

15. It is postulated that in metals, **cations are packed together so that some electrons are mobile**. The ductility is explained by the ease of movement of the cations, and the mobile electrons allow good electric conductivity.

16. **The greater the ionic radii, the weaker the ionic bond.**

17. **The greater the charge the stronger the bond.**

18. Minerals are classified according to **the character of the anion or anion group** which is combined with cations.

19.

	% Weight	Atomic Wt.	Atomic proportions
Fe	46.5 ÷ 55.847		0.8326
S	53.5 ÷ 32.06		1.6687

Consequently the composition of pyrite can be represented by the formula FeS_2.

20. The bonding between the pairs of **S** atoms is likely to be **covalent** in character. The electronegativity values of the **Fe** and **S** are 1.8 and 2.5 respectively and these indicate that the **Fe–S** bond is partly **ionic** in nature.

Chapter 6. Answers to the questions posed in section 6.13

1. There must be **a regularity in the material** such as the pattern of threads in cloth, and there must also be **regularity in the light** which is described in terms of wave motion.

2. In 1912 **Laue was trying to establish the wave nature of X-rays** and he thought that crystals would provide a regular array of atoms with separations of about the same order of magnitude as the postulated wave lengths of X-rays.

3. **K,L,M,N** with **K** representing the inner shell.

4. The **electrons in the K shell have the lowest energies** and **therefore they require an addition of high energy** from the electron beam before they can move from the K shell to higher energy levels. In contrast the electrons in the outer shell, M, have high energies so they do not require much additional energy for their displacement to higher energy levels.

5. After electrons have been displaced from the Cu atoms, **electrons from outer shells move into the vacant lower energy orbits** and **emit quanta of energy which are observed as X-rays with characteristic wavelengths** known as the line spectra.

6. The energy of the emitted quantum of X-radition is inversely proportional to its wavelength according to the Einstein equation, $E = hc/\lambda$. Therefore **the K line spectra have shorter wavelengths, because greater energy is released** when electrons move into the K shell.

7. The $K\alpha_1$ and $K\alpha_2$ doublet is produced by the energy quanta released when electrons move from the L shell to occupy vacant orbits. The $K\beta$ line is produced when electrons move from the M shell into vacant orbits in the K shell and **since greater energy is released, the $K\beta$ line has the shortest wavelength.**

8. **The beam electrons lose energy by collisions** with electrons in the target and because **the energy released may have any magnitude below the maximum value**, a continuous spectrum of X-rays is produced from the shortest wavelength which is related to the maximum electron energy.

9. When X-rays pass through matter they are partly absorbed and the amount of absorption depends on the wavelength

of the X-rays. The long wavelength X-rays do not have sufficient energy to displace electrons from the inner shells of atoms so they tend to be transmitted. **X-rays which have just enough energy to displace electrons from the inner shells of atoms are absorbed,** and **the wavelength at which this pronounced increase in absorption occurs is known as the absorption edge.**

10. **The absorption edge of Ni falls between the Kα and Kβ wavelengths of Cu,** so that the shorter Cu Kβ wavelengths are absorbed. The lower energy, Cu Kα X-rays are largely transmitted, and can be used as a beam of monochromatic X-rays with effectively a single wavelength of 1.54 Å.

11. An electron lying in the path of a beam of monochromatic X-rays vibrates with the frequency of the incident beam. **Secondary X-rays with the same wavelength as the primary beam are radiated in all directions from the excited electron, and this is known as coherent scattering.**

12. When an incident beam of monochromatic X-rays strikes parallel layers of atoms in a crystal structure, coherent scattering occurs. Cooperative interference known as **diffraction will occur in the direction along which the X-rays scattered from the different parallel layers of atoms are in phase because they followed paths which differ in length by a whole number of wavelengths.**

13. The Bragg equation $n\lambda = 2d \sin \theta$, relates the **angle of incidence** θ at which planes of atoms of **spacing** d, will diffract X-rays of **wavelength** λ.

14. The diffracted beam which is produced when $n = 3$, meaning that **there is a path difference equivalent to 3 wavelengths between the X-rays scattered from adjacent layers of atoms,** is known as the 3rd order reflection.

15. The atomic scattering factor is **the ratio of the amplitudes of the X-rays scattered by the atom and by one electron**. In an idealized situation the atomic scattering factor would have the value of the atomic number, the total number of electrons in the atom divided by one.

16. As shown by fig. 6.12, in the position of the 1st order reflection the X-rays scattered from the top and bottom planes of the unit cell have a path difference of λ. However the X-rays scattered from the (002) plane will have a path difference of $\frac{1}{2}\lambda$ compared with the rays scattered from the top and bottom planes of the unit cell, and **almost complete cancellation of the 1st order reflection would occur**.

17. If the atomic scattering powers of the different atoms were similar and there are only half as many of the kind on the (00$\bar{2}$) planes, **the magnitude of the 1st order diffraction beam would be reduced by about one half.**

18. All kinds of waves can be represented by the movement of a point on a reference circle so that movement of the projection of the point on a diameter represents the vibration. **If the motion of the point on the diameter can be defined by an equation of the kind: Displacement** $= r$ **Cos** ψt**, it is called simple harmonic motion.**

19. On the reference circle a phase difference of one wavelength is equivalent to 360° or 2π radians.

20. Angular measurements of phase difference are independent of the wavelength.

Chapter 7. Answers to the questions posed in section 7.13.

1. **The angular positions of the diffracted X-ray beams are recorded** and if possible related to crystallographic features,

so that **the shape and size of the unit cell can be determined**.

2. In the face centred unit cell the central plane of atoms have the indices (**200**). Consequently the X-ray diffraction beam which is produced when there is a **path difference of** λ between the rays scattered from adjacent layers of ions is called **the 200 reflection**.

3. a. The **volume of a mole** of the substance = molecular weight/density.
b. The number of formula units of the substance in a mole is given by **Avogadro's number**.
c. The **volume of the unit cell** can be determined from the X-ray measurements of cell dimensions.
d. **The number of formula units in a unit cell is determined from the simple ratio** bc/a. This is the second major step in the determination of the crystal structure.

4. The third major step is **to postulate a structure** which has the unit cell pattern determined from the reflection angles and contains the appropriate number of formula units. The probable intensities of the reflections from the proposed structure are calculated and compared with the observed intensities. If there is good agreement, the structure determination is accepted.

5. The progressive decrease in the intensities of the higher order reflections indicated that the planes of ions parallel to (100) are similar in character.

6. The alternation of the reflections from the planes parallel to (**111**) suggests that the alternate planes are composed of different ions and this is what is implied by the proposed structure of halite.

7. Sylvite is composed of **K⁺**and Cl⁻ ions which are very similar in mass and scattering power, so that the **111** reflection is completely cancelled out. The **111** reflection appears with low intensi-

ty from halite. Consequently **if sylvite had been studied alone the X-ray reflections would have been interpreted incorrectly** as arising from a primitive cubic lattice.

8. The coordination number indicates **the number of anions which surround a cation** in a crystal structure, or the number of cations adjacent to an anion.

9. Pauling stated that in an ionic crystal structure **the coordination number is determined by the ionic radii ratio of** the cation and anion.

10. When 6 anions surround a cation the group has the form of the **octahedron**.

11. Paulings's second rule states that **the total strength of the valency bonds which reach a cation from the anions surrounding it, is equal to the charge on the cation**, and this is known as the electrostatic valency principle.

12. Minerals which have the same structure type are said to be **isostructural** or **isotypous**.

14. Fluorite is a good illustration of Paulings' second rule, the electrostatic valency principle, which states that **in a stable crystal structure the positive and negative charges are balanced. In fluorite there are twice as many F⁻ ions so if the Ca⁺² ions occur at the face-centred lattice points, the 8 K ions must occur at the centres of the 8 small cubes in order to maintain maximum symmetry and balance the positive charges.**

15. On the plan shown in fig. 7.23 the heights of the ions in the fluorite unit cell are indicated by fractions of the repeat distance in the vertical direction.

16. Fluorite **CaF₂** is said to have **8:4 coordination**, which indicates that the **Ca⁺²** ions are surrounded by 8 F⁻ ions while each **F⁻** ion is surrounded by 4 **Ca⁺²** ions.

17. The structure produced by **cubic close packing** has a face-centred space lattice arrangement.

18. In the structure of halite the ions at the face-centred lattice points are separated to allow the occurrence of oppositely charged ions in the spaces along the ⟨100⟩ lines, so that **the cubic close packing arrangement is retained**.

19. In the halite structure the layers of ions parallel to the {111} planes consist alternately of **Na⁺** and **Cl⁻** ions so that the bonding in the ⟨111⟩ directions of form is **strong**.

20. In the fluorite structure the layers of ions parallel to the {111} planes consist of a sheet of **Ca⁺²** ions flanked by sheets of **F⁻** ions. When these are stacked together sheets of **F⁻** ions are adjacent to each other so that the bonding in the ⟨111⟩ directions of form is weak, and the consequence is that in fluorite **well developed cleavages develop parallel to the {111} form**.

Chapter 8: Answers to the questions posed in section 8.11.

1. If **C** atoms occur at the face-centred lattice points they contribute the equivalent of **4 C** atoms to the unit cell. The other **4 C** atoms are placed within the cell at the centres of four small cubes with a tetrahedral arrangement.

2. The distribution of the **C** atoms in the diamond structure is described by **two interpenetrating faced-centred cubic spaces lattices**, which are related by displacements of ¼ the repeat distance along the triad axes.

3. The diamond structure consists of **covalently bonded, tetrahedral coordination groups**.

4. The coordination tetrahedra are arranged with triad axes parallel to the triad axes of the structure, and with the apices pointing in opposite directions.

5. In the diamond structure the tetrahedral coordination units are

arranged so that **layers of strongly bonded C atoms lie parallel to all the faces of the {111} form**, so that the crystals are usually octahedra.

6. There is a greater distance between the plane of **C** atoms forming the bases of the coordination tetrahedra and the **C** atoms at the apices, so that **the bonding is less strong in the directions perpendicular to the planes of the {111} form**.

7. On a screw axis, **rotation is combined with simultaneous translation parallel to the axis**.

8. The structure of diamond does not possess the tetrad symmetry axes of the holosymmetric class, but **the positions of the C atoms can be related by tetrad screw axes parallel to, but not coincident with the lattice rows** as shown in fig. 8.7.

9. The positions of atoms on either side of a glide plane may be brought into coincidence **by reflection combined with simultaneous translation** parallel to the glide plane.

10. **The group of symmetry operations including screw axes and glide planes** which is used to describe a crystal structure is called a space group, because they do not refer to a single point, the centre of the reference axes.

11. Polymorphism refers to **the occurrence of minerals which have identical compositions but exhibit different crystal structures**.

12. The structure of sphalerite is defined by **two interpenetrating face-centred cubic space lattices which are composed of different atoms, Zn and S**, and the lattices are related by displacements of ¼ the repeat distance on the triad axes.

13. In sphalerite the tetrahedral coordination groups (**ZnS₄**) and (**SZn₄**) are arranged with the triad axes parallel to those of the lattice, but they **point in opposite directions so that the faces of the two coordination groups are parallel**

to the alternative orientations of the tetrahedron within the unit cube.

14. The layers of atoms parallel to the {111} form in sphalerite are composed of **Zn atoms which are prominent on one side** of the the layer and **S atoms which are prominent on the opposite side.**

15. **The opposite sides of the layers of atoms** parallel to the {111} form in sphalerite **will have different chemical characteristics because of the asymmetricial distribution of the Zn and S atoms in the layers.** This asymmetry is reflected in the growth of the tetrahedron crystal form which does not have a symmetry centre.

Chapter 9: Answers to the questions posed in section 9.3.

1. A point group is **the group of symmetry operations** which may include reflection, rotation and rotation combined with inversion, **that describes the symmetry of a crystal about a single point.** The single point is the origin of the reference axes, and it is not repeated by the symmetry operations.

2. Hessel demonstrated in 1820 that there were only **32** different symmetry classes, and each of these is described by a point group.

3. The symmetry classes are distributed in **7 crystal systems** which are characterized by the occurrence of particular symmetry axes.

4. **No,** because the 14 space lattices possess the symmetry of the holosymmetric class of the appropriate crystal systems.

5. (i) **Each point is identical.**
   (ii) Each point has **identical surroundings**.
   (iii) Each point has an **identical orientation.**

6. **The screw axis combines rotation with**

simultaneous translation, and **the glide plane combines reflection with simultaneous translation**, in each case on an atomic scale.

7. A space group is the group of symmetry operations, including translations on a screw axes and glide planes, as well as reflection, rotation, and rotation combined with inversion, which may be used to describe the symmetry of a crystal structure. These **symmetry operations which are used to relate the positions of atoms do not refer to a single point, but extend through the crystal structure** and consequently are called the space group.

8. All the points in a space lattice have identical orientations but **it is appropriate to replace the points by identical asymmetrical shapes which need not have identical orientations within the space lattice**.

9. The identification of the 230 possible combinations of symmetry operations known as the space groups, was made independently by Federov, Schoenflies, and Barlow between **1890** and **1894**. The first crystal structure determinations were made in **1913**.

10. The third major step in crystal analysis is to postulate the positions of the atoms from the observed X-ray reflections. Space group theory limits the possible hypotheses because it identifies the possible arrangements of atoms which possess particular symmetry characteristics.

Chapter 11: Answers to the questions posed in section 11.7.

1. The first stage in the description of a crystal involves the **identification of zones and zone axes,** and the **measurement of the interfacial angles.**

2. The crystallographic axes are **reference**

axes which if possible are placed parallel to axes of symmetry or zone axes, so that **the orientation of the crystal faces may be described in terms of intercept ratios**.

3. A plane parallel to a crystal face, which intersects the crystallographic axes is selected as a reference plane. This reference plane is called the parametral plane because it **determines the ratios of the unit intercepts (parameters) on the crystallographic axes**.

4. Since the parametral plane defines the unit intercepts its indices are (**111**).

5. The form which includes the parametral (unit) plane is called the unit form {111}.

6. **The ratios of the intercepts of the parametral plane** are called the axial ratios.

7. First the **angles between the normal to the parametral plane P and the crystallographic axes are determined** from the stereogram. Then the axial ratios are obtained from the relationship explained in section 11.6.

$$x:y:z = 1 : \frac{\cos \text{XOP}}{\cos \text{YOP}} : \frac{\cos \text{XOP}}{\cos \text{ZOP}}$$

8. **The intercept ratios of each face are divided by the respective axial ratios**, and the simple ratios produced are converted into indices.

9. The indices {110} represents a **tetragonal prism form**, and it is composed of four faces.

10. The indices {111} represent the **tetragonal pyramid form** which has been selected as the unit form, and it is composed of 8 faces.

Chapter 16: Answers to the questions posed in section 16.4.

1. If in a crystal, a plane of atoms ends along a line, the **line defect** is known as an edge dislocation.

2. A lattice layer representing a plane of atoms may spiral about a line steeply inclined to the layer, to join the layer below. **The line around which the lattice layer spirals** is called a screw dislocation.

3. Crystal growth occurs most easily when atoms can fit into a stable position against a step of the kind that occurs on a crystal face due to the presence of a screw dislocation. **During crystal growth the step rotates and spirals about a screw dislocation continuously.**

4. A twinned crystal is **a composite crystal** consisting of two or more parts **in which the lattice orientations are related in certain geometrical ways**.

5. **The plane along which the two parts of twinned crystal are joined** is called the composition plane.

6. A twin is described as a parallel twin **if the twin axis is parallel to the plane across which the two lattice orientations are related**.

7. The Carlsbad twin is a **parallel twin** in which the composition plane is parallel to (**010**) and the twin axis is parallel to z.

8. A twin is described as a normal twin if the **twin axis is normal to the plane across which the two lattice orientations of the twin are related**.

9. **If the two parts of a twinned crystal can be related by reflection across a plane—** this plane is called the twin plane.

10. A crystal may grow with different orientations of the lattice on either side of a composition plane if the change in orientation of the lattice results in only a slightly higher energy state along the composition zone. The structure is continuous through the twin zone and does not necessarily contain dislocations.

Chapter 17: Answers to the questions posed in section 17.16.

1. The most common elements in the earth's crust are **O, Si, Al, Fe, Mg, Ca, Na and K**.

2. **Oxygen** forms about 46% of the weight of the rocks of the earth's crust, and because the oxygen atom is the lightest and the largest of the atoms of the common elements, oxygen atoms occupy about **94%** of the volume.

3. **Silicon** is the second most common element, providing about 27% of the weight, but because of its small size it occupies less than 1% of the volume.

4. Na^+, K^+, Ca^{+2}, Mg^{+2}, Fe^{+2}, Fe^{+3}, Al^{+3}, Si^{+4} and O^{-2}.

5. The **tetrahedral group (SiO_4)** is the most common coordination unit in the rocks of the earth's crust.

6. The electronegativity values of silicon (1.8) and oxygen (3.5) indicate that the bonding in the (SiO_4) tetrahedron is about **50% covalent and 50% ionic**. However, the compositions and structures of the common silicate minerals can be understood by regarding them as ionic structures.

7. **A series of minerals which have similar structures and forms, but which have compositions between those of two end members**, is described as an isomorphous series.

8. The olivine structure is characterized by the occurrence of **separate $(SiO_4)^{-4}$** tetrahedra which are bonded together by Mg^{+2} or Fe^{+2} **cations**.

9. The end members of the olivine isomorphous series are **forsterite Mg_2SiO_4** and **fayalite Fe_2SiO_4**.

10. An olivine with an intermediate composition can be represented by a molecular percentage symbol of the kind $Fo_{60}Fa_{40}$.

11. The radius of the ferrous ion Fe^{+2} (0.74 Å) is larger than the radius of Mg^{+2} (0.66 Å). Consequently the bonds are weaker in fayalite and it has a lower melting temperature (1250°C) than forsterite (1890°C).

12. **The first crystals of olivine will be richer in Mg** which is the component characteristic of the higher melting end member, but as crystallization proceeds the Mg/Fe ratio will decrease.

13. The structure of beryl contains rings of six SiO_4 tetrahedra which share two oxygen atoms. The hexagonal Si_6O_{18} rings are stacked on top of each other and are linked by Be and Al atoms to give hexagonal symmetry.

14. The pyroxenes have structures composed of **single chains** which are aligned parallel to the z axis.

15. The repeat unit of the single chain is $(Si_2O_6)^{-4}$ and the end members of the orthopyroxene series are **enstatite, $Mg_2Si_2O_6$** and **orthoferrosilite, $Fe_2Si_2O_6$**.

16. The composition of diopside is **$CaMgSi_2O_6$** and it crystallizes with **monoclinic symmetry** so it is one of the clinopyroxenes.

17. The single chains lying parallel to z are packed alternately with the bases and apices of the tetrahedra adjacent to each other. In diopside the **Ca^{+2} ions occur between the bases** of adjacent chains, and the **Mg^{+2} ions occur between the apices of the chains**.

18. The pyroxenes are characterized by **two sets of cleavage planes** which are **parallel to the prism form {110} and intersect at about 93°**. The prismatic cleavage planes develop because the bonds between the sides of the single chains and between the apices of the adjacent chains are weak. The angle of 93° between the cleavage and prism faces is determined by the cross section shape of the single chain.

19. The *c* dimensions of all the pyroxenes are similar because they are **determined**

by the length of the repeat unit of the single chains. The **b** dimensions of all pyroxenes are **determined by the width of two adjacent chains**.

20. In the clinopyroxenes the **a** dimension of the unit cell is determined by the height of two adjacent single chains which are stacked so that $\beta = 105°\ 50'$. **In the orthopyroxenes** the **direction of displacement of the single chains in the** z direction alternates so that the repeat distance **a is approximately double that of the clinopyroxenes** and orthorhombic symmetry is attained.

21. The repeat unit of a double chain is (Si_4O_{11}).

22. The **amphiboles** are an important group of silicate minerals which are characterized by double chain structures.

23. The **a** and **c** dimensions of the unit cells of the clinopyroxenes and amphiboles are similar, but the **b dimensions in the double chain structures of the amphiboles are approximately double the b dimensions of the clinopyroxenes**.

24. The amphiboles are characterized by **prismatic cleavages** which are similar to those in the pyroxenes, but because of the width of the double chains **the cleavages in the amphiboles intersect at approximately 124°.**

25. The hydroxyl ion **(OH)**$^-$ is able to fit into the rings of free oxygens in the double chain structure, so that the composition of tremolite is $Ca_2Mg_5Si_8O_{22}(OH)_2$.

26. In the silicate sheet structure the repeat unit has the composition (Si_4O_{10}).

27. The phyllosilicates have a **single, extremely well developed cleavage** and the name is derived from the Greek word for leaf or sheet.

28. The clay mineral kaolinite is composed of the repetition of **a tetrahedral layer** (Si_4O_{10}) **linked to an octahedral layer containing aluminium and additional hydroxyl ions.**

29. **Brucite consists or repetitions of two sheets of (OH)**$^-$ **ions in hexagonal close packing bonded by Mg**$^{+2}$ **cations in 6-fold coordination.**

30. **Gibbsite has a similar structure to brucite but because the Al**$^{+3}$ **is trivalent only $\frac{2}{3}$ of the possible sites are occupied.**

31. The brucite structure is described as a **trioctahedral arrangement** because there are three Mg^{+2} cations. for every six (OH)$^-$ anions. In gibbsite there are only two Al^{+3} ions for each six (OH)$^-$ ions and this is described as a **dioctahedral arrangement.**

32. **Pyrophyllite consists of repetitions of an aluminium octahedral layer which is sandwiched between the apices of the tetrahedra of two silicate sheets.**

33. Talc has a similar structure to pyrophyllite but it contains Mg ions in a trioctahedral layer. Consequently the composition can be derived from that of pyrophllite given in question 32, and it is $Mg_6(Si_4O_{10})_2(OH)_4$.

34. Muscovite is composed of layers similar to those in pyrophyllite but with Al^{-3} occupying $\frac{1}{4}$ of the Si^{+4} sites. **In the muscovite structure these units are linked by the presence of K**$^+$ **ions in 12-fold coordination** between the hexagonal rings at the bases of the silicate sheets.

35. The well developed cleavage parallel to the muscovite sheet structure is regarded as (001). In the structure there are diad axes parallel to y passing through the K$^+$ ions, and there are glide planes parallel to (010). Muscovite crystals have a hexagonal form which reflects the hexagonal nature of the silicate sheets but the overlying sheets are displaced in a zigzag fashion towards -x so that the crystal exhibits **monoclinic symmetry.**

36. The composition of muscovite is represented by the formula $KAl_2(AlSi_3O_{10})(OH)_2$ which indicates that $\frac{1}{4}$ of the Si^{+4} sites are occupied by

Al^{+3} and the additional charge is provided by K^+.

37. The composition of phlogopite is analogous to that of muscovite but with Mg^{+2} ions occurring in place of the Al^{+3} ions in the octahedral layer, so the formula becomes $\mathbf{KMg_3(AlSi_2O_{10})(OH)_2}$.

38. In biotite some ferrous ions Fe^{+2} occur in the Mg^{+2} sites so the composition is $\mathbf{K(Mg,Fe)_3(AlSi_3O_{10})(OH)_2}$.

39. Muscovite, phlogopite, and biotite are members of an isomorphous group known as the micas within the phyllosilicates. **In the micas the negative charges at the bases of the silicate sheets are balanced by the presence of cations such as K^+ and this gives these minerals some strength.**

40. The clay minerals have weakly bonded, composite silicate sheet structures which contain mainly aluminium and water. They are classified on the basis of the repeat distance perpendicular to the layers.

41. When all the oxygen atoms of the SiO_4 tetrahedra are shared, **a framework structure** is produced and the composition is SiO_2.

42. Silica occurs in nature as the minerals **quartz, tridymite, crystobalite** and **opal. Coesite** is a high pressure form found in impact craters.

43. The high temperature form is called β-**quartz** and it exhibits **hexagonal symmetry**. Below 573°C β-quartz changes by a displacive transformation to the less regular structure of α-**quartz** which exhibits **trigonal symmetry** only.

44. In β-quartz the tetrahedra are arranged so that there are **helices of Si—O—Si bonds which are parallel to z and which are either all left handed or all right handed**. This explains the enantiomorphic features of the quartz crystals studied in section 15.7

45. β-quartz is unstable above 870°C and it will tend to change to the more open structure of **tridymite**. Above 1470°C the stable structure is that of **crystobalite**. These two changes in structure are called **reconstructive transformations because they involve the breaking of bonds.**

46. The plagioclase feldspars range in composition from **anorthite $CaAl_2Si_2O_8$**, to **albite, $NaAlSi_3O_8$**.

47. The earliest plagioclase crystals will be **richer in Ca^{+2} and Al^{+3}** than the magma but as crystallization continues some of the Ca^{+2} and Al^{+3} ions will diffuse into the magma and their places will be occupied by Na^+ and Si^{+4} ions from the magma.

48. **Continuous reaction series** was the term used by N. L. Bowen to describe the crystallization of an isomorphous series like the plagioclase feldspars.

49. The composition of the crystals is indicated by the **solidus** and the composition of the liquid in equilibrium with the crystals at a particular temperature is given by the **liquidus**.

50. The low temperature form is called **orthoclase** and it has **monoclinic symmetry** with an ordered distribution of the Al^{+3} ions in the tetrahedral sites. The high temperature form is called **sanadine** and it also has **monoclinic symmetry** but the irregular distribution of Al^{+3} ions is said to be disordered.

51. There is a complete range in composition from high temperature **albite $NaAlSi_3O_8$**, to **sanadine, $KAlSi_3O_8$**, and **the change from the triclinic symmetry of albite to the monoclinic symmetry of sanadine occurs at approximately Ab_{63}.**

52. The crystallization diagram of the alkali feldspars consists of **two pairs of liquidus and solidus curves.** These indicate that when melts richer in K than Or_{30} crystallize, the crystals are richer in K than the melt, but when melts richer in Na than Ab_{70} crystallize the

first crystals are richer in Na than the melt.

53. Alkali feldspar crystals in the composition range $Or_{77}Ab_{23}$ to Or_8Ab_{22} **become unstable at temperatures below about 500°C, and albite is exsolved to form irregular layers in orthoclase**. This texture, composed of two feldspars is called microperthite.

54. Incongruent melting refers to **the transformation of a mineral to another mineral and some liquid**. For example enstatite melts incongruently to form olivine and silica liquid.

55. During the crystallization of magmas, **high temperature minerals become unstable and they may react with the liquid to form another mineral**. For example an olivine may obtain silica from the magma so that an orthopyroxene may crystallize in its place. The two minerals e.g. the olivine and orthopyroxene, are called a reaction pair.

56. N. L. Bowen used the term **discontinuous reaction series** to describe the change from a higher temperature isomorphous series to a lower temperature isomorphous series by reaction with the magma.

Appendix

B *The Construction of the Crystallographic Reference Axes*

The clinographic projection

Consider the reference axes used to describe a cube as shown in fig. 3.10. The units of intercept on the *x,y*, and *z* axes are all of equal length = *a*, and the lengths of the fore-shortened *x* and *y* axes seen in fig. 3.10 can be calculated by viewing the effect of the rotation through an angle θ about the *z* axis as shown in fig. B.1. The result of parallel projection perpendicular to the paper is shown in fig. B.1, where the origin **0** appears on the paper at **A**, the end of the *x* axis appears on the paper at **B** and the end of the *y* axis (−*y*) occurs at **C**. We wish to know the lengths **AB** = *x* and **AC** = −*y*.

It can be seen on fig. B.1 that
$$AB = DO = a \sin θ$$
and $AC = OE = a \cos θ$

Now consider the effect of raising the point of view as shown in fig. 3.11. by studying fig. B.2, which represents a vertical plane parallel to the line of projection now raised through an angle φ. The lines ABC and ODE are both represented by points. The end of the −*y* axis will be projected above the line ABC to the point **P**, while the end

of *x* is projected below the line to the point **Q**. It is necessary to know the distances of the ends of axes from the base line ABC, i.e. **CP** is the vertical distance (upwards) to the end of the −*y* axis, and **BQ** is the vertical distance (downwards) to the end of the +*x* axis in fig. B.2.

$CP = yE \tan φ$ and as $yE = a \sin θ$ in fig. B.1.
$$CP = a \sin θ \tan φ$$
$BQ = xD \tan φ$ and as $xD = a \cos θ$ in fig. B.1.
$$BQ = a \cos θ \tan φ$$

This construction can be simplified by choosing suitable values for the angles θ and φ such as $θ = 18°26'$ ($\tan θ = \frac{1}{3}$), $φ = 9°28'$ ($\tan φ = \frac{1}{6}$). The simplified construction has been given in section 3.3.

When drawing tetragonal and orthorhombic crystals as in chapters 11 and 12, the unit lengths on the axes can be modified to the appropriate axial ratios.

The construction of monoclinic axes

In order to make drawings of a monoclinic crystal such as orthoclase, the orientation of

295

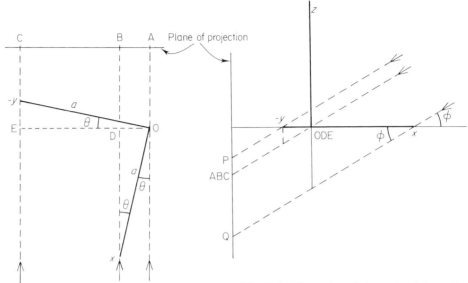

Fig. B.1. The cubic *x* and *y* axes rotated about the *z* axis

Fig. B.2. The point of view raised through angle φ in order to look down on the plane containing *x* and *y*

the *x* axis must be modified to represent the angle of inclination $\widehat{x\ z} = \beta$.

Fig. B.3. is a plan of the construction in the *xz* plane. It can be seen that the perpendicular distance of the end of the *x* axis (*x'*) from the *z* axis $x'M = Ox'$ sin (180° -β) = ON.

Therefore the modified length (ON) of the perpendicular *x* axis on a clinographic projection can be obtained by multiplying the apparent length OX by Sin β.

Fig. B.4 shows the clinographic axes with the perpendicular *x* axis modified in length to **ON**. The vertical distance from **N** to the end of the *x* axis $Nx' = OM = Ox'$ Cos β. The intersection of the line through **N** parallel to the *z* axis and the line through **M** parallel to the *x* axis gives the position of the unit length of the *x* axis inclined at β to the *z* axis. The unit lengths of the *x*, *y*, and *z* axes can be modified to the appropriate axial ratios.

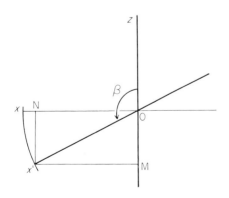

Fig. B.3. The monoclinic *x* axis

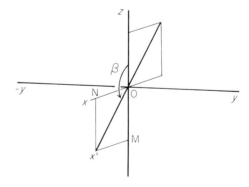

Fig. B.4. Construction of the monoclinic *x* axis on a clinographic projection of the crystallographic axes

The hexagonal crystallographic axes

Fig. B.5 is a plan of the hexagonal axes *x,y,-u*. The axis perpendicular to *y* is produced to T = a tan 60° where *a* is the unit length of the cubic axes.

On a clinographic projection of the cubic axes the position of the *y* axis is retained, but the *x* axis is extended to point T so that OT = *a* tan 60° = 1.73 *a* as shown in fig. B.6. The lines *yT* and *-yT* are drawn and the centre points of these lines give the orientations of the hexagonal axes *x* and *-u*. Finally the unit length of the *z* axis can be modified to the appropriate axial ratio.

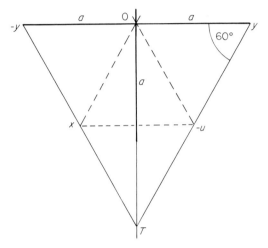

Fig. B.5. The relationship of the hexagonal *x* and *-u* axes to the cubic axes

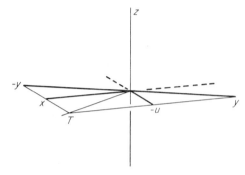

Fig. B.6. Construction of the hexagonal axes in clinographic projection

Appendix

C *The X-ray Powder Method of Mineral Identification*

If a mineral is ground to a powder which can pass through a 325 mesh screen the individual mineral grains are less than 4400 Å in diameter. When a beam of X-rays is passed through the powder of randomly oriented grains there will be sufficient appropriately oriented crystals to produce diffracted X-rays from each set of planes with a characteristic spacing *d*. If the diffraction angles are recorded, the *d* spacings can be calculated and arranged in sequence with the relative intensities of the beams also indicated. Minerals can be identified by the sequence of *d* spacings and intensities, without consideration of the orientation of the sets of planes of atoms.

The cylindrical camera

X-rays diffracted from a mineral powder can be recorded on photographic film. The most widely used photographic method makes use of a cylindrical camera which holds a strip of 35 mm film containing emulsion on both sides, fig. C.1. A commercial cylindrical camera is shown on the left side of fig. C.2. The sample is placed on a fibre which coincides with the axis of the cylinder and it is rotated on this axis during exposure to the X-rays. Two holes are cut in the film so that the X-rays may be concentrated on the sample by means of a tube called a collimator while the primary beam is received in another collimator to reduce scatter. This is known as the Debye–Scherrer method.

The powder is usually mixed with a binder such as collodion, which absorbs little and gives no diffraction patterns, and it is then mounted on a glass fibre. The glass fibre can be rotated in the path of the X-ray beam and this greatly increases the chances of each set of planes being in a position appropriate for diffraction during part of the time.

Each diffraction cone intersects the film as a pair of arcs and the distance between the arcs is proportional to 4θ. The radii of the cylindrical cameras are chosen so that the distance between the arcs is directly related to θ. For example in the camera with a diameter of 114.83 mm, 1 mm = 1°, so the angular measurement between the arcs can be obtained by direct measurement, and θ obtained by division. If it is desired to take into consideration the shrinkage of the film then the distance between the centres of holes equals 180°. In a smaller camera of radius 57.54 mm, 1 mm = 2°. The larger camera separates closely spaced lines more

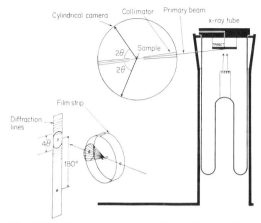

Fig. C.1. The arrangement of the film and recorded diffraction beams in a cylindrical camera

clearly but requires 4 times the exposure because the intensity is proportional to $1/r^2$. Having obtained the diffraction angles from the film the corresponding *d* spacings can be calculated because λ is known. For particular wavelengths, tables are available so the *d* spacings may be read directly from the values of θ or 2θ.

The Diffractometer

The diffractometer consists of a goniometer for measuring the diffraction angles, and it employs a large flat powder sample combined with a parafocusing arrangement to

Fig. C.2. Philips X-ray equipment. An X-ray tube enclosed in a shield stands vertically on the table top of a high voltage transformer. A cylindrical camera with the cover removed is in position at the left side window of the X-ray tube. A diffractometer is in position at the right side window with the cover plate removed so that the specimen plate can be seen at the centre. The proportional counter is in the box at the end of the diffractometer arm and a cable goes to the counter scales and ratemeter. Modern equipment incorporates more elaborate safety devices

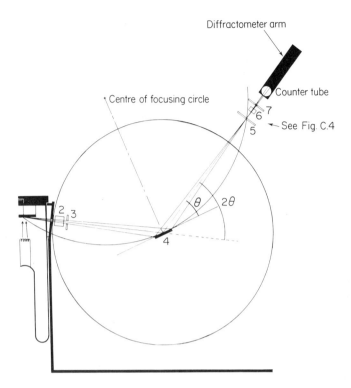

Fig. C.3. The geometrical arrangement of the diffractometer

increase the intensity of diffraction, fig. C.2 and C.3. The parafocusing principle is based on the theorem that all angles standing on an arc are identical and equal to $\frac{1}{2}$ the angle subtended at the centre of the same arc. Ideally the sample should be curved but the diffracted beams are brought to an approximate focus by arranging the sample in the tangent to the focusing circle. This position bisects the directions of the incident beam and the diffracted beam. Consequently the goniometer is constructed so that the sample holder moves at half the rate of the goniometer arm in order to keep the holder, tangent to the focusing circle. In other words as the goniometer moves through 2θ the sample holder moves through θ. Only those crystallites whose lattice planes are parallel to the surface of the specimen holder contribute to the reflection. Consequently the particle size must be small,

600–1000 Å, and the grains must be present in large quantities with good statistical distribution, so the sample area must be large.

The focusing error is kept small by the use of a narrow line focus, produced by viewing the 10 mm × 1 mm line focus at angles of 6° to 3° known as the **take off angle**. The divergence of this line focus parallel to its length is limited by the **divergence parallel slit** assembly, fig. C.4. The beam is allowed to diverge in the direction perpendicular to the goniometer axis but the amount of divergence is controlled by the **divergence slit**. The length of the sample irradiated by the X-ray beam is greatest at low angles of θ so the divergence slit is used to prevent scatter from the sample holder. At high angles the area of the sample covered by the beam decreases, so wider divergent slits are used to get a better statistical cover of the powder and to increase intensities.

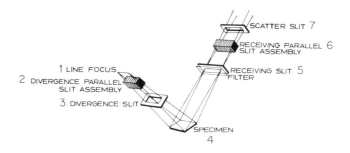

Fig. C.4. The slits used to control the primary and diffracted
X-ray beams in a diffractometer

The diffracted beams are similarly controlled by the **receiving slit** and **receiving parallel slit assembly**. The **scatter slit** is introduced to reduce any scattered X-rays other than the diffracted beams entering the detector.

The diffracted X-rays are detected and measured by the amount of ionization produced in a gas. The detector consists of a gas-filled metal cylinder which has a gas tight window, transparent to X-rays. The tube acts as a cathode and is kept at ground potential. One of the inert gases argon, xenon, or krypton is used to fill the tube (fig. C.5).

When an X-ray quantum enters the tube, it ionizes the atoms of the gas, thus liberating electrons and forming positive ions of the gas. In an electrical field of about 200 volts the ion pairs separate, the electrons move to the wire anode and the positive ions move to the cathode. If the X-ray quanta arrive at a constant rate, a small constant current of the order of 10^{-12} amp passes through the circuit, and this is a measure of the intensity. When the device is operated in this manner it is called an **ionization chamber**.

If the voltage is raised to between 600–900 volts, the electrical field intensity is so high that the electrons produced by the primary ionization are rapidly accelerated towards the wire anode. The electrons acquire enough energy to displace electrons from other gas atoms, so causing further ionization. These electrons continue the ionization process so that the number of atoms ionized by the absorption of a single X-ray quantum is 10^3 to 10^5 times as large as the number produced in the ionization chamber. This process is called gas amplification and the avalanche of electrons which hits the wire, produces an easily detectable pulse of current which is further amplified and then passed on to the counting circuits. The size of the electrical pulse for any given applied voltage is directly proportional to the number of ions formed in the primary ionization process, and this number is in turn proportional to the energy of the X-ray quantum which was absorbed. Consequently this device is called a **proportional counter**, and it is essentially a fast counter because the avalanche of electrons is confined

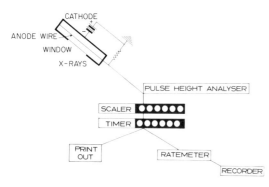

Fig. C.5. The arrangement of the counter tube
and the counting units

to a narrow section of the wire, 0.1 mm or less in length, and it does not spread longitudinally. It can count at rates of up to 10^6 pulses per second. The pulses are very small and have to be amplified to about 1 volt before passing to the counting circuits.

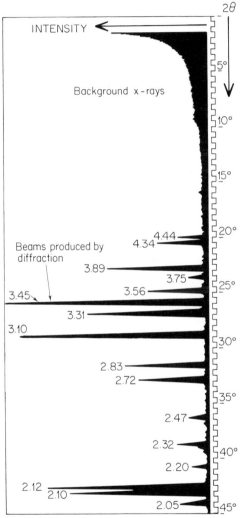

Fig. C.6. A chart of the diffraction peaks produced by the ratemeter. The chart moves upwards under a pen which moves in the horizontal direction in response to the intensity of the X-rays which are diffracted as the 2θ angle increases. The values of the 2θ angles are indicated along the margin of the chart. The numbers on the peaks are the calculated *d* spacings in angstroms, of the sets of planes which produce the diffraction beams

If the operating voltage is very high the electric field is so intense that the electrons cause ionization to spread throughout the tube. As a result, the gas amplification factor reaches a value of 10^7 or more and the electric pulse is comparatively large, 1 to 10 volts. When operated like this the device is called a **Geiger counter** and it usually has a window at the end of the tube. The electric pulses which it produces vary only slightly in size and bear no relation to the energy of the X-ray quanta. Although the electrons reach the wire in less than a microsecond the large positive gas ions take 200 microseconds to reach the cathode and during this time the detector is insensitive to X-ray quanta. Consequently, there is an upper limit to the rate of accurate counting without losses.

It is now necessary to count the number of electric pulses emitted by the detector over a measured period of time, so that the average counting rate is obtained by simple division. The pulses are counted or recorded by a device called a **scaler**. The scaler consists of a number of identical stages connected in series. Each stage passes on only a fraction of the pulses to the next, and in the decade scaler every tenth pulse passes on to the next counter so recording tens, hundreds, etc. The scaler is linked to a **timer** so that the number of counts arriving in certain fixed time, or the time for certain number of counts to arrive may be recorded.

The output of the scaling circuit can be fed into a **ratemeter** which converts the digital information into a continuous electric current which can actuate a strip chart recorder. By changing sets of gears, the goniometer can be made to scan at different rates ranging from 2° per minute to $\frac{1}{16}°$ per minute, but at each half degree a mark is made on the side of the chart so that it is easy to determine **2θ**, as shown in fig. C.6. The chart rate can also be varied so that the trace may be obtained in an uncrowded form.

The identification of a mineral by its diffraction pattern

Consider the diffraction chart shown in fig. C.6, which was produced with X-rays from a copper target. The wavelength of the X-rays from copper is 1.5418 Å so the *d* spacings of the layers of atoms which produced each diffracted beam can be calculated from the Bragg equation $\lambda = 2d \sin \theta$, or can be obtained from specially perpared tables. The 2θ angles and the corresponding *d* spacings and intensities are listed below. The relative intensities of the lines are shown in terms of the most intense line which is labelled 100.

2θ	*d* spacing	Intensity
20.0	4.4	15
20.5	4.3	25
22.85	3.89	50
23.62	3.75	10
24.93	3.56	30
25.8	3.45	100
26.9	3.31	60
28.82	3.10	90
31.59	2.83	40
32.89	2.72	35
36.3	2.47	10
38.79	2.32	15
40.85	2.20	10
42.60	2.12	80
43.10	2.10	75
44.13	2.05	15

In 1919 A.W. Hull demonstrated that every crystalline substance gives a distinctive diffraction pattern and that in a mixture of substances each constituent produces its pattern independently of the other. Hull showed that an unknown crystalline substance could be identified from a diffraction pattern obtained from a minute amount of the material. The unique diffraction pattern can serve as a 'fingerprint' identification of the crystalline phase. However X-ray diffraction analysis would not be a useful method of chemical identification unless the file of X-ray diffraction patterns of known

chemical compounds was maintained and continually increased.

The Joint Committee on Powder Diffraction Standards (JCPDS) is an international organization dedicated to collecting, editing, publishing, and distributing powder diffraction data to serve as standard reference for the identification of crystalline materials from their diffraction patterns. This data compilation, known as the Powder Diffraction File, now includes 35,000 numeric diffraction patterns of crystalline materials and is increasing at the rate of 2000 each year. The committee is based at the headquarters of the American Society for Testing and Materials in Philadelphia.

The Powder Diffraction File search manual is based on an indexing search system designed by J. D. Hanawalt. The Search Manual is constructed by dividing the *d* range from greater than 10.00 Å to 1.00 Å into 45 groups. The *d* value of the strongest line of a pattern determines into which of the 45 main groups the entry will fall. The strongest line on the pattern shown in fig. C.6 is 3.45 Å and this falls in group 3.49–3.40. Part of this group from page 331 of the 1978 Search Manual is shown in fig. C.7. Within the groups the entries are arranged in the order of the decreasing value of the second strongest line in the second column. The next six strongest lines are also listed in the Search Manual and their intensities are indicated by the subscripts, x = 100, 9 = 90 etc. Determine the *d* values of the eight strongest peaks on the pattern shown in fig. C.6 and then identify the substance from the part of the Search Manual shown in fig. C.7.

It will be discovered that the combination of *d* spacings which most closely matches the pattern shown in fig. C.6 is that of $BaSO_4$, the composition of the mineral barytes. Detailed data on each crystal is issued on cards and the card number is shown on the right of the Search Manual entry. The data on barium sulphate is on card 24–1035 which is shown in fig. C.8. Note that the intensities do not match exactly. In some cases the sequence of intensities of the lines obtained experimentally are different from those in the File and it is then necessary to try various combinations such as taking the second line as the most intense. An identification can be made after a small number of attempts and can be confirmed by matching the d spacings of all the lines shown on the appropriate card.

3.49 – 3.40 (+.02)

									File No.
	3.47_5	$3.12x$	3.18_6	3.24_5	1.99_4	3.63_3	2.95_3	1.74_3	$Sr_3(FeF_6)_2$ — 19–1277
	3.46_4	$3.12x$	4.40_5	2.46_4	1.97_4	1.68_4	4.94_2	2.69_2	HBO_2 — 15– 403
*	3.44_8	3.12_6	$3.23x$	3.66_3	5.02_2	2.23_2	1.95_2	1.77_1	$SrSeO_4$ — 15– 363
*	$3.44x$	3.12_4	2.52_4	4.85_3	3.37_3	3.28_3	2.58_3	2.08_2	$FeSO_4.H_2O$ — 21– 925
i	$3.44x$	3.12_7	1.92_7	3.28_6	3.65_6	2.23_6	2.13_6	3.19_5	BaY_2F_8 — 24– 125
	$3.42x$	3.12_8	2.96_8	2.82_8	2.13_8	3.95_6	3.04_6	2.55_6	$Rb_2(CuCl_3)$ — 28– 893
	3.40_3	$3.12x$	3.00_4	3.60_3	2.17_3	2.07_3	1.96_2	1.96_2	$Ba_5F(VO_4)_3$ — 19– 104
i	3.38_8	3.12_7	$4.42x$	2.62_7	2.03_7	1.53_6	6.26_5	1.51_5	$(Mn,Ca,Zn,Mg)Te_2O_5$ — 15– 129
i	3.38_8	$3.12x$	4.03_8	2.52_6	2.47_6	2.29_6	2.20_6	2.02_6	Cs_2NaCrF_6 — 27– 674
	$3.38x$	$3.12x$	$3.46x$	$2.08x$	3.74_5	2.28_5	2.30_4	2.05_4	$RbBeF_3$ — 10– 145
o	3.38_9	3.12_3	$3.29x$	2.16_3	1.97_3	3.00_2	1.82_2	2.62_1	$Pb_9Al_2F_{24}$ — 23–1153
o	3.48_5	$3.11x$	7.06_6	2.00_3	1.74_3	1.94_2	2.48_2	1.69_2	BaU_2O_7 — 13– 76
o	3.48_7	3.11_7	$3.57x$	3.01_5	2.81_4	2.07_4	3.13_4	2.52_4	$BaBi_2S_4$ — 16– 580
i	3.48_7	$3.11x$	3.45_7	3.23_6	3.96_5	3.20_4	2.98_4	2.26_3	$La_2Mo_4O_{15}$ — 23–1146
i	$3.44x$	3.11_4	5.28_3	1.63_2	2.66_2	2.76_1	2.07_1	2.57_1	$(Zn_2V_2O_7)44N$ — 28–1492
	$3.44x$	$3.11x$	1.86_9	3.17_7	4.21_4	2.06_4	5.37_3	3.72_3	$Rb_3Na_3Zr_4F_{22}$ — 10– 128
	3.43_5	$3.11x$	6.15_5	3.60_4	2.83_4	2.71_4	2.48_4	4.12_3	UVO_5 — 19–1388
	3.42_6	$3.11x$	2.76_8	3.21_6	2.99_6	2.59_6	3.29_4	2.97_4	$Ba_5Fe_{14}O_{26}$ — 28– 144
i	$3.42x$	3.11_6	1.91_4	3.27_3	4.18_3	3.63_3	2.22_3	3.17_2	$BaTm_2F_8$ — 27–1036
*	3.41_6	$3.11x$	$2.99x$	3.00_6	2.59_6	1.90_6	1.88_6	7.62_4	$EuGa_2O_9$ — 22– 282
i	$3.41x$	$3.11x$	2.97_7	2.58_5	3.66_5	2.52_4	2.01_3	4.72_3	$LiScGe_2O_6$ — 26–1442
	3.40_6	$3.11x$	3.58_8	3.31_6	3.25_6	2.12_6	2.85_3	2.75_3	Pl_3Bl_3 — 18– 969
	3.40_5	$3.11x$	3.46_8	1.97_4	2.64_4	2.75_3	4.50_3	4.37_3	$Ba_2Mg(PO_4)_2$ — 16– 556
*	3.40_9	3.11_6	$3.22x$	4.73_5	2.68_5	3.60_5	2.80_4	5.43_3	$Pb_5S_2I_6$ — 23– 329
	$3.40x$	$3.11x$	3.14_8	3.07_8	3.05_8	2.78_8	2.31_8	4.46_5	$\beta-Ba_2Cu_7F_{18}$ — 23– 818
	$3.39x$	3.11_7	3.36_5	2.26_5	3.93_4	3.03_4	6.80_3	2.68_3	$KAlGe_3O_8$ — 19– 930
i	$3.48x$	3.10_6	$9.46x$	3.21_6	2.36_5	4.73_4	3.15_3	1.90_3	$CsBiMo_2O_8$ — 26– 361
i	3.47_8	$3.10x$	$3.07x$	2.98_6	4.92_6	4.78_6	3.94_6	3.30_6	$KAlSiO.H_2O$ — 23–1314
i	3.47_8	$3.10x$	$2.80x$	2.77_4	3.41_3	2.64_3	2.71_2	2.90_2	$RbTiOPO_4$ — 25– 740
*	$3.47x$	3.10_8	2.80_8	5.54_4	3.25_3	3.24_3	3.02_3	3.41_2	$TlTiOPO_4$ — 25– 964
i	3.46_8	3.10_8	$3.60x$	3.00_8	5.15_4	3.85_4	3.72_4	4.45_3	$\beta-Cu(IO_3)_2$ — 27– 163
	$3.46x$	$3.10x$	3.20_8	3.59_6	3.17_6	1.63_6	5.20_4	3.68_4	$AlVO_4$ — 18– 74
i	$3.46x$	$3.10x$	$3.06x$	$2.96x$	2.80_8	6.96_7	3.27_7	2.72_7	$KAlSiO.0.72KBr.0.43H_2O$ — 27–1335
*	$3.45x$	3.10_9	2.82_8	2.24_6	2.09_6	2.49_5	4.38_4	3.53_3	O_2BF_4 — 25– 596
*	$3.45x$	$3.10x$	2.12_8	2.11_8	3.32_7	3.90_5	2.84_5	2.73_5	$BaSO_4$ — 24–1035
c	$3.45x$	$3.10x$	2.12_7	2.11_6	3.32_6	3.90_5	2.84_5	2.73_5	$BaSO_4$ — 24– 20
i	$3.45x$	$3.10x$	$1.98x$	$1.97x$	2.92_6	2.25_8	2.20_8	4.75_7	$Pb_3Al_2F_{12}$ — 22– 380
i	3.44_8	3.10_8	$6.90x$	2.82_6	2.49_4	2.30_4	1.79_4	1.73_4	$K_2Te_4O_9.4H_2O$ — 27– 442
*	3.43_5	$3.10x$	3.33_6	3.08_5	3.02_4	3.46_3	3.19_3	2.90_3	$Sr_2P_2O_7$ — 13– 194
*	3.40_7	$3.10x$	2.71_8	3.25_7	2.50_6	2.97_5	2.16_5	8.35_4	$Na_3Mg_4AlSi_8O_{22}(OH)_2$ — 20– 386
	3.39_4	$3.10x$	6.16_7	2.48_4	3.31_2	2.15_2	2.41_2	1.82_2	$RbBe_2F_5$ — 10– 143
	3.39_7	3.10_6	$3.47x$	3.01_5	6.76_3	5.04_3	3.04_3	2.09_3	$Pb_5B_8O_{17}$ — 20– 573
	$3.38x$	$3.10x$	$4.49x$	4.71_8	4.07_8	2.93_8	2.80_8	2.75_8	$Li_2B_{10}O_{16}.H_2O$ — 14– 674
*	$3.50x$	3.09_4	3.83_3	3.62_3	4.31_3	3.82_3	3.59_3	2.25_2	$Cs_2Cr_2O_7$ — 18– 350
*	$3.49x$	3.09_8	8.45_3	7.67_7	3.56_7	3.86_6	2.88_5	2.74_5	$NaH_2PO_2.H_2O$ — 25– 832
	3.48_4	$3.09x$	2.96_6	2.07_4	3.67_2	3.32_2	2.79_2	2.72_2	$K_3PO_4.V_2O_5$ — 14– 530
o	$3.47x$	$3.09x$	$2.05x$	1.90_9	3.13_8	2.97_7	2.43_5	5.55_4	$Mg_6B_4O_{10}(SO_4)_2.9H_2O$ — 14– 639
	$3.45x$	$3.09x$	9.45_5	2.50_3	2.68_2	2.38_2	2.15_2	4.76_2	$\beta-CuLuMo_2O_8$ — 28– 318
*	3.44_9	$3.09x$	$2.10x$	3.39_7	3.32_6	3.06_6	2.82_5	1.73_5	BaS_2O_7 — 21– 86
	3.43_4	$3.09x$	1.96_6	2.02_4	3.59_3	3.15_3	2.23_3	2.91_3	$SrAlF_5$ — 20–1183

Fig. C.7. Part of the 1978 Powder Diffraction File Search Manual produced by the Joint Committee on Powder Diffraction Standards (JCPDS). This shows part of the 3.49–3.40 group from page 331. The d spacings of the families of planes which produce the eight strongest diffraction lines from a substance are listed, and the number of the data card in the file is shown on the right side. (Reproduced by permission of JCPDS—International Centre for Diffraction Data)

24 - 1 0 3 5

d	3.45	3.10	2.12	5.58	BaSO₄
I/I_1	100	95	80	2	Barium Sulfate

24-1034

★

(Barite)

Rad. CuKα₁ λ 1.54056 Filter Mono. Dia.
Cut off I/I_1 Diffractometer I/I cor.= 2.6
Ref. NBS Monograph 25, Sec. 10 (1972)

Sys. Orthorhombic   S.G. Pbnm (62)
a₀ 7.1565 b₀ 8.8811 c₀ 5.4541 A 0.8058 C 0.6141
a   β   γ   Z 4   Dx 4.472
Ref. Ibid.

εα   nωβ   εγ   Sign
2V   D   mp   Color Colorless
Ref. Ibid.

BaCO₃ was treated with H₂SO₄. Excess H₂SO₄ was then
fumed off and the sample was annealed overnight at
690°C.
Pattern at 25°C.

See following card

FORM T-2

B

d A	I/I_1	hkl
5.58	2	110
4.440	16	020
4.339	30	101
3.899	50	111
3.773	12	120
3.577	30	200
3.445	100	021
3.319	70	210
3.103	95	121
2.836	50	211
2.735	15	130
2.729	45	002
2.482	13	221
2.447	2	112,131
2.325	14	022
2.305	6	310
2.282	8	230
2.211	25	122
2.169	3	202
2.121	80	311,140

d A	I/I_1	hkl
2.106	75	212,231
2.057	19	041
1.9486	1	222
1.9317	7	132
1.8575	18	330
1.7889	4	400
1.7616	8	103
1.7584	10	312,331
1.7540	8	410
1.7284	4	113
1.7239	5	150
1.6823	8	023
1.6741	14	142
1.6699	11	411
1.6596	2	420
1.6440	3	151
1.6378	8	123
1.6258	1	340
1.5944	8	213
1.5906	6	250

Fig. C.8. The JCPDS card numbered 24–1035. (Reproduced by permission of JCPDS—International Centre for Diffraction Data)

The pulse height analyser

When a diffractometer is used, another method is available for reducing fluorescent radiation from the sample which was discussed in sections 6.6 and 6.7. The amplitude or size of the electric pulses (measured in volts) passed from the proportional counter to the scaler, depends on the wavelength (i.e. energy) of the X-ray quanta which produced the ionization of the gas. An instrument for determining the size of the pulses passed by the counter, and for passing on only a selected range of these pulses to the scaler, is known as a **pulse-height analyser**. The instrument has an adjustable lower voltage limit, called the **lower level**, and a channel or **window** of adjustable width. The analyser will pass only those pulses whose amplitudes lie between the lower level and the lower level + window width. For example, if the lower level is set at 1 volt and the window is set at 0.5 volts, then only pulses ranging between 1 and 1.5 volts will be passed on by the analyser.

Consider the diagram fig. C.9, which shows the lower level voltage on the baseline, and the number of pulses recorded at the side. The diffractometer is set on a prominent quartz or silicon peak for **Cu Kα** radiation, and the window is set at 1% of the full range which happens to be 5 volts, so the window is 0.05 volts. The number of pulses received in the scaler over a set period of time, e.g. 10 seconds, are recorded for different settings of the lower level. The first reading gives the number of pulses ranging from 0 to 0.05 volts, since the lower level is 0 and the window is 0.05 volts. The lower level setting is moved and the second reading gives the number of pulses with energies between 0.05 and 0.10 volts, and so on. The resulting curve shows that most of the pulses produced by copper radiation occur within a relatively narrow voltage zone.

If an **iron sample** is placed in the diffractometer in a position where no diffraction from Cu Kα radiation occurs, then the above procedure can be repeated to deter-

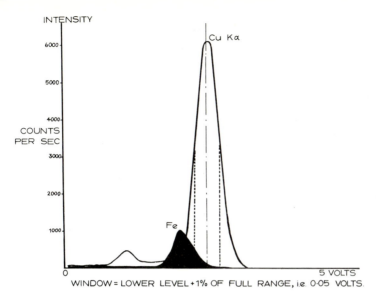

Fig. C.9. Counts of electric pulses of different voltage produced by X-rays of different energies from a Cu Kα primary beam and from iron fluorescent radiation

mine the size of the pulses produced by **Fe fluorescent radiation**. The second curve shows that the voltage of the pulses produced in the proportional counter by the **Fe** X-rays are significantly different from the voltage of the pulses produced by **Cu Kα**.

This graph can be used to set the lower level and the window so that most of the pulses produced by **Fe** flourescent radiation are excluded by the analyser and not passed on to the scaler. This reduces the background and improves the peak to background ratio.

Although the continuous spectrum is reduced by the filter, a large proportion is passed onto the counter and adds to the background. These rays have a peak intensity near 0.5 Å and their energy is very different from **Cu Kα** so the pulse height analyser will reduce this also. Strongly radioactive samples also give high voltage background pulses and these too are reduced by the analyser. The energy of **Cu Kβ** is too close to that of **Cu Kα** for these to be separated by the analyser so the **Ni filter** is still essential. The incoherently scattered X-rays which have wavelengths slightly greater than **Cu Kα** cannot be separated by the analyser so these form the minimum background.

Fig. C.10 is a diffraction pattern produced from biotite without the use of a **Ni filter** and the peaks from the **Cu Kα** and **Cu Kβ** X-rays are labelled. When a **Ni filter** is used the Cu Kβ lines are absorbed as shown in fig. C.11, and the intensities of the X-rays are reduced. The high background in fig. C.11 is due partly to fluorescent X-rays from the iron in the biotite. With **pulse height selection** and a **Ni filter** a greatly improved peak to background ratio is obtained as shown in fig. C.12.

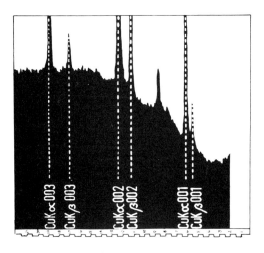

Fig. C.10. A diffraction chart of biotite produced without a Ni filter or pulse height analyser

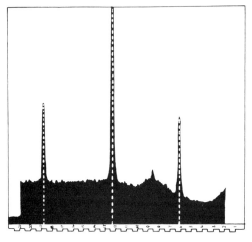

Fig. C.11. A diffraction chart produced with a Ni filter to suppress the Kβ lines

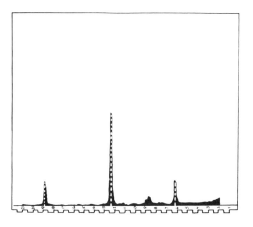

Fig. C.12. A diffraction chart produced with a Ni filter and pulse height selection to suppress Fe fluorescent radiation

Appendix

D The Classification and Characteristics of Some Common Minerals and Important Ore Minerals

The objectives of this appendix are listed below.

1. To illustrate the criteria on which minerals are classified.
2. To provide brief notes on common minerals and important ore minerals, so that the reader may become familiar with the names and main characteristics of these minerals. There are notes on 144 minerals and the 77 most common and important are indicated by heavy type.
3. The classification may be used to confirm an identification of a mineral made from the tables in Appendix E.

The classification of minerals

The classification of minerals was introduced in section 5.14. Minerals are first arranged into **classes** according to the nature of the anion or anion group which is present, and the classes used here are listed in the table.

Class	Anion
1. Native elements	
2. Sulphides	S^{-2}
3. Arsenides	As^{-2}
Tellurides	Te^{-3}
4. Oxides	O^{-2}
Hydroxides	$(OH)^{-1}$
5. Halides	Cl^{-1}, F^{-1}
6. Carbonates	$(CO_3)^{-2}$
7. Nitrates	$(NO_3)^{-1}$
8. Borates	$(BO_3)^{-3}$
9. Phosphates	$(PO_4)^{-3}$
10. Sulphates	$(SO_4)^{-2}$
11. Vanadates	$(VO_4)^{-3}$
Tungstates	$(WO_4)^{-2}$
Molybdates	$(MO_4)^{-2}$
12. Silicates	

12a. Silicate structures with **independent** (SiO_4) or **paired** (Si_2O_7) tetrahedra, *Orthosilicates.*

12b. **Ring** silicates, (Si_6O_8). *Cyclosilicates.*

12c. **Single chain** silicates, (Si_2O_6). *Inosilicates.*

12d. **Double chain** silicates, (Si_4O_{11}). *Inosilicates.*

12e. **Sheet** silicates, (Si_4O_{10}). *Phyllosilicates.*

12f. **Framework** silicates, $(AlSi)_2O_4$. *Tektosilicates.*

The classes listed in the table are the same as those used in Dana's *System of Mineralogy* (7th edition, 1944, John Wiley & Sons), with the exception of class 3. The **sulphides** of class 2 consist of a metal (e.g. FeS_2) or a semi-metal (e.g. AsS) combined with sulphur. In Dana's system class 3 consists of **sulphosalts** in which a semi-metal takes the place of sulphur in the structure as in arsenopyrite FeAsS. In the classification used here sulphosalts are included with the sulphides, and class 3 contains arsenides and tellurides.

The classes are subdivided into **subclasses** on the basis of chemical or structural characteristics. The **native elements** of class **1** can be divided into the semi-metals and non-metals which have covalent bonding, and the metals which are characterized by metallic bonding. The **carbonates** included in class **6** fall into subclasses characterized by trigonal, orthorhombic, and monoclinic symmetry. The **silicates** of class **12** have been divided into subclasses on the basis of their structures which were considered in chapter 17.

Within the subclasses, minerals are classified into **groups** which contain all the minerals that are closely related chemically and structurally, for example the **feldspar group**. The feldspar group is further subdivided into the plagioclase and alkali feldspar **isomorphous series**.

Some groups of minerals are related only by having similar modes of occurrence. The **feldspathoid group** contains minerals which have compositions similar to those of the feldspars but with a lower ratio of silica, so that they crystallize instead of the feldspars in magmas which are alkali-rich and silica deficient. Consequently the feldspathoids such as leucite, nepheline and sodalite never occur with primary quartz.

The **zeolite group** contains minerals which have distinct compositions and structures, but they are hydrated silicate minerals which can lose some or all of the water when

heated. The name zeolite is derived from the Greek word *to boil*. The zeolite minerals crystallize from the residual, low temperature, water rich fluids and are usually found within amygdales and other cavities in lavas.

Terms used in the tables to describe the rocks in which minerals occur, or the situation in which minerals crystallized

Igneous rocks are named on the basis of their primary minerals and the texture. The minerals which compose the main coarse-grained igneous rocks in which the crystals have diameters of 5 to 10 mm are listed below.

Granite	Quartz, K feldspar $>$ plagioclase.
Granodiorite	Quartz, K feldspar $<$ plagioclase.
Granitic	General term referring to granite, granodiorite and other closely related igneous rocks.
Syenite	Feldspars. No quartz.
Gabbro	Pyroxene, Ca-plagioclase, sometimes olivine.
Peridotite	Olivine dominant.
Kimberlite	Peridotite containing mica.
Ultrabasic	General term including rocks in which olivine and/or pyroxenes are dominant.

Basalt is composed of the same minerals as gabbro but with fine-grained or porphyritic textures.

Pegmatite refers to veins or patches of larger crystals commonly exceeding 2 cm in length, but in places exceeding several metres in length. Pegmatites are formed from the residual, low viscosity, water-rich magmatic fluids.

Hydrothermal refers to hot aqueous solutions. Quartz, calcite and sulphides are common minerals which have crystallized from hydrothermal solutions in fractures to form hydrothermal veins.

Syngenetic mineral deposits were formed at the same time as the rocks that enclose them, e.g. diamonds in kimberlite pipes and chromite rich layers in peridotites of the Bushveld.

Epigenetic mineral deposits formed later than the rocks within which they occur. e.g. gold and sulphide minerals in hydrothermal quartz veins intruded into older rocks.

Hypogene ore minerals were deposited during the original (**primary**) period of mineralization. The examples of syngenetic and epigenetic mineral deposits given above are all hypogene.

Secondary minerals crystallize as the result of the breakdown of primary minerals, e.g. feldspar breaks down and forms kaolinite.

Supergene minerals form as a result of the alteration of primary ore minerals. This term has a wider meaning than secondary because it includes the assemblage of minerals formed during weathering processes as well as the local alteration of primary minerals.

Supergene sulphide enrichment may occur below the oxidized zone of weathering under certain circumstances. The first requirement is that primary sulphides including the iron sulphides are present. In the oxidized zone pyrite changes to ferrous and ferric sulphates, ferric hydroxide and sulphuric acid. The ferric hydroxide gives rise to haematite and limonite which frequently forms a capping known as **gossan** over sulphide mineral deposits. The ferric sulphate attacks other primary sulphide minerals to yield soluble sulphates which trickle down through the oxidized zone to the water table. If conditions are suitable, sulphide minerals such as covellite are precipitated as supergene sulphide minerals which enrich the primary (hypogene) sulphide zone.

Placer deposits are concentrations of heavy, stable minerals in certain places in sand and gravel deposits as the result of flowing water.

The descriptions of the minerals

The descriptions of the minerals are arranged in the order of the **classes** and **subclasses** listed above. Within all the classes and subclasses except the silicates of class 12, the **individual minerals are described in order of increasing specific gravity** because this property is directly related to the chemical composition and degree of packing of the atoms. In the subclasses of the silicates the minerals are arranged according to their structures.

The arrangement of the information in the brief description of each mineral is shown below and this should be studied carefully because headings are not repeated.

NAME.   Composition.   **Specific gravity.**   Hardness.
Colour. Streak. Lustre.
System, form, habit. Cleavage indicated by indices of form. Fracture.
Observations.

1. Native elements

SULPHUR S Yellow. Streak white. Resinous. Orthorhombic, pyramidal habit. Conchoidal fracture. Deposited near volcanic vents.	**2.05–2.09**	**2**
GRAPHITE C Black. Streak black. Greasy. Hexagonal, massive, platey habit. {0001} perfect. Most commonly occurs in metamorphic rocks, fig. 1.12.	**2.2**	**1**
DIAMOND C Colourless or pale yellow when pure. Adamantine. Cubic, octahedra, plate 7, fig. 1.12. {111} perfect. Occurs in kimberlite pipes and alluvial deposits.	**3.5**	**10**
COPPER Cu Copper red. Streak metallic copper. Metallic when fresh, dull when tarnished. Cubic, usually irregular masses. Malleable, ductile. Found in oxidized zones with the carbonates malachite and azurite, plate 3.	**8.9**	**2½–3**
SILVER Ag Silver white, tarnishes brown or black. Streak silver white. Metallic. Cubic, usually irregular masses or curved fibres. Malleable, sectile. Occurs in hydrothermal veins and alteration zones. Associated with other silver minerals, or with Co, Ni minerals, or with the oxide uraninite.	**10.5** (8–10.5)	**2½–3**
GOLD Au Gold yellow, paler when alloyed with silver. Streak similar colour. Metallic, plate 5. Cubic, octahedra, irregular masses. Malleable, sectile. Occurs mainly in hydrothermal quartz veins and placer deposits.	**19.3** (15–19.3)	**2½–3**
PLATINUM Pt Steel grey to dark grey. Streak grey. Metallic. Cubic, usually in small grains. Malleable. Rare metal occurring in ultrabasic igneous rocks.	**21.45** (14–21.45)	**4**

2. Sulphides

REALGAR AsS Red. Streak red to orange. Resinous. Monoclinic, prismatic or granular. {010}, sectile. Occurs in veins with ore minerals of lead, silver, and gold, associated with orpiment.	3.48	$1\frac{1}{2}$	
ORPIMENT As_2S_3 Yellow. Streak yellow. Resinous. Monoclinic, prismatic usually foliated masses. {010} perfect, sectile. Rare mineral usually associated with realgar.	3.49	$1\frac{1}{2}$	
SPHALERITE ZnS Brownish black to amber. Streak brown. Resinous, plate 11. Cubic, tetrahedra, fig. 2.39, usually massive. {110} perfect. Most commonly occurs in hydrothermal veins and as replacement masses in limestones. Most important ore of Zn.	3.9–4.1	$3\frac{1}{2}$–4	
CHALCOPYRITE $CuFeS_2$ Brass yellow. Streak greenish black. Metallic, plate 12. Tetragonal, commonly pseudo-tetrahedral habit or massive.{011} Most widespread copper mineral, important ore found in hydrothermal veins and as disseminated grains.	4.1–4.3	$3\frac{1}{2}$	
ENARGITE Cu_3AsS_4 Black to grey. Streak similar. Metallic. Orthorhombic, columnar, bladed, massive. {100} perfect. Rare, but an ore of copper found in vein and replacement deposits.	4.43–4.45	3	
STANNITE Cu_2FeSnS_4 Grey to black. Streak black. Metallic. Tetragonal, massive. Rare, a minor ore of tin.	4.4	4	
STIBNITE Sb_2S_3 Black to lead grey. Streak similar. Metallic. Orthorhombic, tabular of bladed, fig. 1.25. {010} perfect. An important antimony mineral found in low temperature hydrothermal veins.	4.52–4.62	2	
PYRRHOTITE $Fe_{1-x}S$ Brownish bronze. Streak black. Metallic. Hexagonal, usually massive. Common accessory mineral in igneous rocks. Magnetic.	4.58–4.65	4	
COVELLITE CuS Indigo blue. Streak lead grey or black. Metallic. Hexagonal, platey habit or massive. {0001} perfect. Secondary mineral found in zone of sulphide enrichment over altered copper deposits.	4.6–4.76	$1\frac{1}{2}$	

PENTLANDITE	(Fe, Ni)S	4.6–5.0	$3\frac{1}{2}$

Light bronze yellow. Streak light brown. Metallic.
Cubic, usually in grains. No cleavage.
Almost always associated with pyrrhotite and pyrite. The
main ore mineral of nickel.

TETRAHEDRITE	$(Cu,Fe,Au,Ag)_{12}Sb_4S_{13}$	4.6–5.1	$3–4\frac{1}{2}$

Grey to black. Streak black or brown. Metallic.
Cubic, tetrahedra. No cleavage.
Occurs in low temperature hydrothermal veins with Cu, Pb, Zn and Ag
minerals. An important ore mineral for silver and copper.

MOLYBDENITE	MoS_2	4.62–4.73	1

Lead grey colour. Streak greenish grey. Metallic.
Hexagonal, foliate habit. {0001} perfect, sectile, greasy.
Accessory mineral in certain granitic rocks and veins associated
cassiterite, scheelite, wolframite and fluorite. An ore of molybdenum.

MARCASITE	FeS_2	4.89	6

Pale bronze yellow. Streak grey black. Metallic.
Orthorhombic, tabular but commonly radiating fibres
forming nodules.
Twinning produces spear head shaped crystals,
fig. 16.19. {101}.
Paler coloured and less stable than pyrite, formed in low
temperature veins, alteration zones and in sedimentary beds.

PYRITE	FeS_2	5.02	$6–6\frac{1}{2}$

Pale brass-yellow. Streak greenish or brownish black. Metallic.
Cubic, striated cubes, pyritohedra, plates 6 and 18. {100} poor.
The most common and widespread sulphide mineral. Often
occurs in large amounts with other sulphides containing
Cu,Pb,Zn,Au. Pyrite alters to iron sulphate which forms limonite and the
brown coloured capping known as gossan over sulphide mineral deposits.

BORNITE	Cu_2FeS_4	5.06–5.08	3

Brownish bronze on fresh surfaces, tarnishes to variagated purples
and blue. Streak grey black. Metallic, plate 12.
Cubic, usually massive. {111} poor.
A hypogene copper sulphide ore mineral found in gabbroic and
granitic rocks and veins, and in adjacent metamorphosed rocks.

MILLERITE	NiS	5.5	3

Pale brass yellow. Streak greenish black. Metallic.
Trigonal, usually occurs as fibres. {10$\bar{1}$1}, {01$\bar{1}$2}, perfect.
Occurs in low temperature hydrothermal deposits, often as an
alteration of other Ni minerals. Minor ore of nickel.

CHALCOCITE	Cu_2S	5.5–5.8	$2\frac{1}{2}$

Lead grey tarnishes black. Streak black. Metallic.
Orthorhombic, usually fine grains and massive. {110} poor, slightly
sectile, conchoidal.
An important copper ore mineral occurring mainly in supergene
enriched zones.

BOURNONITE $PbCuSbS_3$		5.8–5.9	$2\frac{1}{2}$

Steel grey to black. Streak similar. Brilliant metallic.
Orthorhombic, short prismatic crystals, massive. {010} poor.
Occurs in hydrothermal veins, often associated with galena.
An ore mineral of copper, lead, and antimony.

PYRARGYRITE Ag_3SbS_3		5.85	$2\frac{1}{2}$

Deep red. Streak red. Translucent, adamantine.
Trigonal, prismatic, rhombohedral, massive. {10$\bar{1}$1} poor.
Occurs in hydrothermal veins and is an ore of silver. Proustite
Ag_3AsS_3 is a similar mineral but lighter in colour.

ARSENOPYRITE $FeAsS$		6.07	$5\frac{1}{2}$

Silver white. Streak grey black. Metallic.
Orthorhombic, crystals common, showing combinations of prisms
and domes giving rhomb shaped cross-sections. {110}.
An ore of arsenic, occurs in high temperature hydrothermal
veins often associated with gold or tin minerals.

COBALTITE $CoAsS$		6.33	$5\frac{1}{2}$

Silver white may be reddish. Streak grey black. Metallic.
Cubic, striated cubes, and pyritohedrons. {100} perfect.
An ore of cobalt found in veins with nickel sulphides and
disseminated in metamorphic rocks.

ARGENTITE Ag_2S		7.2–7.4	2

Black. Streak black and shining. Metallic.
Cubic, octahedral, massive or encrustations. Sectile.
Important primary ore mineral of silver found in low temperature
hydrothermal veins.

GALENA PbS		7.4–7.61	$2\frac{1}{2}$

Black. Streak lead grey. Metallic.
Cubic, cubes and octahedra, fig. 1.25 {100} perfect.
Common and widespread, occurs in veins, sedimentary beds and as
replacement masses. The source of lead.

CINNABAR HgS		8.10	$2\frac{1}{2}$

Red. Streak scarlet. Adamantine to dull.
Trigonal, rhombohedral, usually granular or in encrustations.
{10$\bar{1}$1} perfect.
An important mercury ore mineral found in volcanic rocks and
near hot springs.

3. Arsenides and Tellurides

NICCOLITE — NiAs Pale copper red, quickly alters to green 'nickel bloom'. Streak brownish black. Metallic. Hexagonal, usually massive. A minor nickel ore mineral found in basic igneous rocks and associated veins.		7.78	$5-5\frac{1}{2}$
SYLVANITE — $(Au,Ag)Te_2$ Silver white. Streak grey. Brilliant metallic lustre. Monoclinic, prismatic, usually bladed or granular. {010} perfect. An ore mineral of gold and silver. Occurs in veins with pyrite, other sulphides and tellurides.		8.16	$1\frac{1}{2}-2$
CALAVERITE — $AuTe_2$ Brass yellow to silver white. Streak yellowish to greenish grey. Metallic. Monoclinic, usually granular, no cleavage. A telluride gold ore mineral distinguished from sylvanite by the absence of cleavage.		9.35	$2\frac{1}{2}$

4. Oxides and Hydroxides

BAUXITE — **Aluminium hydroxides** White, grey, reddish brown when stained by haematite. Streak brown. Dull. A mixture of minerals. Often shows pisolitic structure, fig. 1.45. Produced by weathering in tropical climates where Si and other elements have been leached.		**2.0–2.55**	**1–3**
BRUCITE — $Mg(OH)_2$ White, sometimes greenish. Streak white. Pearly, waxy or vitreous. Trigonal, usually foliate // (0001). {0001} perfect. Found associated with serpentine as a decomposition product of Mg silicates.		2.39	$2\frac{1}{2}$
DIASPORE — $HAlO_2$ White, grey, rarely brown, pink. Streak white. Vitreous, brilliant. Orthorhombic, platey, foliate, massive. {010} perfect. Widespread occurrence as fine-grained massive material in bauxites, also occurs with corundum in pegmatites and crystalline limestone. Used as a refractory.		3.3–3.5	$6\frac{1}{2}-7$
LIMONITE — $FeO(OH)_n, H_2O$ Dark brown. Streak yellow brown, pigment is known as yellow ochre, plate 1. Dull. Amorphous. No cleavage. Secondary mineral found in oxidized zones often associated with goethite.		**3.6–4.0**	**5**

SPINEL	**MgAl$_2$O$_4$**	**3.6–4.0**	**8**

White, reddish when pure. Fe spinel, *pleonaste* is greenish black;
Cr spinel, *picotite* is greenish brown. Streak white or grey. Resinous.
Cubic, octahedra. No cleavage.
The spinel group consists of isomorphous series in which there
may be almost complete substitution of Mg by Fe, Zn or Mn
and Fe, Cr may substitute for Al. Common in metamorphic rocks and as
an accessory mineral in gabbros and peridotites.

CHRYSOBERYL	**BeAl$_2$O$_4$**	3.65–3.8	8$\frac{1}{2}$

Yellow, may be green. Streak white. Vitreous.
Orthorhombic, tabular but twinning gives pseudohexagonal
appearance. {110}.
Rare mineral, occurs in granitic pegmatites. Used as a gemstone,
name means golden beryl.

CORUNDUM	**Al$_2$O$_3$**	4.02	9

Variable, grey, blue, pink, red. Vitreous.
Trigonal, columnar often with tapering hexagonal prisms,
plates 7 and 8. {0001} poor.
Accessory mineral in metamorphic rocks and in igneous rocks
poor in silica.

RUTILE	TiO$_2$	4.18–4.25	6–6$\frac{1}{2}$

Reddish brown to black. Streak pale brown. Adamantine.
Tetragonal, prismatic. {110} good.
Accessory mineral in granitic and metamorphic rocks. Ore of
titanium. *Anatase* (tetrag.) and *brookite* (ortho.) are polymorphs.

MANGANITE	MnO(OH)	4.3	4

Black to steel grey. Steak dark brown. Metallic.
Orthorhombic, long prismatic crystals or radiating bundles. {010} perfect.
Occurs in veins associated with granitic rocks. Minor ore of manganese.

GOETHITE	**HFeO$_2$**	4.37	5

Brown. Streak yellow brown. Dull
Orthorhombic, usually massive, foliated. Bog iron ore has a porous
texture. {010} perfect, distinguishes it from limonite.
A common secondary mineral.

CHROMITE	**FeCr$_2$O$_4$**	**4.6**	5$\frac{1}{2}$

Black. Streak dark brown. Metallic to submetallic.
Cubic, octahedral, usually massive or in grains. No cleavage.
Occurs as an accessory mineral in ultrabasic rocks such as
peridotites. The only chromium ore mineral.

ILMENITE	**FeTiO$_3$**	**4.7**	5$\frac{1}{2}$–6

Black. Streak black. Metallic.
Trigonal, tabular, massive. {0001} poor.
Occurs as an accessory mineral in gabbroic rocks, as large
masses in metamorphic rocks and in some beach sands.
An ore of titanium.

PSILOMELANE $(Ba, H_2O)_2Mn_5O_{10}$		4.7	5–6
Black. Streak brownish black. Dull to submetallic. Orthorhombic, massive as reinform encrustations, also earthy. No cleavage. Secondary mineral formed under surface conditions, associated with pyrolusite, goethite, and limonite. Found in large residual deposits, an ore of manganese.		(3.7–4.7)	
PYROLUSITE MnO_2		**4.75**	**1–2**
Black. Streak black. Metallic. Tetragonal, Usually granular or powdery. {100} perfect. A secondary mineral often forming dendritic growths on fracture surfaces. Nodular masses found on the sea bottom, and in sedimentary rocks. The main ore mineral of manganese.			
FRANKLINITE $(Fe,Zn,Mn)(Fe, Mn)_2O_4$		5.15	6
Black. Streak reddish brown. Metallic. Cubic, octahedra, massive. No cleavage. Weakly magnetic. The main Zn ore mineral in crystalline limestones at Franklin, New Jersey, plates 16 and 17.			
MAGNETITE Fe_3O_4		**5.15**	**6**
Black. Streak black. Metallic. Cubic, octahedra, massive. No cleavage. Strongly magnetic iron ore mineral. Occurs as an accessory mineral in igneous rocks, in large beds in metamorphic rocks, and in beach sands.			
HAEMATITE Fe_2O_3		**5.26**	$5\frac{1}{2}$–$6\frac{1}{2}$
Steel grey to black. Streak red. Metallic to dull, plate 1. Trigonal, rhombohedral, radiating fibres, reniform masses. {0001} poor. Widespread mineral. The most important ore of iron mined from rocks mainly of sedimentary origin.			
ZINCITE ZnO		5.68	4–$4\frac{1}{2}$
Orange to deep red. Streak orange yellow. Sub-adamantine. Hexagonal, usually massive. {10$\overline{1}$0} perfect, conchoidal fracture. Rare, occurs mainly at Franklin with franklinite and willemite, plate 16.			
CUPRITE Cu_2O		**6.1**	$3\frac{1}{2}$
Red. Streak brownish red. Submetallic. Cubic, octahedra, fig. 2.33.{111}. Occurs in the oxidized zone of copper deposits, often coated by malachite.			
CASSITERITE SnO_2		**6.8–7.1**	**6–7**
Brown to black, sometimes yellowish. Streak white. Translucent, submetallic to adamantine. Tetragonal, fig. 11.12, {100} perfect, {110} poor. Occurs in veins of quartz and pegmatite near granites, associated with fluorite, topaz and tourmaline. Main ore of tin often mined from placer deposits.			

URANINITE (pitchblende) UO_2 Black. Streak brownish black. Submetallic, pitch-like lustre. Cubic, octahedra, usually massive and botryoidal. Radioactive. Occurs as a primary mineral in some granites, pegmatites and hydrothermal veins, and also in certain sedimentary rocks.	**10.95** (8–10.95)	$5\frac{1}{2}$

5. Halides

SYLVITE KCl Colourless or white, shades of blue or red from impurities. Streak white. Vitreous. Cubic, cube and octahedron forms. {100} perfect. Salty taste more bitter than halite, occurs in evaporite deposits. Used as a fertilizer. Main source of K.	1.99	2
HALITE **NaCl** Colourless or white, may be yellow, red, or blue when impure. Streak white. Vitreous. Cubic habit, fig. 1.19, {100} perfect. Rock salt occurs in evaporite deposits.	**2.16**	$2\frac{1}{2}$
FLUORITE CaF_2 Colourless and wide range of colours, plate16. Streak white. Vitreous. Cubic, cubes, {111} perfect, plate 15. Common mineral found in veins with metallic ores and in limestones. Accessory mineral in granitic rocks. Used as a flux in steel manufacture.	**3.18**	4
CERARGYRITE AgCl Grey on fresh surfaces, may be greenish or brownish. Streak white. Translucent, wax like. Cubic, usually massive. Sectile. An important supergene ore of silver in enriched zones.	5.51	$2\frac{1}{2}$

6. Carbonates

Trigonal subclass

CALCITE $CaCO_3$ Colourless, white may show wide range of colours. Streak white. Vitreous. Trigonal, hexagonal columnar habit, rhombohedra, massive, figs. 1.22–1.24, {$10\bar{1}1$} perfect with interfacial angle = 74° 55′. Widespread mineral, forming extensive beds of limestone. Effervesces in cold HCl.	**2.72**	3

DOLOMITE $CaMg(CO_3)_2$		2.85	$3\frac{1}{2}$–4
Usually pink tinted, may be colourless, white, brown. Streak white. Vitreous. Trigonal, rhombohedra with curved faces, massive, $\{10\bar{1}1\}$ perfect with interfacial angle $= 73°45'$ Usually found in sedimentary deposits.			
MAGNESITE $MgCO_3$		3.0–3.2	$3\frac{1}{2}$–5
White, grey, yellowish. Streak white. Vitreous. Trigonal, usually massive. $\{10\bar{1}1\}$ perfect. Occurs in veins and patches formed by the alteration of serpentine, olivine, or pyroxene. Does not effervesce in cold HCl. SiO_2 as an impurity produces chert like appearance and greater hardness.			
RHODOCHROSITE $MnCO_3$		3.45–3.6	$3\frac{1}{2}$
Pink. Streak white. Vitreous. Trigonal, usually massive, stalactitic. $\{10\bar{1}1\}$ perfect. Occurs in veins with ore minerals of silver, lead and copper.			
SIDERITE $FeCO_3$		3.83–3.88	$3\frac{1}{2}$
Brown. Streak white. Vitreous. Trigonal, rhombohedra with interfacial angle $= 73°$. Usually granular. $\{10\bar{1}1\}$ perfect. Widespread in sedimentary beds associated with clay, used as iron ore. Also occurs in hydrothermal veins.			
SMITHSONITE $ZnCO_3$		4.35–4.4	$4\frac{1}{2}$–5
Dirty brown, dry bone ore may be colourless or white. Streak white. Translucent. Vitreous. Trigonal, rhombohedral, usually botryoidal or stalactitic. $\{10\bar{1}1\}$ perfect. Secondary mineral usually associated with Zn deposits; in limestones. Effervesces in acid.			

Orthorhombic subclass

ARAGONITE $CaCO_3$		2.95	$3\frac{1}{2}$–4
Colourless, white, yellowish. Streak white. Vitreous. Orthorhombic, prismatic, tabular, pseudohexagonal twins, figs. 1.42 and 16.17. $\{101\}$ and $\{110\}$ imperfect. The less stable form of $CaCO_3$ tends to be precipitated from warmer water, and hot springs. The pearly layer of many shells is aragonite.			
STRONTIANITE $SrCO_3$		3.7	$3\frac{1}{2}$
White, grey, yellow, green. Streak white. Translucent. Vitreous. Orthorhombic, prismatic. $\{110\}$ good. Occurs in low temperature hydrothermal veins associated with barytes, celestite and calcite. Effervesces in HCl.			

WITHERITE $BaCO_3$		4.3	$3\frac{1}{2}$
White, grey. Streak white. Vitreous. Orthorhombic, always twinned on {110} forming pseudohexagonal pyramids, fig. 16.18. {010} good. Found in veins associated with galena, effervesces in HCl.			
CERUSSITE $PbCO_3$		6.55	3
Colourless, white, grey. Streak white. Translucent, adamantine. Orthorhombic, crystals common, tabular, twinning on {110} gives pseudohexagonal forms. {110} good, {021} poor. Widespread supergene lead mineral, an important ore. Effervesces in warm nitric acid.			

Monoclinic subclass

MALACHITE $Cu_2Co_3(OH)_2$		3.9–4.03	$3\frac{1}{2}$
Bright green. Streak pale green. Dull. Monoclinic, usually massive, botryoidal, plate 2. Common secondary mineral associated with copper deposits.			
AZURITE $Cu_3(CO_3)_2(OH)_2$		3.77	$3\frac{1}{2}$
Azure blue to deep blue. Streak light blue. Vitreous, plate 3. Monoclinic, complex crystals or massive. {011} perfect. Secondary mineral in copper deposits, effervesces in HCl.			

7. Nitrates

NITER (saltpeter) KNO_3		2.09–2.14	2
White. Streak white. Vitreous. Orthorhombic usually in thin encrustations. {011} perfect. Found in some soils, used as source of nitrogen.			
SODA NITER (chile saltpeter) $NaNO_3$		2.29	2
White, sometimes brown. Streak white. Vitreous. Trigonal, usually massive. {10$\bar{1}$1} perfect. Strongly deliquescent, found only in arid regions.			

8. Borates

BORAX $Na_2B_4O_7. 10H_2O$		1.7	$2–2\frac{1}{2}$
White. Streak white. Translucent, Vitreous. Monoclinic, prismatic, massive encrustations. {100}, {110} perfect. The most widespread borate formed by evaporation of salt lakes. Used for cleansing, as a solvent and a flux.			

9. Phosphates

TURQUOISE	$CuAl_6(PO_4)_4(OH)_8, 2H_2O$	2.6–2.8	6
Pale blue, bluish green. Streak white or greenish. Waxlike, plate 2. Triclinic, usually cryptocrystalline and massive. A secondary mineral found in veins in altered volcanic rocks, used as a gemstone.			
AMBLYGONITE	$LiAlFPO_4$	3.0–3.1	6
White, grey, pale greenish, or bluish. Streak white. Vitreous to greasy. Triclinic, large masses like plagioclase, but different cleavages. {100}, {110}. Occurs in granite pegmatites with spodumene and lepidolite. A lithium ore mineral.			
APATITE	$Ca_5(F,Cl,OH)(PO_4)_3$	3.15–3.2	5
Green, sometimes brown or blue. Streak white. Vitreous or sub-resinous. Hexagonal, prismatic, columnar, fig. 1.10. {0001} poor, conchoidal fracture. Common accessory mineral. Phosphate rock formed by the precipitation of apatite or by the accumulation of bones. Used as a fertilizer.			
AUTUNITE	$Ca(UO_2)_2(PO_4)_2. 10–12H_2O$	3.15–3.2	2
Yellowish green. Streak yellow. Vitreous. plate 16. Tetragonal, tabular parallel to {001} foliate aggregates. {001} perfect. Secondary mineral formed in the oxidation zone of uranium mineral deposits. Fluorescent, plate 17.			
MONAZITE	$(Ce,La,Y,Th)PO_4$	4.62–5.3	5
Reddish brown to yellow. Streak white. Resinous. Monoclinic, usually granular often in sands. {100} poor. This phosphate of rare earth elements may contain up to 20% ThO_2 and is the main source of thorium, a radioactive element.			
PYROMORPHITE	$Pb_5Cl(PO_4)_3$	7.04	$3\frac{1}{2}$
Green, yellow or brown. Streak white. Resinous. Hexagonal, prismatic groups, globular reniform masses, encrustations. Uneven fracture. Secondary mineral occurring in the oxidized zone of lead deposits.			

10. Sulphates

CHALCANTHITE $CuSO_4 . 5H_2O$ Azure blue. Streak white. Translucent. Vitreous. Triclinic, prismatic, reniform. {110} poor, conchoidal. Occurs in the surface oxidized zones of copper deposits in arid regions. Often deposited on iron in copper mines. Minor ore of copper.	2.12–2.3	$2\frac{1}{2}$
GYPSUM $CaSO_4 . 2H_2O$ Colourless. Streak white. Transparent. Vitreous. Monoclinic, fig. 1.17, sometimes fibrous. fig. 1.9 {010} perfect. Extensive sedimentary beds. Normally the first salt deposited by evaporation of the sea. Also found around volcanic vents and in mineral deposits.	2.32	2
ANHYDRITE $CaSO_4$ White, may be pink, brown, or bluish. Streak white. Vitreous to pearly. Orthorhombic, usually massive. {100}, {010}, {001} tabular cleavage blocks. Usually occurs in evaporite deposits. Anhydrite changes to gypsum by absorption of water and increase in volume.	2.9–3.0	$3–3\frac{1}{2}$
ANTLERITE $Cu_3SO_4(OH)_4$ Green to black. Streak pale green. Vitreous. Orthorhombic, slender prismatic crystals, reniform, massive. {010} perfect. An ore of copper found in oxidized zones of Cu deposits. *Brochantite* $Cu_4SO_4(OH)_6$ is very similar.	3.9	$3\frac{1}{2}$
CELESTITE $SrSO_4$ White, colourless. Streak white. Vitreous. Orthorhombic, crystals resemble those of barytes. {001} perfect {210} good, intersecting at 75°58′. Occurs disseminated in sedimentary rocks, associated with gypsum, anhydrite or halite, also in cavities and veins. Main source of strontium.	3.95–3.97	$3–3\frac{1}{2}$
BARYTES $BaSO_4$ Colourless, white. Streak white. Vitreous. Orthorhombic, tabular, figs. 1.20, 1.21, and 12.1. {001} perfect {210} good intersecting at 78°28′. Occurs associated with metallic mineral veins, sometimes as a cement in sandstones.	4.5	3
ANGLESITE $PbSO_4$ Colourless to white, adamantine when pure, otherwise dull. Orthorhombic, crystals similar to barytes. {001}, {210}. Common supergene lead mineral formed by the oxidation of galena. Does not effervesce in nitric acid.	6.2–6.4	3

11. Vanadates, Tungstates, Molybdates

CARNOTITE $K_2(UO_2)_2(VO_4)_2\,3H_2O$ Yellow. Streak yellow, a strong pigment. Dull. Orthorhombic, usually occurs as a powder. {001} perfect. A secondary mineral found in the oxidized zone, an ore of vanadium and uranium.		4.1	1
SCHEELITE $CaWO_4$ Colourless to white, yellow or brownish. Streak white. Vitreous. Tetragonal, dipyramidal habit, commonly massive. {101}. An ore of tungsten. Occurs in pegmatites and thermally metamorphosed rocks associated with cassiterite, topaz, and wolframite.		5.9–6.1	$4\frac{1}{2}$
WOLFRAMITE **(Fe,Mn)WO₄** Brownish black. Streak reddish or brownish black. Submetallic. Monoclinic, prismatic or tabular. {010} perfect. The main tungsten ore mineral found in granitic pegmatites and quartz veins.		**7.1–7.5**	**4–5**
WULFENITE $PbMoO_4$ Yellow to orange, may be green or brown, Streak white. Vitreous. Tetragonal, square tabular crystals. {101}. Occurs in the oxidized parts of lead bearing veins. Minor source of molybdenum.		6.5–7.0	3

12(a) Silicate structures with independent (SiO_2) tetrahedra or pairs (Si_2O_7) (*Orthosilicates*)

OLIVINE **(Mg,Fe)₂SiO₄** Green. Streak white. Translucent, vitreous, plate 13. Orthorhombic, tabular, figs. 12.14 & 12.15. {010} poor. Occurs in igneous rocks rich in Mg and Fe, structure explained in section 17.5.		**3.27–4.37**	**$6\frac{1}{2}$**
WILLEMITE Zn_2SiO_4 White when pure, yellow, green, red. Streak white. Vitreous. Trigonal, prismatic or rhombohedral, usually massive or granular. {0001} good. Occurs in crystalline limestones. An important ore of zinc.		3.9–4.2	$5\frac{1}{2}$
ZIRCON **ZrSiO₄** Brown, may be colourless, green or blue. Streak white. Vitreous. Tetragonal, prismatic, fig. 11.1. {110} poor. Common accessory mineral in granitic rocks.		**4.68**	**$7\frac{1}{2}$**
SPHENE $CaTiSiO_5$ Brown. Streak white. Adamantine. Monoclinic, wedge shaped crystals. {110} distinct. Common accessory mineral in granitic rocks.		3.4–3.55	$5–5\frac{1}{2}$

GROSSULARITE $Ca_3Al_2(SiO_4)_3$	3.58	7
Pale green. Streak white. Vitreous to resinous.		
Cubic, icositetrahedra or rhombdodecahedra. No cleavage.		
A garnet occurring in metamorphosed impure limestones.		
ALMANDINE $Fe_3Al_2(SiO_4)_3$	3.58–4.3	7
Dark red to black or pale reddish. Streak grey. Vitreous.		
Cubic, icositetrahedra or rhombdodecahedra. No cleavage.		
The commonest garnet, found mainly in metamorphic rocks.		
The garnet group includes a number of isomorphous series.		
Pyrope $Mg_3Al_2(SiO_4)_3$ is similar in appearance to almandine.		
IDOCRASE $Ca_{10}(Mg,Fe)_2Al_4(SiO_4)_5(Si_2O_7)_2(OH)_4$	3.35–3.45	7
Brown, may be green, yellow, blue. Streak white. Resinous.		
Tetragonal, prismatic crystals often terminated by {001}. {110} poor.		
Found in thermally metamorphosed impure limestones.		
SILLIMANITE Al_2SiO_5	3.23	6–7
Grey, sometimes brownish or greenish. Streak white. Vitreous.		
Orthorhombic, slender prisms or fibres. {010} perfect.		
Occurs in high grade Al-rich metamorphic rocks.		
ANDALUSITE Al_2SiO_5	3.16	$7\frac{1}{2}$
Grey, brown sometimes greenish. Streak white. Vitreous.		
Orthorhombic, prismatic with nearly square cross section,		
fig. 1.44. {110}		
Usually found in thermally metamorphosed shales. Chiastolite is		
a variety with diagonal planes of inclusions.		
KYANITE Al_2SiO_5	3.63	5–7
Blue, grey. Streak white. Vitreous.		
Triclinic, bladed or tabular. {100} perfect.		
Occurs in schists and gneisses.		
TOPAZ $Al_2(SiO_4)(F,OH)_2$	3.4–3.6	8
Colourless, white, yellow or pale blue. Streak white. Vitreous.		
Orthorhombic, prismatic, fig. 1.11. {001} perfect.		
Occurs in pegmatites associated with granitic rocks.		
STAUROLITE $(Fe^{+2},Mg)_2(Al,Fe^{+3})_9O_6(SiO_4)_4(O,OH)_2$	3.7–3.8	7
Red-brown or brownish black. Streak grey. Vitreous.		
Orthorhombic, prismatic, $(110)\widehat{\ }(\bar{1}10) = 50°$ often showing		
cruciform twins, fig. 16.20. {010} distinct.		
Occurs in schists and gneisses.		
HEMIMORPHITE $Zn_4(Si_2O_7)(OH)_2H_2O$	3.4–3.5	5
White. Streak white. Translucent. Vitreous.		
Orthorhombic, tabular parallel to {010}, massive, stalactitic.		
{110} perfect.		
Occurs in the oxidized zone of Zn deposits often associated with		
smithsonite. A minor ore of Zn.		

EPIDOTE $Ca_2Fe^{+3}Al_2(Si_2O_7)(SiO_4)O(OH)$	3.35–3.45	6–7

EPIDOTE $Ca_2Fe^{+3}Al_2(Si_2O_7)(SiO_4)O(OH)$ 3.35–3.45   6–7
Yellowish green to greenish black. Streak grey. Vitreous.
Monoclinic, prismatic. {001} perfect.
Common in low grade metamorphic rocks. Clinozoisite does
not contain Fe and is a common secondary mineral formed by the
alteration of plagioclase in gabbroic rocks.

12(b) Ring silicates (*Cyclosilicates*)

BERYL $Be_3Al_2(Si_6O_{18})$ 2.75–2–8 $7\frac{1}{2}$
Green, yellowish green. Streak white. Vitreous.
Hexagonal columnar habit, plate 14. {0001} poor.
Occurs in granitic pegmatite veins. Gemstone.

CORDIERITE $Mg_2Al_3(AlSi_5O_{18})$ 2.6–2.66 $7–7\frac{1}{2}$
Blue, grey, yellowish or brown. Steak white. Vitreous.
Orthorhombic usually as grains. {010} poor.
Occurs in metamorphic rocks and in some granites, resembles
quartz, but has cleavage.

TOURMALINE $Na(Mg,Fe)_3Al_6(BO_3)_3Si_6O_{18}(OH)_4$ 3.0–3.2 $7\frac{1}{2}$
Black, colourless when free of Fe. Li varieties are pink,
blue or green. Streak white. Vitreous.
Trigonal, prismatic usually with rounded triangular cross
sections. {10$\bar{1}$1} poor.
Occurs in granite pegmatites and also as an accessory mineral
in metamorphic rocks.

AXINITE $Ca_2(Fe,Mn)Al_2(BO_3)(Si_4O_{12})(OH)$ 3.27–3.35 $6\frac{1}{2}$
Brown. Streak white. Vitreous.
Triclinic, wedge shaped crystals with sharp edges. {100} good.
Occurs in high temperature metamorphic rocks near
granite intrusions.

12(c) Single chain silicates, (*Inosilicates*)

ENSTATITE $Mg_2Si_2O_6$ 3.2–3.5   6
White, grey to pale green. Streak white. Vitreous.
Orthorhombic pyroxene, prismatic or massive. {110} good,
intersecting at 87°.
Occurs in basic and ultrabasic rocks. The most pure varieties
are found in meteorites. Belongs to an isomorphous
series with hypersthene.

HYPERSTHENE $(Mg,Fe)_2Si_2O_6$	3.5–3.9	6
Brownish green to black. Streak grey. Vitreous. Orthorhombic pyroxene, prismatic or massive. {110} good, intersecting at 87°. Occurs in basic and ultrabasic igneous and metamorphic rocks. See enstatite.		
DIOPSIDE $CaMgSi_2O_6$	3.2–3.3	6
White to greenish grey. Streak white. Vitreous. Monoclinic pyroxene, prismatic, fig. 13.16. {110} good intersecting at 87°. Most common in metamorphic rocks rich in Ca. Diopside is the end member of an isomorphous series ranging in composition to hedenbergite $CaFe(Si_2O_6)$.		
AUGITE $Ca,(Mg,Fe,Al)(Al,Si)_2O_6$	3.2–3.4	5–6
Dark green, black. Streak white or grey. Vitreous, plate 13. Monoclinic pyroxene, prismatic, fig. 13.16. {110} good intersecting at 87°, fig. 17.11. The most common pyroxene particularly in igneous rocks rich in Ca, Mg, Fe.		
AEGIRINE $NaFeSi_2O_6$	3.4–3.55	6
Dark green, brownish or nearly black. Streak white. Vitreous. Monoclinic pyroxene, prismatic or grains. {110} good with angles 87°. Found in Na rich, Si poor rocks such as nepheline syenites.		
JADEITE $NaAlSi_2O_6$	3.25–3.5	$6\frac{1}{2}$
Green. Streak white. Vitreous. Monoclinic pyroxene, usually fibrous and compact. {110} intersecting at 87°. Occurs in metamorphic rocks associated with serpentine. Jade ornaments are made from jadeite and the amphibole nephrite, a compact form of tremolite.		
SPODUMENE $LiAlSi_2O_6$	3.1–3.2	$6\frac{1}{2}$
White, grey, violet. Streak white. Vitreous. Monoclinic pyroxene, prismatic. {110} intersecting at 87°, {100} good. Occurs in pegmatites with lepidolite and amblygonite. A source of lithium.		
WOLLASTONITE $CaSiO_3$	2.8–2.9	5
White to grey, vitreous to pearly lustre. Triclinic, tabular crystals or massive. {001} and {100} perfect. Occurs in thermally metamorphosed impure limestones.		

12(d) Double chain silicates (*Inosilicates*)

TREMOLITE $Ca_2Mg_5Si_8O_{22}(OH)_2$ White, may be greenish. Streak white. Vitreous. Monoclinic amphibole, prismatic, bladed. {110} perfect intersecting at 56°. Isomorphous series with actinolite. Occurs in metamorphosed impure limestones. A compact variety of tremolite called nephrite is used to make jade ornaments.	3.0–3.3	5–6
ACTINOLITE $Ca_2(Mg,Fe)_5Si_8O_{22}(OH)_2$ Green. Streak white. Vitreous. Monoclinic amphibole, prismatic, bladed. {110} perfect, intersecting at 56°. Occurs in low and medium grade metamorphic rocks containing some iron. See tremolite.	3.35	6
HORNBLENDE $NaCa_2(Mg,Fe,Al)_5(Si_8O_{22})(OH,F)_2$ Dark green to black, sometimes brownish. Streak white to grey. Vitreous. Monoclinic amphibole, prismatic, bladed. {110} good intersecting at 56°, fig. 13.9. Common in igneous and metamorphic rocks.	3.2	5–6
GLAUCOPHANE $Na_2(Mg,Fe,Al)_5Si_8O_{22}(OH)_2$ Pale blue. Streak white to blue grey. Vitreous, silky in fibrous varieties. Monoclinic amphibole, prismatic habit. {110} perfect with angles of 58°. Glaucophane and riebeckite form the end members of an isomorphous series. Glaucophane occurs in low grade schists.	3.0–3.2	6
RIEBECKITE $Na_2(Mg,Fe,Al)_5Si_8O_{22}(OH)_2$ Dark blue in Fe rich minerals. Streak blue grey. Vitreous. Monoclinic amphibole, prismatic. {110} perfect, with angles of 56°. Riebeckite occurs in metamorphic rocks and Na rich granitic rocks. See glaucophane.	3.41	6

12(e) Sheet silicates. (*Phyllosilicates*)

KAOLINITE $Al_4(Si_4O_{10})(OH)_8$ White, grey. Streak white. Dull. Monoclinic, minute clay size plates. {001} perfect. A common secondary clay mineral formed by the alteration of feldspars.	2.6–2.63	$2–2\frac{1}{2}$
SERPENTINE $Mg_6(Si_4O_{10})(OH)_8$ Green, often mottled. Streak white. Vitreous, waxy or greasy lustre. Monoclinic, fibrous variety is called *chrysotile*, platey variety is called *antigorite* {001}. Secondary mineral formed by alteration of olivine, pyroxene or amphibole. Chrysotile is used as asbestos. Infusible.	2.2–2.65	2–5

TALC $Mg_3(Si_4O_{10})(OH)_2$ Green to grey. Streak white. Greasy lustre and feel. Monoclinic, silicate sheet structure. {001} perfect. Sectile, fig. 1.8. Secondary mineral formed from magnesium silicates. Used in wide range of products.	2.8–2.82	1
MUSCOVITE $KAl_2(AlSi_3O_{10})(OH)_2$ Colourless. Streak white. Vitreous, pearly. Monoclinic, silicate sheet structure, {001} perfect, flexible elastic sheets. Fig. 1.16. Common rock forming mineral. Used for electrical insulation.	2.76–3.1	$2\frac{1}{2}$
GLAUCONITE $K_2(Mg,Fe)_2Al_6(Si_4O_{10})_3(OH)_{12}$ Green. Streak green. Dull. Monoclinic, usually granular. {001}. Dioctahedral mica, authigenic grains in marine sediments.	2.5–2.8	2
BIOTITE $K(Mg,Fe)_3(AlSi_3O_{10})(OH)_2$ Brown to black. Streak white to grey. Vitreous, pearly. Monoclinic silicate sheet structure. {001} perfect. Common mica. A paler coloured Mg mica is called *phlogopite*.	2.8–3.2	$2\frac{1}{2}$–3
LEPIDOLITE $K_2Li_3Al_3(AlSi_3O_{10})_2(OH,F)_4$ Pink or red, may be colourless, grey, yellow. Streak white. Vitreous. Monoclinic, silicate sheet structure. {001} perfect. Occurs in granitic pegmatites, source of lithium.	2.8–3.0	$2\frac{1}{2}$–4
CHLORITE $Mg_3(Si_4O_{10})(OH)_2Mg_3(OH)_6$ Green. Streak white to pale green. Vitreous to pearly lustre. Monoclinic, silicate sheet structure. {001} perfect. Common secondary mineral.	2.6–2.9	2
PREHNITE $Ca_2Al_2Si_3O_{10}(OH)_2$ Light green to white. Streak white. Vitreous. Orthorhombic, usually in rounded groups of tabular crystals. {001} good. Occurs in cavities in basaltic volcanic rocks.	2.8–2.95	6

12(f) Framework silicates. (*Tektosilicates*)

QUARTZ SiO_2 Colourless, white, coloured by impurities, plate 4. Streak white. Vitreous. Trigonal, hexagonal columnar habit, fig. 1.1, massive, conchoidal fracture, fig. 1.15. Common rock and vein forming mineral. Massive cryptocrystalline variety is called *chert* or *flint*, fig. 1.14. Laminated cryptocrystalline variety is called *agate*, plate 4.	2.65	7
JASPER SiO_2 with haematite inclusions. Red. Streak pinkish. Vitreous to dull. Usually massive and fine grained. Conchoidal fracture. Often found associated with pillow lavas.	2.65	7

OPAL $SiO_2.\,nH_2O$ Colourless, white sometimes with play of colours. Streak white. Vitreous, pearly. Amorphous. Conchoidal fracture. Deposited in cavities from hot aqueous solutions.		1.9–2.2	5–6
CHRYSOCOLLA $CuSiO_3.\,2H_2O$ Green, bluish brown when impure. Streak light blue. Vitreous to earthy. Cryptocrystalline, conchoidal fracture. Secondary mineral in oxidized zones over copper deposits. A minor Cu ore.		2.0–2.4	2–4
ORTHOCLASE **KAlSi$_3$O$_8$** White, grey sometimes pink, Streak white. Vitreous. Monoclinic, tabular, fig. 13.1. {001} perfect {010} good, fig. 1.18. Common rock forming mineral particularly in granitic rocks.		**2.57**	**6**
PLAGIOCLASE **Albite NaAlSi$_3$O$_8$** **Anorthite CaAl$_2$Si$_2$O$_8$** White, grey sometimes pink. Streak white. Vitreous. Triclinic, tabular, fig. 1.40. {001} perfect, {010} intersecting at 86° 24′, fig. 14.2. Plagioclase is the most common rock forming mineral group.		**2.62–2.76**	**6**
NEPHELINE **(Na,K)(AlSiO$_4$)** White to grey. Streak white. Greasy lustre and feel. Hexagonal, prismatic, {10$\bar{1}$0}. Feldspathoid found in plutonic and volcanic rocks. Used in glass and ceramics.		**2.55–2.65**	**5$\frac{1}{2}$–6**
PETALITE $LiAlSi_4O_{10}$ White to grey. Streak white. Pearly. Monoclinic, platy habit. {001} perfect, brittle. Found in pegmatites with other Li bearing minerals such as lepidolite. An important lithium ore.		2.41	6
LEUCITE $K\,(AlSi_2O_6)$ White to grey. Streak white. Dull. Cubic, icositetrahedra. Rare feldspathoid found in lavas in which there is insufficient silica to combine with K to form feldspar.		2.45–2.5	5$\frac{1}{2}$
SODALITE $Na_4(AlSiO_4)_3Cl$ Blue. Streak white. Vitreous. Cubic, rhombdodecahedra, usually in grains. Rare feldspathoid. Transparent crystals found in lavas of Vesuvius with other feldspathoid minerals.		2.15–2.3	5$\frac{1}{2}$
LAZURITE (lapis lazuli) $(Na,Ca)_4(AlSiO_4)_3(SO_4,S,Cl)$ Azure blue. Streak white. Vitreous. Cubic, usually massive. A rare mineral occurring in thermally metamorphosed limestones. Commonly contains pyrite as in plate 2.		2.4–2.45	5–5$\frac{1}{2}$

ANALCIME	**Na(AlSi$_2$O$_6$)H$_2$O**	**2.27**	**5**
White. Streak white. Vitreous.			
Cubic, icositetrahedra.			
Usually found in cavities in basaltic lavas, associated with zeolites.			
NATROLITE	Na$_2$(Al$_2$Si$_3$O$_{10}$), 2H$_2$O	2.25	5
White. Streak white. Translucent. Vitreous.			
Monoclinic, prismatic, fibre crystals radiating from centres.{110}.			
A zeolite found in amygdaloidal lavas.			
CHABAZITE	(Ca,Na)$_2$(Al$_2$Si$_4$O$_{12}$) . 6H$_2$O	2.05–2.15	4–5
White, pinkish. Streak white. Vitreous.			
Trigonal, rhombohedron form with angles almost 90°.			
Poor rhombohedral cleavage.			
Zeolite found in amygdaloidal basalts, fig. 1.41.			

Appendix

E *Procedure and Tables for the Identification of Common Minerals and Important Ore Minerals*

The X-ray powder method for the identification of minerals is based on the spacings of the planes of atoms in mineral structures, and consequently the identification procedure is often straightforward and precise. However variations in the composition of a particular mineral lead to small changes in the *d* spacings and X-ray intensities so that the index leads to a number of possible identifications. The final identification is then made on the basis of the other physical or chemical characteristics. The identification of a hand specimen of a mineral on the basis of its physical properties is less precise, and consequently it is common to arrive at a number of possibilities from which a final identification can be made only after more detailed examination.

The first characteristics which most observers note are the **colour and lustre** of the specimen, so the simplified tables for mineral identification provided here are based on these features. Minerals which commonly exhibit more than one colour are repeated in the appropriate lists. The minerals are arranged in order of increasing **hardness** and

the large number of colourless, white, or grey minerals are divided into three categories on the basis of hardness $<2\frac{1}{2}$, $2\frac{1}{2}$–6, >6. The tables for mineral identification are listed below.

1. Black, metallic.
   Black, non-metallic.
2. Blue.
3. Green.
4. Red, metallic.
   Red, non-metallic.
5. Brown, metallic.
   Brown, non-metallic.
6. Yellow, metallic.
   Yellow, non-metallic.
7. Colourless, white or grey, metallic
   Colourless, white or grey, non-metallic.
   $H < 2\frac{1}{2}$
   Colourless, white or grey, non-metallic,
   $H\ 2\frac{1}{2}$–6
   Colourless, white or grey, non-metallic,
   $H > 6$

The entries in the tables are abbreviated usually to one line and they consist of the following information.

Hardness	Specific gravity	Diagnostic characteristics	Mineral name	Class or subclass

The steps in the procedure which should be followed when using the tables are outlined below.

1. **Determine the colour**. Dark grey is included with black, light grey with white. Very pale metallic yellow is included with yellow minerals.
2. **Determine the lustre**. The most important division is between metallic and non-metallic lustre.
3. **Determine the hardness** on the basis of the three categories.

   $H < 2\frac{1}{2}$—can be scratched by a fingernail (2) or copper coin (3).

   H $2\frac{1}{2}$–6—can be scratched with increasing difficulty by a knife (5) or file $(6\frac{1}{2})$.

   $H > 6$—mineral will scratch window glass.

   If necessary a more precise determination of the hardness can be made later by comparison with a Mohs set of minerals.
4. **Turn to the appropriate table of colour and lustre, and** to the part of the table which lists minerals in the indicated **hardness range**. There will be a number of possible identifications, so it is necessary to try to match the characteristics of the specimen with those of the minerals in the appropriate part of the table. Proceed as follows.
5. **Scan the specific gravity values** of the possible minerals, and note if there are any large differences. With a little experience it will be possible to judge large differences in specific gravity of large specimens by holding in the hand. Most common non-metallic minerals has specific gravities near quartz (2.65) so with experience it is possible to judge whether a mineral has a much higher specific gravity such as barytes (4.5) or anglesite (6.2). Graphite (2.2) is significantly lighter than the very similar black sulphide, molybdenite (4.62). If a pure,

compact fragment of the specimen can be obtained, a careful measurement of the specific gravity may lead to a precise identification.

6. **Scan through the diagnostic features** of the number of possible minerals and note any particularly significant features which might confirm or eliminate a possible identification. The special features include the following properties.
7. **Determine the streak**. This may lead to an identification, for example molybdenite is distinguished from graphite by its greenish black streak, and almost black sphalerite may be identified by its pale brown streak as shown in plate 11.
8. **Determine the habit** and if possible the **symmetry**. This may indicate the crystal system to which the mineral specimen belongs, or at least it may indicate a system to which the mineral does not belong; prismatic crystals do not belong to the cubic system.
9. **Determine whether the specimen possesses cleavage planes**. If there are a number of cleavages determine the angles between them and their symmetry. In the tables, cleavages are indicated by indices of form followed by the angle of intersection if this is significant. Cleavage is particularly useful for the identification of the common rock forming minerals including, the micas, pyroxenes, amphiboles, feldspars, and trigonal carbonates.
10. **Comparision of the physical features** of a specimen of a common mineral with those of the closely similar minerals listed in the appropriate table will often result in a particular identification or at least in a small number of possibilities. **Note the class numbers, mineral names and S.G. values** of the minerals indicated by the identification tables.
11. **Turn to the appropriate class or subclass, in Appendix D** and locate the description of the mineral by using the

S. G. value. In the silicate subclasses the minerals have been listed on the basis of their structural characteristics so in these subclasses it will be necessary to search for the mineral name.

12. In some cases **simple chemical tests** may be useful to make a final identification. For example cerussite (6.55) may be distinguished from angelsite (6.2) because if small fragments are placed in a test tube containing warm nitric acid, cerussite will effervesce. The most common chemical test is the use of cold dilute HCl to distinguish certain carbonate minerals by the effervescence of carbon dioxide.

Abbreviations used in the tables.

st.—streak.
tetrag.—tetragonal
ortho.—orthorhombic
mono.—monoclinic
tri.—triclinic
hex.—hexagonal
trig.—trigonal

1. Black minerals, metallic lustre

H	S.G.	Diagnostic features	Name	Class
1	2.2	St. black, greasy, sectile.	**GRAPHITE**	1
1	4.62	St. greenish grey, greasy, sectile.	**MOLYBDENITE**	2
2	4.52	St. lead grey, ortho. bladed, fig. 1.25.	**STIBNITE**	2
$2\frac{1}{2}$	7.4	St. lead grey, cubic, {100}, fig. 1.25.	**GALENA**	2
2	7.2	St. shining black, cubic, easily sectile.	ARGENTITE	2
$2\frac{1}{2}$	5.8	St. steel grey, ortho. short prismatic.	BOURNONITE	2
$2\frac{1}{2}$	5.5	Lead grey. st. black, ortho. slightly sectile.	**CHALCOCITE**	2
3	4.43	Ortho. bladed, {100} perfect	ENARGITE	2
4	4.6	Grey. st. brownish black, tetrahedra.	TETRAHEDRITE	2
4	4.4	Grey. st. black, tetrag.	STANNITE	2
4	21.46	Steel grey, malleable	PLATINUM	1
4	4.3	St. brown, ortho. prismatic, radiating, {010}.	MANGANITE	4
$4\frac{1}{2}$	7.1	St. reddish or brownish black, mono. {010}.	**WOLFRAMITE**	11
$5\frac{1}{2}$	5.26	St. red, trig. rhombohedra, prismatic, plate 1.	**HAEMATITE**	4
$5\frac{1}{2}$	4.7	St. black, trig. tabular {0001}, poor.	**ILMENITE**	4
$5\frac{1}{2}$	4.6	St. dark brown, octahedra, no cleavage.	CHROMITE	4
6	5.18	St. black. octahedra, no cleavage, magnetic.	**MAGNETITE**	4
6	5.15	St. reddish brown, octahedra, no cleavage.	FRANKLINITE	4

Black, non-metallic lustre

H	S.G.	Diagnostic features	Name	Class
1	4.75	St. black, dull may be metallic, dendritic on fractures.	PYROLUSITE	4
$2\frac{1}{2}$	2.8	St. white to grey. vitreous. {001} mica.	**BIOTITE**	12e
$3\frac{1}{2}$	3.9	St. pale brown, resinous, cubic, {110}, plate 11.	**SPHALERITE**	2
5	4.71	St. brownish black, reniform encrustations, earthy.	PSILOMELANE	4
$5\frac{1}{2}$	10.5	Pitch-like lustre, radioactive.	URANINITE	4
6	6.8	Brownish, st. white, adamantine, tetrag. {100}.	**CASSITERITE**	4
5	3.0	Greenish, st. white, mono., prismatic. {110} 56°.	**HORNBLENDE**	12d
6	3.5	Brownish. st. grey, ortho., prismatic. {110} 87°.	HYPERSTHENE	12c
6	3.02	Greenish, st. grey, mono., prismatic. {110} 87°.	**AUGITE**	12c
$7\frac{1}{2}$	4.32	Reddish, cubic forms	**ALMANDINE**	12a
$7\frac{1}{2}$	3.0	Trig. prismatic, rounded triangular cross-sections.	**TOURMALINE**	12b

2. Blue minerals

H	S.G.	Diagnostic features	Name	Class
$1\frac{1}{2}$	4.6	Metallic, indigo blue, st. black, {0001} flexible.	**COVELLITE**	2
3	5.06	Metallic, bronze with blue tarnish, plate 12.	**BORNITE**	2
$2\frac{1}{2}$	2.12	Azure blue, st. white, vitreous, tri., soluble in water.	CHALCANTHITE	10
$3\frac{1}{2}$	3.77	Deep blue crystals, st. light blue, effervesces, plate 3.	**AZURITE**	6
2–4	2.0	Greenish, st. light blue, conchoidal fracture.	CHRYSOCOLLA	12f
4	3.18	Cubic, cubes, {111}, plate 16.	**FLUORITE**	5
$5\frac{1}{2}$	2.15	Deep blue, st. white, cubic, crystals in lavas.	SODALITE	12f
5	2.4	Azure blue, st. light blue, contains pyrite, plate 2.	LAZURITE	12f
6	2.6	St. white or greenish, waxy, plate 2.	TURQUOISE	9
6	3.0	Pale blue, st. grey, mono., prismatic, {110} 58°.	GLAUCOPHANE	12d
6	3.4	Dark blue, st. blue grey, mono. {110} 56°.	RIEBECKITE	12d
5–7	3.63	Patchy blue, st. white, tri. bladed {100}.	**KYANITE**	12a
7	2.6	Dark blue to grey, st. white, vitreous, ortho. {010}.	**CORDIERITE**	12b
9	4.02	Trig. tapering hexagonal columns, plate 7.	**CORUNDUM**	4

3. Green minerals

H	S.G.	Diagnostic features	Name	Class
1	2.8	Greyish, greasy lustre and feel, sectile.	**TALC**	12e
2–5	2.2	Mottled, st. white. platey or fibrous.	**SERPENTINE**	12e
$2\frac{1}{2}$	2.39	Pale green. trig. {0001} foliate, sectile.	BRUCITE	4
2	2.5	Dark green grains in sandstones, mica {001}.	**GLAUCONITE**	12e
2	2.6	Micaceous {001}, folia not elastic.	**CHLORITE**	12e
2	3.1	St. yellow, foliate, fluorescent, plate 16.	AUTUNITE	9
2–4	2.0	St. light blue, conchoidal fracture.	CHRYSOCOLLA	12f
$3\frac{1}{2}$	3.6	Bright green. st. pale green, mono. effervesces, plates 2 and 3.	**MALACHITE**	6
$3\frac{1}{2}$	3.9	Green to black, st. pale green. ortho. prismatic, reniform {010}.	ANTLERITE	10
$3\frac{1}{2}$	7.04	Yellowish, st. white, hexag. encrustations.	PYROMORPHITE	9
4	3.18	Cubic, cubes, {111}, plate 16.	**FLUORITE**	5
5	3.15	Pale green, hex. columnar, {0001} poor.	**APATITE**	9
6	3.0	Greenish black, mono, prismatic, {110} 56°.	**HORNBLENDE**	12d
6	3.35	Green, mono. prismatic, {110} 56°.	ACTINOLITE	12d
6	3.5	Greenish black, ortho. prismatic, {110} 87°.	HYPERSTHENE	12c
6	3.2	Greenish black, mono. prismatic, {110} 87°.	**AUGITE**	12c
6	3.2	Pale green, mono. prismatic, {110} 87°.	**DIOPSIDE**	12c
6	3.4	Dark green, mono. prismatic, {110} 87°.	AEGIRINE	12c
$6\frac{1}{2}$	3.2	Green, mono. fibreous, compact.	JADEITE	12c
$6\frac{1}{2}$	2.8	Pale green, ortho. fibre groups in amygdales.	PREHNITE	12e
$6\frac{1}{2}$	3.22	St. white, vitreous, granular, plate 13.	**OLIVINE**	12a
7	3.59	Pale green, cubic forms	GROSSULARITE	12a
7	3.3	Greenish black, mono. prismatic, {001}.	EPIDOTE	12a
$7\frac{1}{2}$	3.0	Greenish black, trig., prismatic with rounded triangular cross-sections.	TOURMALINE	12a
$7\frac{1}{2}$	2.75	Pale green, hex. columnar, plate 14.	**BERYL**	12b
8	3.6	Greenish black, cubic, octahedra.	SPINEL	4

4. Red minerals, metallic lustre

H	S.G.	Diagnostic features	Name	Class
$2\frac{1}{2}$	8.9	Copper red colour and streak, malleable.	**COPPER**	1
$3\frac{1}{2}$	6.1	Ruby red, st. brownish, metallic to adamantine.	CUPRITE	4
5	7.78	Pale copper red, st. black, alters to green 'nickel bloom'.	NICCOLITE	3

Red, non-metallic lustre

H	S.G.	Diagnostic features	Name	Class
1–3	2.0	Haematite pigment, dull, powdery, pisolitic, fig. 1.45.	**BAUXITE**	4
$1\frac{1}{2}$	3.48	St. orange red, resinous, prismatic or granular.	REALGAR	2
$2\frac{1}{2}$	8.10	St. scarlet, trig. $\{10\bar{1}1\}$, granular.	**CINNABAR**	2
$2\frac{1}{2}$	5.85	Deep red, st. red, adamantine. trig, prismatic, rhombohedral.	PYRARGYRITE	2
$2\frac{1}{2}$	2.8	Pink or red, st. white, $\{001\}$ mica.	LEPIDOLITE	12e
$3\frac{1}{2}$	3.45	Rose pink, st. white. trig. $\{10\bar{1}1\}$.	RHODOCHROSITE	6
$3\frac{1}{2}$	2.85	Pinkish, st. white. trig. $\{10\bar{1}1\}$, curved faces.	**DOLOMITE**	6
4	5.68	Orange to deep red, st. yellow. $\{10\bar{1}0\}$.	ZINCITE	4
$5\frac{1}{2}$	5.26	St. red, dull, powdery, strong pigment, plate 1.	**HAEMATITE**	4
6	2.57	Mono. $\{001\}$, $\{010\}$, pink haematite pigment.	**ORTHOCLASE**	12f
6	2.62	Tri. $\{001\}^\wedge\{010\}$ 86°, pink haematite pigment.	**PLAGIOCLASE**	12f
7	2.65	Massive, conchoidal fracture, quartz with red haematite pigment.	JASPER	12f
$7\frac{1}{2}$	4.32	Cubic forms, plate 10.	**ALMANDINE**	12a
9	4.02	Trig. tapering hexagonal columns, plate 8.	CORUNDUM	4

5. Brown, metallic lustre

H	S.G.	Diagnostic features	Name	Class
3	5.06	Bronze colour, st. grey black, purple tarnish, plate 12.	**BORNITE**	2
4	4.58	Bronze, st. black, magnetic.	PYRRHOTITE	2

Brown, non-metallic lustre

H	S.G.	Diagnostic features	Name	Class
$2\frac{1}{2}$	2.8	Brown to black, vitreous, mica, {001}.	**BIOTITE**	12e
$3\frac{1}{2}$	3.9	St. brown, resinous, cubic, {110}, plate 11.	**SPHALERITE**	2
$3\frac{1}{2}$	3.83	Pale brownish, vitreous. trig. {10$\bar{1}$1}	SIDERITE	6
$4\frac{1}{2}$	4.35	Trig. {10$\bar{1}$1}, reniform 'dry bone' ore, effervesces.	SMITHSONITE	6
5	3.6	St. yellow, amorphous, dull.	**LIMONITE**	4
5	4.37	St. yellow, dull. ortho. {010} cleavage.	**GOETHITE**	4
5	4.62	Reddish brown to yellow, st. white, resinous, granular.	MONAZITE	9
5	3.4	St. white, mono, wedge shaped.	SPHENE	12a
$6\frac{1}{2}$	3.27	Tri, wedge shaped crystals, sharp edges, vitreous.	AXINITE	12b
$6\frac{1}{2}$	3.35	Tetrag. prismatic often with {001} faces, st. white, resinous.	IDOCRASE	12a
$6\frac{1}{2}$	4.18	Reddish, st. pale brown, adamantine, tetrag. prismatic.	RUTILE	4
$6\frac{1}{2}$	6.8	Yellowish, st. white. adamantine. tetrag. {100}.	**CASSITERITE**	4
7	3.7	Reddish, st. grey, ortho. (110) (1$\bar{1}$0) = 50°, cruciform twins.	**STAUROLITE**	12a
$7\frac{1}{2}$	3.16	Ortho. prismatic, nearly square cross-sections.	**ANDALUSITE**	12a
$7\frac{1}{2}$	4.68	Tetrag. prismatic.	**ZIRCON**	12a
8	4.0	Octahedra.	SPINEL	4

6. Yellow, metallic lustre

H	S.G.	Diagnostic features	Name	Class
$2\frac{1}{4}$	19.3	Malleable, plate 5.	**GOLD**	1
$2\frac{1}{2}$	9.35	St. yellowish to greenish grey, no cleavage.	CALAVERITE	3
$3\frac{1}{2}$	4.1	St. greenish black, tetrag. {011}.	**CHALCOPYRITE**	2
$3\frac{1}{2}$	4.6	Bronze colour and st. cubic, {111}.	PENTLANDITE	2
3	5.5	Trig. fibres, {10$\bar{1}$1}, {01$\bar{1}$2}	MILLERITE	2
6	5.02	Pale brass yellow, striated cubes and pyritohedra.	**PYRITE**	2
6	4.89	Paler than pyrite, st. black, ortho., radiating fibres.	MARCASITE	2

Yellow, non-metallic lustre

H	S.G.	Diagnostic features	Name	Class
1	4.1	Dull, strong pigment after uranium minerals.	CARNOTITE	11
$1\frac{1}{2}$	3.49	Resinous, foliate masses, sectile.	ORPIMENT	2
$2\frac{1}{2}$	2.05	Resinous, ortho., no cleavage.	**SULPHUR**	1
2	3.1	Yellowish green, foliate, fluorescent, plate 16.	AUTUNITE	9
3	6.5	Yellow to orange, st. white, tetrag. tabular // (001).	WULFENITE	11
4	3.18	Cubes, {111}, plate 16.	**FLUORITE**	5
5	3.6	Dull, yellow ochre, amorphous.	**LIMONITE**	4
8	3.4	Pale, vitreous. ortho, {001}, fig. 1.11.	TOPAZ	12a
$8\frac{1}{2}$	3.65	Ortho. pseudo-hex. columns.	CHRYSOBERYL	4
10	3.5	Pale, adamantine, octahedra.	**DIAMOND**	1

7. Colourless, white or grey, with metallic lustre

H	S.G.	Diagnostic features	Name	Class
$1\frac{1}{2}$	8.16	Silver white, st. grey, mono, bladed, {010}.	SYLVANITE	3
$2\frac{1}{2}$	10.5	St. silver, malleable, sectile.	SILVER	1
$5\frac{1}{2}$	6.33	Silver white, reddish, st. black, striated cubes, {100} perfect.	COBALTITE	2
$5\frac{1}{2}$	6.07	Silver white, st. black, ortho. prisms and domes.	ARSENOPYRITE	2
6	4.89	Yellowish, st. black, fibrous.	MARCASITE	2

Colourless, white or grey, non-metallic, $H < 2\frac{1}{2}$

H	S.G.	Diagnostic features	Name	Class
1	2.7	Grey, greasy lustre and feel, sectile.	**TALC**	12e
$2\frac{1}{2}$	2.39	Trig. {0001} foliate, sectile.	**BRUCITE**	4
$2\frac{1}{2}$	2.76	Mica {001}, flexible, elastic sheets.	**MUSCOVITE**	12e
2	2.6	Dull, {001} clay.	**KAOLINITE**	12e
2	2.32	Mono. tabular, fibrous {010} fig. 1.17, vitreous.	**GYPSUM**	10
$2\frac{1}{2}$	2.16	Cubic, {100}, salty, fig. 1.19, vitreous.	**HALITE**	5
2	1.99	Cubic, {100}, salty, bitter, vitreous.	SYLVITE	5
2	5.5	Grey, tarnishes brown, waxlike, sectile.	CERAGYRITE	5
1–3	2.0	Powdery, pisolitic, fig. 1.45.	**BAUXITE**	4
2	2.09	Encrustions on soils, cooling taste.	NITER	7
2	2.29	Strongly deliquescent.	SODA NITER	7
2	1.7	Encrustations {100}, {110}, easily fusible.	BORAX	8

Colourless, white or grey, non-metallic lustre, *H* 2½–6

3	2.72	Trig. {10Ī1} fig. 1.24. vigorous effervescence in HCl.	**CALCITE**	6
3½	2.85	Pinkish. trig. {10Ī1}, curved surfaces.	**DOLOMITE**	6
3–5	3.0	Trig. {10Ī1}, does not effervesce in HCl.	MAGNESITE	6
3½	2.95	Ortho. tabular, fig. 1.42. pseudo-hex. fig. 16.17.	**ARAGONITE**	6
3½	3.7	Ortho. {110}, effervesces in HCl.	STRONTIANITE	6
3½	4.3	Ortho. pseudo-hex. fig. 16.18. {010} effervesces in HC1.	WITHERITE	6
			CERUSSITE	6
3½	6.55	Ortho. {110}, effervesces in warm HNO₃.		
3	2.9	Ortho. tabular, rectangular cleavage blocks.	ANHYDRITE	10
3	4.5	Ortho. {001}, {210} intersecting at 78° 20′.	BARYTES	10
3	3.95	Ortho. {001}, {210} intersecting at 75° 58′.	CELESTITE	10
3	6.2	Ortho. {001}, {210} does not effervesce in HNO₃.	ANGLESITE	10
4	3.18	Cubic, {111}, plate 15.	**FLUORITE**	5
4	2.05	Trig. rhombohedron faces at nearly 90°.	CHABAZITE	12f
4½	5.9	Tetrag. dipyramidal habit, {101}.	SCHEELITE	11
4½	3.4	Ortho. tabular, {110}.	HEMIMORPHITE	12a
5½	3.9	Trig. {0001}.	WILLEMITE	12a
5	2.8	Tri. {100}, {001}.	WOLLASTONITE	12c
5	1.9	Amorphous, sometimes shows play of colours.	OPAL	12f
5	2.27	Icositetrahedra, in amygdules.	ANALCIME	12f
5½	2.45	Icositetrahedra, in lavas.	LEUCITE	12f
5	2.25	Radiating fibres in amygdales.	NATROLITE	12f
5½	2.55	Greasy grey lustre. trig. {10Ī0}.	**NEPHELINE**	12f
5	3	Mono. prismatic, {110} 56°.	TREMOLITE	12d

Colourless, white or grey, non-metallic, *H* > 6

6	2.57	Mono. {010}, {001} fig. 1.18.	**ORTHOCLASE**	12f
6	2.62	Tri. {010} {001} = 86°, fig. 14.2.	**PLAGIOCLASE**	12f
6	3.0	Tri. {100}, {110}, fusible with red flame.	AMBLYGONITE	9
6	3.2	Mono. prismatic, {110} 87°.	**DIOPSIDE**	12c
6	3.1	Mono. prismatic, {110} 87°, {100}.	SPODUMENE	12c
6	3.2	Ortho. prismatic, {110} 87°.	ENSTATITE	12c
6	2.8	Ortho. prismatic, rounded groups in amygdales.	PREHNITE	12e
6	2.4	Mono. {001}, foliate masses in pegmatites.	PETALITE	12f
7	2.65	Trig. conchoidal fracture, vitreous.	**QUARTZ**	12f
7	2.6	Grey, {010} poor.	CORDIERITE	12a
$7\frac{1}{2}$	3.16	Ortho. prismatic with nearly square end sections. fig. 1.44.	**ANDALUSITE**	12a
$6\frac{1}{2}$	3.77	Ortho. slender prisms, {010} perfect.	**SILLIMANITE**	12a
$6\frac{1}{2}$	3.3	Ortho. {010}, fine bladed crystals in bauxite.	DIASPORE	4
$7\frac{1}{2}$	4.68	Tetrag. prisms. fig. 11.1.	ZIRCON	12a
8	3.4	Ortho. {001} fig. 1.11.	TOPAZ	12a
8	3.6	Octahedra.	SPINEL	4
9	4.02	Trig. tapering hexagonal columns.	CORUNDUM	4
10	3.5	Octahedra, adamantine.	DIAMOND	1

Index